Arming
the Luftwaffe

Arming the Luftwaffe

The German Aviation Industry in World War II

Daniel Uziel

McFarland & Company, Inc., Publishers
Jefferson, North Carolina, and London

LIBRARY OF CONGRESS CATALOGUING-IN-PUBLICATION DATA

Uziel, Daniel, 1967–
Arming the Luftwaffe : the German aviation
industry in World War II / Daniel Uziel.
      p.   cm.
Includes bibliographical references and index.

ISBN 978-0-7864-6521-7
softcover : 50# alkaline paper ∞

1. Aeronautics, Military — Germany — History — 20th century.
2. Germany. Luftwaffe — Equipment and supplies — History — 20th cenuty.
3. Germany. Luftwaffe — History — World War, 1939–1945.
    I. Title. II. Title: German aviation industry in World War II.
UG635.G3U95 2012   338.4'7623746094309044 — dc23   2011038151

BRITISH LIBRARY CATALOGUING DATA ARE AVAILABLE

© 2012 Daniel Uziel. All rights reserved

*No part of this book may be reproduced or transmitted in any form
or by any means, electronic or mechanical, including photocopying
or recording, or by any information storage and retrieval system,
without permission in writing from the publisher.*

Front cover image: Production of Me 109 fuselages inside a cramped
wooden hut at the Gauting Forest factory (courtesy of NARA);
cover design by David K. Landis (Shake It Loose Graphics)

Manufactured in the United States of America

*McFarland & Company, Inc., Publishers
Box 611, Jefferson, North Carolina 28640
www.mcfarlandpub.com*

Acknowledgments

This research was made possible through the support of several organizations and individuals. Foremost among them are the Smithsonian Institution and the U.S. National Air and Space Museum. The core of this research came into being during a twelve-month fellowship at that magnificent museum. Besides providing access to a large amount of resources, the museum, as well as other researchers' forums of the Smithsonian Institution, was an intellectually stimulating and supportive environment and I fully commend them for that.

A special thanks to Michael Neufeld and Dik Dasso — two historians and curators that agreed to be my advisors and supported this research in various ways. The staff of the museum's archives and library was also extremely helpful in locating material and providing me photos for this book.

During my term at the National Air and Space Museum, I spent long hours and days at the U.S. National Archives' College Park facility. Researching in this impressive modern archive was particularly pleasant and fruitful. Mr. Larry McDonald of the reference room was particularly friendly and helpful. Thanks to his efforts I located some materials I would have never found on my own.

The German Museum in Munich and the head of its research department Professor Helmut Trischler, were kind enough to host me for a couple of weeks in 2006 and allow me to work at their archives. I would like to thank in particular Mr. Wolfgang Schinhan, an archive worker who not only served me in an exemplary way, but also was extremely pleasant to me and to other visitors researching there at that time.

As it was during my past researches, the German Federal Archives proved again to be a bottomless pit of information and knowledge. As usual, its staff was also very helpful.

Several researchers and scholars shared with me their thoughts and offered me documents from their own collections. Richard Egger and Manfred Boehme were particularly helpful in this regard. Frank Bajohr, Russell Lee and Florian Schmalz also kindly provided me copies of some important documents.

Two distinguished scholars supported this research from the beginning: Professor Moshe Zimmermann, my mentor and teacher in the field of German history, and Professor Martin van Creveld, both of the Hebrew University.

Professor Bernd Wegner of the Bundeswehr University in Hamburg allowed me to try some of the main ideas of this research on his colleagues and students and get their feedback.

A big "thank you" also to several friends and colleagues who supported me along the way: Tobias Bütow, my sister Yeal Goshen and my brother-in-law Gil, Beate Meyer, Miriam Moschytz, Christine Schmidt, Joe White and Kim Wünschmann.

I would also like to mention Neil Maher, Shai Kalansky and Teasel Muir-Harmony, who provided me some fascinating and enjoyable glimpses into the American way of life. Stephan Zellmeyer, a great Swiss, provided me much-needed spiritual support both in Washington and in Munich.

I guess the best part of this research was the opportunity to get some hands-on aviation experience. While flying with fellow aviation researcher Allen Meyer in the little Citabria aerobatic plane I learned anew the real meaning of an adrenaline rush. These flights also gave me some insight into the way some World War II aircraft were flown.

Table of Contents

Acknowledgments v

Introduction 1

1. The Aviation Industry at War — 7
2. The Aviation Industry and the Air War — 51
3. Reorganization of Aircraft Production — 71
4. From Technological Expertise to Slave Labor — 144
5. On the Production Lines — Daily Life in the Factories — 194
6. The "People's Fighter" as Case Study of a Late-War Program — 236

Conclusion 263
Chapter Notes 269
Bibliography 293
Index 299

Introduction

In 2002 a private association in the city of Rostock in former East Germany opened a new exhibition dedicated to famous aircraft designer and producer Ernst Heinkel. The exhibition intended to present Heinkel's technical innovation and competence to the younger generations. It also sought to present the social and economical benefits his firm brought to Rostock. Heinkel's association with the Nazi regime was mentioned, but not in the forefront of the exhibition. This negligence caused a public outcry and debate, which eventually resulted in the temporary closure of the exhibition. A special scientific committee was established and was asked to advise on how to reorganize the exhibition in order to make it historically balanced.

The entire debate had practically nothing to do with Heinkel's skills as an aircraft designer and as a trailblazer of new aviation technologies. The organizers of the exhibition wanted to pay tribute to the man whose technological enterprise deeply influenced Rostock in many respects. He was simply viewed as a great German engineer and entrepreneur, who cooperated only mildly with the Nazi regime.

The critics of the exhibition argued that Heinkel not only provided the Nazi regime with important and sophisticated aerial weapon systems, but that he was willingly involved in some of its darkest war crimes.[1]

The debate around the Heinkel exhibition dully demonstrated the fact that aircraft development and production, particularly in the Nazi German context, has broader significance, far beyond the purely technical and organizational aspects. The city of Rostock, for instance, grew almost threefold as a result of Heinkel's expansion during the twelve years of the Third Reich.

The debate highlighted some moral issues: Heinkel's cooperation with the Nazi war machine, the early employment of forced labor in his factories, and finally his "whitewashed" postwar biography, which largely served as the basis for the historical narrative of the exhibition.

Heinkel was not the only aeronautics professional who came under fire in recent years. Even Hans von Ohain, the inventor of the German jet turbine, and coincidently Heinkel's protégé, came into public crossfire when it was intended to name a new terminal at the Rostock-Laage airport after him. His opportunism was cited as the main reason behind the protest.[2]

These and other debates underscored the crucial role of the aviation industry in the history of Nazi Germany. During World War II the aviation industry became the largest branch of the German armaments industry. American intelligence officers and scholars pointed in a postwar report to the sheer magnitude of the German aviation industry: "The story of the German aircraft industry is inseparable from that of the German Air Force. During the war years, aircraft, together with air force equipment, represented approximately 40 percent of total German war production, and their manufacture, in all its phases involved the employment of 2,000,000 men."[3] In terms of net product value, the aviation industry even surpassed the traditionally dominant munitions industry. The annual value of its products went well over 10 billion reichsmarks,[4] and as German records show, in September 1944 over 3 million people were involved in one way or another in its production programs.[5] At the time of Germany's surrender the Junkers firm alone employed approximately 147,000 workers; around 67,000 of them, or 46 percent, were concentration camp inmates, prisoners of war (POW) and foreign workers.[6] All these people worked in a network of at least 96 Junkers-owned main and sub-locations, mostly in central Germany.[7]

Therefore, the aviation industry was not only the leading branch of the German war industry; its factories became the place where millions experienced World War II. The number of people employed in production programs of the *Luftwaffe** is staggering when compared with the numbers of those producing for the army and the navy. On 20 July 1944, Armaments Minister Albert Speer reported to Hitler that "1,940,000 [workers] are engaged in production for the army; 2,330,000 are engaged in production for the air force; 530,000 are engaged in production for the navy." Furthermore, "3,180,000 are engaged in trade, banks, insurance and catering, 1,450,000 are engaged in domestic work."[8]

These figures suggest that in 1944 the aviation industry was not only the largest branch of the armaments industry, but was also one of the largest German employers. Therefore the sheer size of this industry and its central role within the contemporary German economy makes it an important research topic.

It seems that the focal point of the debates around Heinkel, von Ohain and others was the apparent incompatibility between an industrial branch representing modernity and technical excellence and one of the worst and most radical political systems in history. This incompatibility looks, however, less obvious when considering what the Germans termed "modern technology" at that time. As historian Michael T. Allen pointed out: "Almost all Germans agreed that technological modernity meant mechanization, standardization, mass production and the centralized direction of hierarchically managed organizations."[9] Streamlining and industrial efficiency were also perceived through the narrow prism of increased productivity. The Nazi regime inherited a country where these ideals had prevailed already during the nineteenth century. It was quick to grasp the benefits technology and industrialization offered it in fulfilling its political and ideological aims. As part of the mobilization of the industry it sought to modernize, but instead of modernizing the industry through the introduction of modern production technologies, it made it increase its productivity through other means. The industry was happy to work along these lines because output increase through reorganization and standardization demanded less long-term investments and cost them less money.[10]

A common belief in the primacy of community in all aspects of life was another notion that bound technological modernism and Nazi ideology together:

**For the sake of readability and because of the widespread use of common German terms from the Nazi era, many are not italicized beyond their first use in this book.*

German engineers — the greatest enthusiasts of industrial modernization in Germany — advocated a vision of factory community along with mechanization. Their advocacy did not stop at the factory floor but widened to include all of society.... Many histories dwell upon distinctions between Nazi fanaticism characterized by Hitler or the SS, on one hand, and the "rational pragmatism" of "normal" engineers or "technocrats" like Fritz Todt, on the other. But these distinctions can be maintained only if we consider machines themselves or modern management to have no social or cultural history, as if "pragmatic" choices between different factory systems did not inherently involve choices among different visions of "community" and "society."[11]

As a result, the factories of the aviation industry became an important element of the Nazi effort to create a national community (*Volksgemeinschaft*).[12] Modern technology and industry were therefore synonymous with some central social and ideological goals of National Socialism, besides providing it the military hardware it needed.

This background is important when considering the development of the German aviation industry during World War II. This industry represented perhaps the climax of modern technology and workmanship in the broader context of creating a national community. Today the aviation industry is associated above all with the hi-tech industry. This industry is characterized by cutting-edge products, heavy reliance on information technologies, decent employment conditions for its employees, tidy work environment, unique technical know-how, use of the newest technologies and materials, and of course giant budgets. These elements also characterized the German aviation industry and therefore placed it in the same social and economic position of today's high-tech industry. This was the situation at least until the outbreak of World War II.

Towards the end of World War II almost all the social and cultural ideals associated with the aviation industry were eroded by the demands of total war and a looming defeat. This industry became something completely different from the modern and national community-creating industry it strove to be before the war. It became widely dispersed in more-or-less improvised conventional and underground facilities. It also became an important employer of slave labor. These changes happened at a time when the industry geared up for the mass production of aircraft using the most modern technologies available at that time — in fact, groundbreaking aviation technologies in every respect. At the same time, the aviation industry went through a process of demodernization, characterized mainly by the changing nature of its workforce.

Astonishingly, on the eve of total defeat and in the midst of extreme changes, in 1944 the Germans reached unprecedented aircraft production rates.

This research will try to explain the discrepancies between the prewar industry and its final form. It will also try to show how and why this industry achieved the highest production rates at the height of its demodernization process.

Hermann Göring, the Reich's Air Minister and supreme commander of the Luftwaffe, gave the following answer when asked by his interrogators after the war about wartime changes in aircraft production: "All these things are so deep-rooted that you would have to go back some ten to fifteen years."[13] He was absolutely right. In order to understand the story of the aviation industry in the later parts of World War II, it is necessary indeed to look back at the early war period and even at the prewar era. Some processes that reached their climax and culmination in 1944, like the general dispersal, the use of slave labor and aircraft type reduction, started much earlier. Therefore, although this research focuses on the events of 1943–1945, large portions of it go back in time in search of the historical roots of some key developments.

The story of the Luftwaffe is still one of the most popular topics in the history of World War II. This popularity is even traceable in popular culture. American rock band Blue Öyster Cult dedicated a song to the Messerschmitt Me 262 jet fighter, and a photo of the aircraft appeared on the cover of its gold album *Secret Treaties*. Even Homer Simpson was once caught watching a documentary about the Luftwaffe on the History Channel — a clever gag pointing at the popularity of the topic on this popular TV channel.[14] A simple search in databases of libraries and online bookstores using the keyword "Luftwaffe" will bring up several hundred results. However, surprisingly little was written of a scholarly nature about the German aviation industry. Especially the late-war period was left largely untouched. Most of the literature about German World War II aviation technology is of popular nature. Some of these books include phrases like: "Production of day fighters amounted to 1,900 in September 1944, yet in spite of repeated Allied bombing raids this figure jumped to 3,300 fighters per month in November of that year. This, incidentally, must be one of the most outstanding feats ever carried out by a developed industrial nation."[15] Simplistic descriptions of this kind are often found in popular books and films dealing with the Luftwaffe and its aircraft. The above citation fails to put the "outstanding feat" in its historical context: the fact that this "feat" was achieved mainly through brutal use of slave labor, that in late 1944 Germany was far from being a "developed industrial nation," and that in November 1944 output was decreasing and not increasing. It also fails to mention the primitive conditions prevailing in most of the factories at that time, the fanatic push for higher production rates at any cost, and finally the fact that this "feat" was completely useless because the Luftwaffe was a defeated force by 1944 and was unable to make efficient use of these aircraft.

This research survey will deal therefore with the scholarly researches written about the German aviation industry, although some of the more serious popular works are used when they include important and reliable information.[16] Richard Overy and Edward Homze were among the first scholars to deal with the German air armament and its problems in the 1970s. Their groundbreaking works concentrated on the prewar phase (Homze) and on the early war years (Overy) and described some of the problematic aspects of this important industry.[17]

During the 1990s several important researches were published — mainly in Germany. The most comprehensive work on this topic was written and published in this decade. Lutz Budrass's massive book about the German aviation industry from 1918 to 1945 is based on extensive archival research and concentrates mainly on social and business aspects.[18] Unfortunately, due to deadline pressures, the late-war period was largely left out of this important book.

The development and use of jet aircraft in Germany proved to be a particularly attractive and popular topic. Popular histories and memoirs of former Luftwaffe generals helped to create several misconceptions about it. They also helped to create the impression that the production of jet planes formed the bulk of German aircraft production towards the end of World War II. It was not until 1993 that the record was put straight by a German historian.[19] Ralf Schabel's book about the development and production of German jets and air-defense missiles countered several myths. Most importantly, he proved that the prolonged development of these technologies was inherent to their complexity and that their operational use could not have been better.

The problems of aviation factories' dispersal and the use of slave labor attracted the attention of several local German historians, who wrote several important histories of factories and labor camps located in their region.[20]

During the 1990s several German firms started hiring historians to write their histories during the Third Reich. This trend resulted in several publications dealing with specific firms associated with the aviation industry.[21]

Primary sources used in this research are mostly of 3 types: World War II German records, Allied intelligence reports — mostly composed immediately after the war — and eyewitness reports. The most important surviving German records are those of the Reich's Air Ministry (*Reichsluftfahrtministerium*— RLM), of the Armaments Ministry, of the Luftwaffe and of different firms.

Immediately after the war the Allies were highly interested in finding out what happened on the other side of the hill. Numerous military and civilian intelligence and survey teams were tasked with researching almost every aspect of the wartime German economy. Most important among them were the well-known United States Strategic Bombing Survey (USSBS) teams, which produced a large series of detailed reports. Less known teams operated under the guises of the Combined Intelligence Objectives Sub-Committee (CIOS), the U.S. Naval Technical Mission to Europe, the Field Intelligence Agency, Technical (FIAT) teams and the British Bombing Survey Unit (BBSU). The Allies were mainly interested in putting their hands on new technologies developed by the Germans, finding out which technologies and weapons were delivered to the Japanese, and measuring the effectiveness of their strategic bombing campaign.[22] The work of these teams resulted in a dazzling array of reports, covering almost everything from instant coffee manufacture to ballistic missile technology. Besides the published reports, most of the primary sources used by these teams, including interrogation transcripts, can be found in the USSBS records group at the U.S. National Archives.

The third group of sources, the eyewitness reports, is mostly of Holocaust survivors who worked in the German aviation industry. Besides providing information about their own harsh experience, many of them also provide insights into the way large parts of this industry operated from 1943 until the end of the war. These accounts are, however, primarily personal testimonies on the daily life and miseries prevailing in this important industrial branch. They highlight the fact that there is a strong human aspect to the predominantly business and technological façade of the German aviation industry.

1

The Aviation Industry at War

The German aviation industry went through a series of changes during World War II. Some of them resulted from wartime developments, but a large number of wartime changes started before the war. Since the Nazis came to power the aviation industry had produced mostly for the military, but because it was a peacetime industry, it used only a small portion of its potential capacity. The shift from peacetime production to wartime production was not as dramatic as it could have been. We would have expected that after the outbreak of World War II the aviation industry would fully gear up in order to deliver the increased amount of hardware required by a country waging a major war. This gear-up happened only to a limited extent.

As it finally became obvious to the German leadership that Germany was involved in a total war, it sought to reform the organization of aircraft production. The reform took place in two main areas: management and production methods. Procurement policies remain practically the same all through the war and led to several spectacular and traumatic acquisition fiascos.

Some of the basic problems the Germans encountered with aviation production in later stages of World War II evolved before the war or shortly after it started. Therefore, in order to understand the transformation of the aviation industry and its structure, we need to take a look at the earlier history of this branch. This chapter deals with several key aspects and problems of the aviation industry before the outbreak of World War II and afterwards.

Organization of the Aviation Industry

Until Hitler came to power in 1933, Germany's aircraft industry was composed of several small firms manufacturing limited series of mostly light planes. The Treaty of Versailles forbade Germany from producing or possessing military aircraft and practically killed a large industrial sector which had produced thousands of aircraft for Germany and its allies during World War I.[1] Although several important civilian aircraft were developed in Germany in the 1920s and the early 1930s, production runs were limited. Total German aircraft production in 1932 was 36 and the aircraft and aero-engine industry employed in January 1932 a total of only 3,988 workers.[2] This industrial sector could hardly be viewed as a meaningful

Aviation enthusiasts Hitler and Göring meet jubilant crowd during a visit to the Luftwaffe's flight test center at Rechlin, 1939 (courtesy Yad Vashem Archives).

branch of German industry at that time. The Nazi seizure of power brought a dramatic change. In April 1933 Hitler appointed Hermann Göring as Air Minister and established for him a completely new ministry: the *Reichsluftfahrtministerium* (RLM). The RLM immediately instructed the active aircraft firms to increase the production of current types and to begin designing new types. At that time Germany was still restricted by the Treaty of Versailles, but the RLM started preparing plans for aerial rearmament and for a massive expansion of the aviation industry.

In May 1933 the aviation-conscious Minister of Defense, General von Blomberg, transferred the army's small air-defense department (*Luftschutzamt*— Air Defense Office) to the RLM. This move meant a significant expansion of the RLM, which now consisted of two large departments: the military Luftschutzamt and the civilian *Allgemeines Luftamt* (General Air Office). State Secretary Erhard Milch, former general director of national airline Lufthansa, was placed in direct control of the Luftschutzamt. In September 1933 reorganization was undertaken to reduce duplication of effort between the departments. The primary changes were to move the staffing and technical development organizations out of the civilian *Allgemeines Luftamt*, and to create out of them new dedicated departments. The ministry was thus composed of six departments, one of which, the *Technisches Amt* (Technical Office), dealt with development and production. In 1934 an additional department was created, the *Luftzeugmeister* (Aerial Equipment), in order to deal with logistics.[3] This department later expended and played a central role in the management of production. At a time when no

German air force existed, this reorganization was an important change, which practically placed under the supposedly civilian ministry both civilian and military aviation matters and made it Germany's top aviation organization.

On 26 February 1935, Hitler ordered Göring to recreate the German air force, the Luftwaffe, and shortly afterwards its existence was openly acknowledged, thus formally ending Germany's commitment to the Versailles armament restrictions. Although all operational matters were now the responsibility of the Luftwaffe's high command, matters of design, research and development, production, and procurement stayed in the hands of the RLM. This division of responsibilities was trouble-prone and later caused some severe problems. While the Luftwaffe was supposed to submit the characteristics of the equipment dictated by its doctrine, the RLM controlled the development and production of this equipment. As a result the Luftwaffe received several times something completely different from what it had ordered. In several cases it even got products it never ordered or needed. The majority of aircraft flown by the Germans in World War II were conceived along lines set by the RLM in 1935–1936 and were consequently designed and developed by the aircraft manufacturers. The Luftwaffe was represented in this process by numerous liaison officers allocated permanently or temporarily to the RLM. However, even this liaison system failed to prevent some embarrassing procurement fiascos.

Unlike most other parts of the German industry under Nazi rule, the RLM sought to centrally plan and manage the production of aircraft.[4] The two most important persons involved in the central management of military aircraft production were the aforementioned State Secretary and later Field Marshall Erhard Milch and Colonel Ernst Udet. Göring pulled out Milch in 1933 from his position as general director of national airline Lufthansa and made him his deputy in the RLM. Udet was appointed in 1936 to head the Technical Office under the title *Generalluftzeugmeister*. Milch was sidelined at that time and Udet took over from him the technical departments.[5] Udet was a former World War I ace. Afterwards he became an internationally acclaimed stunt pilot and co-owned a small aircraft production firm in the 1920s. Unlike Milch, who accumulated organizational and managerial experience during his term with Lufthansa and in other positions, Udet lacked the capabilities required to manage large and complicated organizations. Furthermore, although he was a capable aviator, he lacked formal technical training and was ignorant in most technical matters he dealt with as head of the Technical Office. Most members of his staff were trained technicians and professional engineers, but their impressive collective skills did little to compensate for their boss's shortcomings.[6] Udet's initial influence was immense, and he was a key figure behind the development and procurement of tactical dive-bombers, which played a crucial role in Germany's blitzkrieg success. On the long term, however, his appointment to this important position proved disastrous.[7]

One of the main problems facing the German armaments industry during the 1930s was financing. The industry was dependent on imports of most raw materials, so lack of foreign exchange could have severely limited Germany's industry on the long term and was arguably an important motivating force behind Hitler's decisions to push towards war. These restrictions influenced also the rearmament of the Luftwaffe and were one of the reasons for its inability to create a strategic bomber force.[8] The RLM was the principal provider of funds for aerial rearmament and for the expansion of the aviation industry, either directly or through bank credits guaranteed by the RLM. These arrangements were liberal enough to allow close cooperation between the government and the firms, making it possible to repay loans quickly and thus build up the ownership of the expended factories. This system brought a quick succession of takeovers of several aviation firms. During the 1930s the RLM became the owner of Arado and Junkers, Saxony's State Bank took over Erla, the Allgemeine Elektrizität Gesellschaft controlled Focke-Wulf, the Mitteldeutsche Stahlwerke consortium controlled ATG, and steel giant Krupp controlled Wesser. The management of the state-owned concerns, however, continued operations with little interference from the government.[9] From a scattered group of small private and independent companies directed mainly by their chief designers, some aviation producers turned into large firms owned by the state or by giant concerns. The booming aviation industry attracted other firms, like locomotive producer Henschel, train car producer Gotha and shipbuilding giant Blohm & Voss, which were not associated before with aviation production. These firms recognized the business potential offered by this booming sector and entered the aviation business.[10]

During these years of expansion and takeovers, some firms stayed under the direct control of their original owners. One of them was Heinkel. Ernst Heinkel, a stubby and energetic Swabian entrepreneur and aircraft designer, originally owned in the 1920s a single small factory in Warnemünde on the Baltic coast. Even before the rearmament, his firm won a solid reputation for its fast airliners and mail planes. Soon Heinkel was able to leverage his expertise in this field in order to win a significant military contract. In 1936 the RLM chose his design as Germany's new medium bomber. This contract brought Heinkel capital and state support, which enabled a massive expansion of the firm. Soon afterwards he established two new large factories in the Hanseatic city of Rostock, not far from his original factory, and in Oranienburg, north of Berlin. Besides expanding his firm's production capacity, Heinkel also started expanding its research and development activities. Since he was highly interested in high-speed flight, he approved and supported research of alternative propulsion technologies. Towards the end of the 1930s his sharp instincts and business-oriented approach, combined with enthusiasm for aviation technologies, made his firm the most important developer of jet and rocket engines. Therefore, at the outbreak of World War II Heinkel was not only a leading aircraft designer and industrialist, but also a pioneer of groundbreaking aviation technologies, whose test pilots already flew prototype jet and rocket planes.

Heinkel's success and expansion after the Nazis came to power exemplifies one of the main characteristics of the German aviation industry: It expanded to an unprecedented extent within a short time. The general mobilization plan prepared by the RLM in summer 1935 foresaw a Luftwaffe with strength of 2,370 aircraft by April 1938. Such strength meant production of around 18,000 aircraft in order to cover attrition, provide reserves and cover other needs. This plan turned the aviation industry almost overnight into the leading industry of the Third Reich's economy.[11] The following table demonstrates the massive expansion of this industrial branch during the '30s:

Year	Number of aircraft produced
1931	13
1932	36
1933	368
1934	1,968
1935	3,183

Top: Leaders and future antagonists Erhard Milch (second from left) and Ernst Udet (second from right) during the "Winter Feast of Aviators" on 16 December 1933 (courtesy Yad Vashem Archives). *Bottom:* A historical day. Ernst Heinkel (center), wearing a Nazi Party pin on his lapel, test pilot Erich Warsitz (left) and jet engine developer Hans von Ohain (right) celebrating in Rostock the successful first flight of the He 178 experimental jet plane on 27 August 1939 (courtesy National Air and Space Museum, Smithsonian Institution, SI 80-1894).

Year	Number of aircraft produced
1936	5,112
1937	5,606
1938	5,235
1939	8,295

Annual aircraft production in Germany, 1931–1939.[12]

It should be noted that the low figure for 1938 is misleading. Series production of several aircraft of the "second generation" started in 1938, and while their production was still being geared up, production of older types was terminated. As a result production figures dropped. This drop was therefore a sign of modernization and not of regression. The high figure of 1939 signifies the full production run of the "second generation" aircraft as well as the fruition of the huge investments in this industry since 1933.

Just before the outbreak of World War II, Udet sought to streamline aircraft production in order to expand production. Among others, aircraft firms were licensed to produce aircraft developed by other firms. It was done in order to make use of available idle production capacity of firms whose designs were not purchased in large numbers. Arado, for example, was licensed in 1938 to produce the He 111 bomber, the Me 109 fighter and the He 59 seaplane, producing a total of 346 planes designed by other firms.[13] Fieseler was also a rather minor manufacturer that produced mostly aircraft developed by other firms. Out of the total of 768 aircraft it produced in 1939 only 225 were of its own Fi 156 liaison plane; the rest were FW 58 and Kl 35 utility planes, and Me 109 fighters.[14]

The licensing scheme proved crucial during the war, as the number of types was steadily reduced while production rates gradually increased. An ever-increasing number of firms, including some general engineering firms, produced almost solely aircraft designed by others.

During the early years of World War II the basic economic and manpower limitations affecting Germany's aviation industry were aggravated by Udet's mismanagement. Germany entered World War II with a surprisingly low production rate. Throughout the first 12 months of the war, average monthly output of the aviation industry was around 800 aircraft.[15] At the same time England — then Germany's main enemy — fully geared up its aircraft production. While in 1939 Great Britain produced 7,940 aircraft compared to Germany's 8,295, in 1940 Great Britain produced 15,049 aircraft while Germany produced only 10,826.[16] These figures are quite surprising taking into account Great Britain's size, limited supply of raw materials and obstructive trade unions and bureaucracy. Germany's low rates reflected on one hand the limitations imposed by the economic reality of that time. On the other hand, although the military prepared itself for a long war, the extra production capacity of the aviation industry remained largely unused because of the blitzkrieg mentality prevailing in the Reich's leadership.

Another important and less-known factor was the general manpower shortage affecting Germany's wartime economy. Unlike the Allies, Germany's unemployment problem was practically solved by 1939, so no meaningful manpower reserves existed. The expanded draft aggravated this shortage. While in Great Britain the same problem was solved through massive employment of women and heavy cuts in the commodities industry, Germany failed to take similar measures. As suggested by Richard Overy, contrary to some basic assumptions, women employment in the German economy was much higher than thought, especially in the immediate prewar years. However, their redistribution at the beginning of the war favored other sectors than the armaments industry, like agriculture, although armament firms repeatedly requested more female workers.[17]

Therefore, largely due to manpower shortage, even after the outbreak of World War II, German aviation factories kept working on a one-shift-a-day basis, forty work hours a week.[18] The manpower shortage proved to be a crucial restrictive element in the entire German war effort. Ever-expanding army and industry, alongside inadequate allocation and redistribution policies, overstretched Germany's manpower reserves almost to the breaking point shortly after the beginning of the war. As wartime losses mounted and higher output was demanded, some unorthodox solutions were sought and found in order to solve this problem.

The general optimism of the German leadership concerning the outcome of the war (the blitzkrieg mentality) began to change only after the Battle of Britain and following the first months of the war with the Soviet Union. By that time the Luftwaffe's commitments had meaningfully expanded while production rates stayed painfully low, as the following figures show.[19]

Year	Total Aircraft Production
1939	8,295
1940	10,826
1941	11,776

The marginal difference between the figures for 1940 and 1941 clearly indicates the extent of the problem the Germans began to face. Considering the fact that by the end of 1941 the Luftwaffe was committed to operations over Great Britain, over the Atlantic Ocean, in the Soviet Union, over the Mediterranean and in North Africa, Germany failed to produce enough aircraft to support these commitments. Furthermore, production failed to keep up with attrition rates and almost all flying units operated in 1941 far below authorized strength. Projected production plans created by Udet's office remained unfulfilled and in official reports to the RLM they were corrected downward in order to hide the failure of production. The extent of the looming crisis is exemplified by the degraded status of the Luftwaffe's bomber force by the end of 1941. In December 1941 the Luftwaffe possessed only 47.1 percent of its authorized bomber strength, and of these only 51 percent were in commission. This statistic means that on 6 December 1941, on the eve of the United States' entering the war, the Luftwaffe had only 468 serviceable operational bombers, or only 24 percent of the authorized strength of this type of aircraft.[20]

Beside the failure to produce enough aircraft to support the expanding commitments of the Luftwaffe, the development and procurement of several new aircraft types went badly wrong. As part of the early streamlining drive, Udet and his staff strove to focus most of the production capacity in the production of four main types: the Me 109 as single-engine fighter; the Ju 88 as medium bomber; the He 177 as heavy bomber; and the Me 210 as twin-engine multi-role fighter. The plan was largely shattered as the last two aircraft proved to be costly failures.

The RLM awarded Heinkel a contract to produce the He 177 bomber in 1939. The aircraft flew for the first time on 20 November 1939. It was supposed to enter service in 1941, but due to serious technical problems with the prototypes, including two crashes, the

RLM halted work on the aircraft in early 1940. The main reason for the troubles was the effort to take a heavy bomber and make it a "jack of all trades." It turned out to be master of none. In addition to normal level bombing, it was also supposed to dive-bomb, perform high-altitude bombing, deploy guided weapons and even serve as a "destroyer"—a heavy fighter-bomber.[21] Another source of problems was the aircraft's unique coupled engines arrangement, which caused heating and engine fires. Heinkel redesigned the plane and submitted a new design several months later. The RLM reapproved the procurement of the aircraft, but the redesigned version was still unsatisfactory and its flight testing proceeded extremely slowly. Even after low-rate preseries production started in early November 1941, some 170 design changes were incorporated in order to rectify different issues.[22] Only in mid–1942 did first-series He 177s start leaving the production lines and even then in small numbers. It entered large-scale production only in late 1943. The He 177 and its troubled history are going to appear again later in this book.

The Me 210 was an even bigger failure. The RLM ordered 1,000 aircraft of this type even before it flew for the first time, but flight testing soon revealed grave problems with its flight characteristics. The RLM made another mistake by authorizing production of the aircraft before the problems were ironed out. Messerschmitt even optimistically licensed three other firms to produce the plane, but it soon became clear that the program was in deep trouble. The RLM suspended the production of the Me 210 on 14 April 1942 after only 184 aircraft were produced, using as a pretext Messerschmitt's failure to provide an improved version of the aircraft on time. Messerschmitt's losses due to this failure were estimated at around 38 million RM[23]; total costs of the failed program were estimated at around 200 million RM. Around 1,000 trainloads of 300 almost complete aircraft, manufacturing tools, parts and metal cut for 800 additional planes were sent to the scrap yard and the Luftwaffe lost almost 2,000 aircraft. Following the Me 210 debacle, Milch removed Willy Messerschmitt from his position as chairman and general director of his company; he nevertheless stayed chief designer and retained influence because of his prestige as a genius designer.

The He 177 and Me 210 affairs clearly demonstrated the complexity of manufacturing modern military aircraft and the cost of mismanaging their procurement.[24] As a result of these and other failures, German airpower in 1941 suffered a blow from which it never recovered.

The crisis of 1941 was the most important turning point since the reorganization of the aviation industry following the Nazi takeover. Although the industry was supposedly under the centralized control of the RLM long before 1941, in practice Udet failed to exercise this control and the producers were mostly left to do whatever they wanted. Waste of valuable raw materials was one problem that can be blamed directly on lack of tight control. Messerschmitt factories, for example, used valuable aluminum to build tropical shelters and ladders for use in vineyards.[25]

The crises in aircraft production came to a head shortly before the invasion of the Soviet Union, as Udet's office could no longer hide the growing disparity between planned and actual production. Göring reinstituted Milch's status and placed him above Udet with the responsibility to generally supervise and reorganize production and supply. Göring and Milch gradually sidelined Udet and he gradually lost his power. This humiliation, together with the realization of the chaos he had created, led to his suicide on 17 November 1941.

Meanwhile Milch made plans to fulfill Göring's order to quadruple aircraft production. In September 1941, even before Udet's death, he submitted a plan to produce 50,000 aircraft by March 1944. In order to achieve this goal he decided to impose tighter control on the industry, forbidding the development of new types (although contrary to common belief, he authorized continued development of rocket and jet aircraft), put a stop to raw material waste by the factories, and strictly obliged factory mangers to provide their allocated quotas.[26]

Centralization and streamlining became the order of the day and things began to improve slowly. However, as the main postwar American report on the German aviation industry pointed out, under Milch's reign tight control sometimes caused even more chaos. One of the most spectacular examples of this state of affairs is the changing of production plans for Henschel's factory in Schönefeld near Berlin. In 1940–1941 it produced Ju 88 bombers under license from Junkers. In 1942 the RLM authorized it to produce Henschel's own Hs 129 ground-attack aircraft. When production tooling was about 50 percent completed, the RLM canceled the program and the factory was ordered to prepare tooling for the production of the Ju 188 bomber (a modified version of the Ju 88). This retooling was completed, but before production started, the RLM decided that the Me 410 multi-role fighter (successor of the failed Me 210) should be produced at Schönefeld at a rate of 400 planes a month. The story was not over yet. After 8 months, the tooling up was 80 percent complete, but the order was canceled and Henschel was ordered to produce the Ju 388 bomber (yet another Junkers medium bomber). Altogether some 3,000,000 to 4,000,000 man-hours had been expended on tooling for this project when bomber production was canceled in summer 1944. The factory ended the war producing wings for the night-fighter version of the Ju 88 — the very same aircraft it produced under Udet.[27]

Milch spent most of 1942 trying to repair the damage caused by Udet and the rogue industry under his authority. In retrospective he fought a losing battle, but at that time the damage looked reversible, mainly due to the availability of large amounts of unused capacity within the aviation industry and due to emerging revolutionary aviation technologies that started to appear as flying prototypes. Many crucial decisions which had an impact on late war production were made at that time, but these could not change the fate of German airpower. As airpower historian Williamson Murray remarked: "Between July 1940 and December 1941, the Germans lost the air war over Europe for 1943 and 1944."[28]

The Aviation Industry and German Society

The expansion of the German aviation industry did not take place in an empty space. It took place within a society under a new political system and at times of important social and economic changes. It is important to describe here the prewar social character of the aviation industry in the Nazi era in order to understand the deep changes it went through during the war, and especially in its later phases.

As a rather new industry, which boomed in a matter of a few years, the aviation industry influenced German society in several respects. Generally, it started expanding into areas stricken by unemployment and suffering from a relatively low standard of living during the economical crisis of the Weimar Republic. The process of growth brought immediate economic, demographical and social changes to the communities around the factories. Furthermore, this industry quickly developed a unique self-consciousness. It viewed itself as an elite industrial sector, producing cutting-edge modern machines.²⁹ As such, it can be viewed as equivalent to today's high-tech industries, and as many firms of this branch do, the aviation industry also sought to implement new ideas concerning working environment, standard of living, special benefits, employer-employee relations and workplace design.

Immediately before the outbreak of World War II there were 27 aircraft factories and 16 aero-engine factories in Germany. Most of these factories were located in the proximity of cities and within their public transportation network.³⁰ Originally the main factories were located in specific cities. Messerschmitt's main plants were located in Bavaria, in Augsburg and Regensburg; Focke-

Ideal worker. A prewar image of an aviation industry worker as portrayed in a Junkers publication (courtesy U.S. National Archives and Records Administration).

Wulf's center was in Bremen; Junkers' in Dessau, Fieseler's in Kassel and Heinkel's in Rostock. The expansion of the aviation industry meant that in a matter of several years aviation firms became leading employers in these cities. From an insignificant total of 3,988 workers in January 1933, the number of people employed in the aircraft and aero-engine industry jumped to 293,000 by October 1938.³¹ Some firms expanded rapidly from humble workshops or small design bureaus into industrial giants. Messerschmitt AG in Regensburg had in 1937 only 530 employees on its payroll. Two years later and after winning a contract to produce the Luftwaffe's new single-engine fighter, the firm employed 4,580 workers. In 1944 Messerschmitt's Regensburg complex alone employed 14,508 wage earners and salaried employees.³² Erla, a minor engineering firm in Leipzig, had in 1934 exactly 111 workers on its payroll. One year later, after the company entered the aviation business, this number increased to 698. By 1939 Erla employed 5,821 workers in its two factories in Leipzig, producing the Me 109 fighter under license. This expansion continued during the war and by 1943 the firm employed a total of 24,991 men and women; 8,959 of them were Germans, and the rest were foreigners, POWs and concentration camp inmates.³³

The story of Taucha, a town in Saxony, exemplifies the industry's growth trend. The population of Taucha, as registered on 1 April 1933, numbered 7,388 people, of which 2,019 were unemployed. When the Mitteldeutsche Motorenwerke (MIMO) firm opened an aero-engine

Ideal factory. Junkers' Dessau main factory before World War II. Note the Bauhaus design of the building (courtesy U.S. National Archives and Records Administration).

factory near the town in 1936, it created 1,544 new jobs. This number grew to 4,789 by 1939. The pace of expansion was so quick that during 1938 the number of workers in the factory doubled from 2,391 to 4,573 and at the end of the year some 7.6 percent of the workers of the entire aero-engine industry lived in Taucha. Yet, the factory was still short of skilled workers. The management therefore launched in 1939 a training scheme for skilled and unskilled personnel. In early 1939 some 33 percent of the factory's workers were skilled, 27 percent were trainees and 25 percent were unskilled. The training courses retrained almost one third of the workforce — around 1,300 men — from other professions.[34] It should be taken into account that the expanding Erla aircraft factory in nearby Leipzig also hired workers from Taucha, thereby further underlining the dominance of the aviation industry in this region. The advantages the MIMO factory offered to the unemployment-stricken Taucha are obvious, even though large numbers of the employees came from other places or were brought in by the firm from its other branches.[35] Furthermore, just like the present day's high-tech industry, the aviation industry functioned in Taucha and other places as a locomotive pulling the rest of the local economy. Therefore, MIMO and Erla contributed to the local economy also in an indirect way.

The aviation industry also started to shape the physical face of its associated cities, and Taucha also exemplifies this trend. The expanding industry stretched the housing capacity of Taucha to its limit, so in 1936 MIMO acquired 250 homes and 220 furnished rooms for its employees. The RLM provided most of the funds for these acquisitions, and MIMO complemented the rest from its own budget. In order to prepare the ground for further expansion, MIMO started its own housing projects. A total of 1,116 homes were constructed in Taucha specifically for the employees of the MIMO factory between 1936 and 1939.[36] MIMO planned to construct 527 more homes in 1939, but due to the outbreak of the war, only 22 of them were completed and 29 were left uncompleted. Bombing raids on the Taucha factory destroyed 48 firm-owned homes by the end of 1944.[37]

Housing projects initiated by the RLM and the aviation industry appeared all over Germany. Between 1935 and the end of 1940 the RLM authorized the construction of 34,150 homes in towns and cities associated with the aviation industry. The ministry's housing construction budget for the fiscal year 1939–1940 projected the investment of 40,000,000 RM in the construction of 15,000 homes and houses. The biggest housing projects of 1935 were Junkers' settlement in Dassau with 307 homes and 205 cottages, and Heinkel's complex of 3-story apartment buildings with a total of 588 apartments in Rostock.[38]

Junkers, another firm that expanded rapidly at that time, also exemplifies this wave of prosperity. This aircraft producer — one of the few that had prospered in the Weimar Republic — became a giant of the German aviation industry in 1936 after merging with its aero-engine-producing sister firm Junkers Motorenwerk (Jumo). Junkers' original factories were grouped to the south of the Magdeburg metro area. The expansion of the aviation industry, and particularly of Junkers, was the main cause of a 222 percent growth in employment in the Saale-Elbe-Erzgebirge region between 1933 and 1939. In 1936 Junkers employed in the region between Magdeburg and Leipzig almost the same number of people as IG Farben — the well-established giant petrochemical concern. Wherever Junkers factories expanded, their neighboring cities and towns profited in terms of reduced unemployment, influx of new residents, construction of homes and apartment buildings, and general economic prosperity.[39]

Sometimes efficient public transportation systems made it unnecessary to relocate workers or to bring new ones from other places. Henschel's two new aircraft manufacturing factories near Berlin, for example, relied on the region's public transportation network and on Berlin's expanding S-Bahn network for the daily commuting of its workers from around the region. Although some workers spent each day a total of around 4 hours on the road, in most cases public transportation in this region was quite efficient. The Berlin public transportation system proved so successful in saving the need to construct new housing that other bomber producing firms moved their production facilities to greater Berlin. One of them was Heinkel with its new factory in Oranienburg, north of the capital.[40]

Heinkel stood not only at the forefront of cutting-edge aviation technologies and expanding business. The firm also led a trend of designing a new style of factories and creating a friendly working environment for the workers. Heinkel's original main factory at the Marienehe suburb of Rostock employed around 400 workers when Hitler came to power. During its rapid expansion in the following years this plant was rebuilt and expanded as a model plant.[41] Heinkel hired Herbert Rimpl, a young and unknown architect, who was willing to implement his boss's visions regarding the design of the factories of the future.

Opposite: Ideal living environment. Scenes from a housing project Junkers constructed for its workers in Dessau (courtesy U.S. National Archives and Records Administration).

Straße in einer Junkers-Siedlung

Gesunder Nachwuchs

Heimstätten im Reihenhausbau

Eigenheime im Doppelhausbau

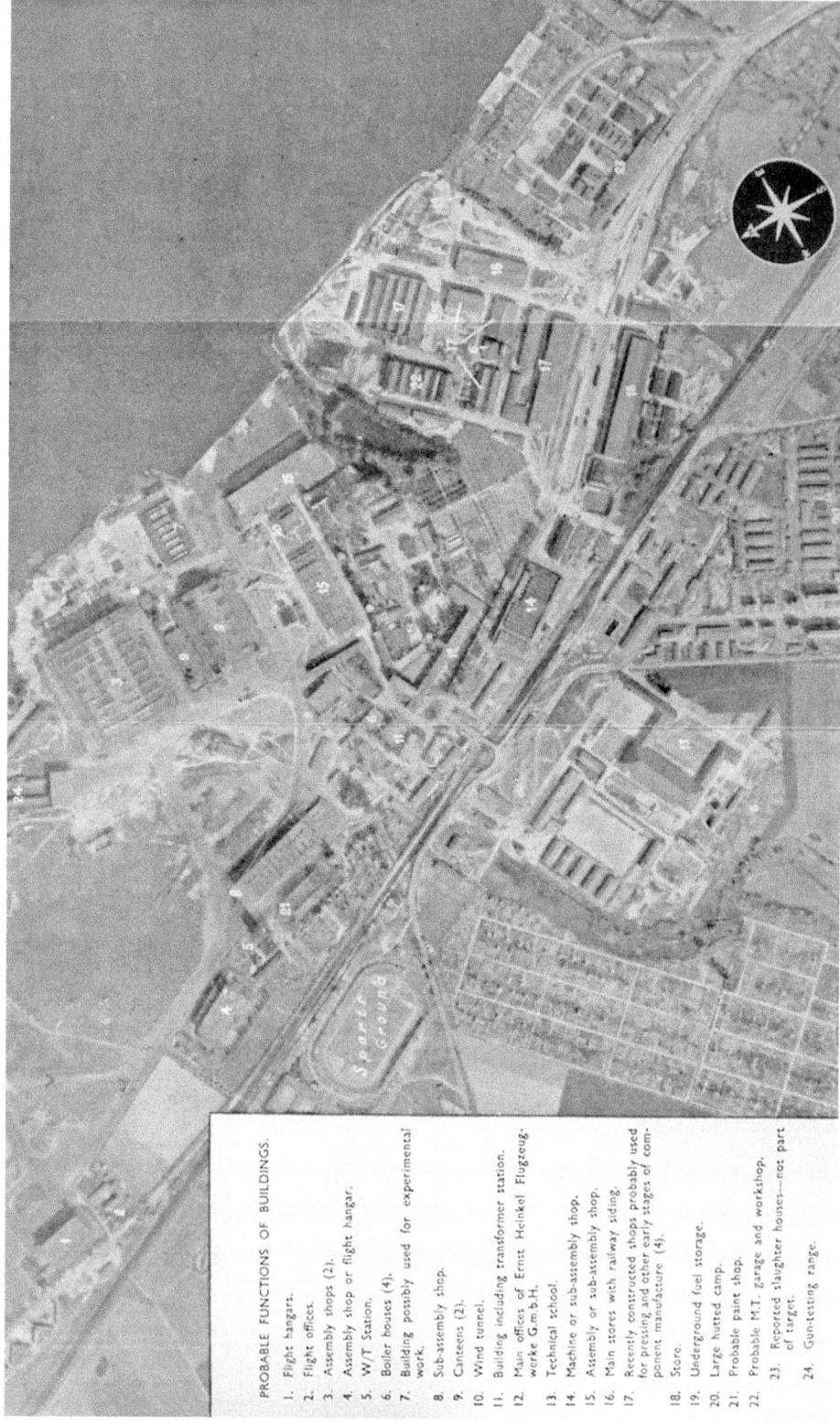

Heinkel was generally interested in health, particularly in kneippism (hydrotherapy), and he wanted to provide his "plant community" (*Betriebsgemeinschaft*) different health-care services. Therefore, the new Marienehe complex included a health-care and recreation center, and an assortment of other health-care facilities. A factory doctor was hired and functioned also as the plant's hydrotherapist. The plant offered medical and dental plans to its employees and their families. A health education program was also included in the firm's health-care plan. In order to enable the plant cantina to serve healthy food, Heinkel bought extra land nearby, and established an organic farm. It even produced its own honey. A flock of sheep was used both for supplying the cantina with organic meat and for grazing the grass of the factory airfield. Following his initiative for healthy nutrition education, the employees' mess hall at Marienehe served a healthy vegetarian meal twice a week. Idyllic apartment buildings and houses were constructed near the factory. These houses and their surroundings represented Heinkel's vision of healthy modern living: a mix of modern architecture with countryside tranquility.[42] Heinkel's management style was markedly paternalistic, especially when it came to welfare and education. He was highly interested in social matters and regularly read reports about different social aspects of his firm. The education department at the firm's headquarters in Rostock was placed directly under him.[43]

After Heinkel won the contract for the production of the Luftwaffe's medium bomber in 1935, he decided to establish a new factory in Oranienburg. The new plant included features of the renewed Marienehe plant and much more. Besides modern production and assembly halls, using Bauhaus design features, it offered its workers a health center with swimming pool and spa, sport field, gym hall and a modern employees' restaurant. Once again, an agriculture farm was established near the plant in order to supply it with fresh and healthy organic food products. Two housing complexes were constructed for the workers with social welfare and administration offices nearby. The plant included a special campus for trainees, complete with modern dorms. Looking today at the idyllic pictures of the pre-war Oranienburg plant one cannot avoid thinking of today's high-tech industry and its ultra modern complexes.[44]

Other firms soon followed Heinkel's example, especially by establishing their own modern housing projects, health-care schemes and recreation facilities. Focke-Wulf constructed in 1936 in Bremen some 95 cottages for its workers and in the following year another 79. By 1941 it bought or constructed 1,059 homes and offered them to its employees at affordable prices. Focke-Wulf's housing projects were constructed in a country style and were surrounded by parks and gardens. The firm also offered its workers subsidized leisure activities and tickets for cultural events, some of which took place inside the plant. An employees' orchestra was established and performed on many occasions.[45] The factory established its own medical department in 1939, which offered medical services to its "community." In 1944 some 100 people worked in this department. In addition to the medical center in the Bremen factory, Focke-Wulf also owned a recreation facility in Tirol (purchased in 1941), a health-care center in North Germany and a hospital in the eastern provinces.[46]

Opposite: **Housing for the masses. An Allied aerial photograph of Heinkel's Rostock-Marienehe factory taken on 29 July 1943. Visible south of the sports ground (left edge) is the suburban housing project Heinkel constructed for its workers. To its right is an area populated initially by foreign workers and then by slave workers (courtesy U.S. National Archives and Records Administration).**

The good life. Workers at the recreation center in Messerschmitt's Augsburg factory (courtesy U.S. National Archives and Records Administration).

State-owned Junkers also fully incorporated elements of modern factory design into its older and newer plants. It offered its workers a wide range of health-care programs and recreation opportunities. Its new expanded factories were neatly landscaped and the workers received hot and healthy meals in the cantinas. Junkers was also engaged in a massive housing construction scheme around its factories in order to provide its workers cheap and convenient housing close to their workplace.[47]

Sport was an important element in these new model factories. Junkers organized sport groups and clubs in all its factories. Although all kinds of sports were available, it seems that soccer was most popular: in the main Dessau complex alone there were 41 soccer teams by 1937.[48]

Focke-Wulf established its own sport club in June 1936, and the firm's first Sport Day took place in July 1938 in Bremen. Besides sports, this club also organized outdoor hikes for the workers.[49] The firm continued to encourage its employees to take part in the firm's sport activities well into the war. The central organization of the firm's sport activities was dispersed along with the Focke-Wulf factories in 1941. From this stage each factory established its own sport department, which was independent from the main department in Bremen. It seems, however, that as the war progressed the employees cared less about taking part in organized sport events. This negligence has led the management of the Sorau factory to publish on 28 June 1943 a leaflet, in which the employees received the following message:

"Dear work comrades [addressed both to men and women]! We can not help but emphasize expressly that the *establishment of regular sport activity at the Sorau plant* is not just for certain groups of 'active' and 'competitor' sportman, but meant *for all workers!*" [emphasis in original].[50] The employees were told that it was their duty to do more sport as workdays were becoming longer and therefore physically more demanding. The workers were therefore ordered to participate at least once a week in evening sport activities organized by the factory. These activities started usually at 8 PM and took place in the gym hall or at the sport field. Even foreign workers were allowed to take part in special sport sessions organized for them once a week.

Organized sport activities were used not only to keep the workers fit for physical work. They also fulfilled a social function in a broader sense. One of the most important functions of the aviation industry in the Nazi era was social integration into the new social and political system. Subsidized housing, organized leisure activities, and workplace-based health care formed important parts of this function. These new measures were not forced upon the aviation industry from above, but seem to have been fully adopted by most of its chiefs and carried out enthusiastically on their own initiative. These measures, at the same time, served the firms' own corporate interests, because through implementation of new ideals, factories sought ways to increase the motivation of their workers, leading to better productivity.

Junkers was one of those firms heavily engaged in social activities and social "engineering." Heinrich Koppenberg, Junkers' general director, emphasized the need to create a strong feeling of "plant community" among the workers. From 1935 Junkers' directorate sought to create this spirit by organizing festivals, factory parties, internal public relations campaigns and exhibitions. The purpose was to create a "Junkers spirit," intended to give the workers a new feeling of belonging and to emphasize the unique value of their work. An important organ of this effort was the firm's official newspaper, the *Propeller*. Besides normal firm news and announcements, it published articles about issues concerning workplace and lifestyle, and it published the social activities of the firm and the way it was taking care of its workers.[51] Another firm magazine dealing with similar issues was titled *Junkers Nachrichten*. The social goals of Junkers since its expansion began in 1933 were described in an article titled "Mensch und Werk" ("Human and Work") in 1939:

> It was necessary, in the first place, to create a set of regulations, the *Betriebsordnung*, which may be regarded as a codification of duties and rights of the two parties to the agreement: employer and employees. It was the first document of this kind in Germany to be officially acknowledged and legally approved. It is based on three fundamental conditions to be fulfilled in every working community: honor, loyalty and team spirit. It welds the employer and his *Gefolgschaft* (following) into a single unit.[52]

The new order at Junkers reflected the new national-socialist order in Germany with its "national community" ideal and leadership principles. Junkers viewed this system and regulations as a new dawn in the way factories and firms worked:

> Our firm community is made of a *Betriebsführer* (factory leader) and *Gefolgschaft* (following). The earlier conflicts between the so-called *Arbeitgeber* (employer) and *Arbeitnehmer* (employee) are eliminated when *Führer* (leader) and *Gefolgschaft* are reaching hands for co-operation and creating an *Arbeits- und Betriebsgemeinschaft* (work and factory community). Through this provision we transfused the basic ideas of national-socialist *völkischen Kampf- und Lebensgemeinschaft* (folkish struggle and life community) into our firm."[53]

Nazification. Signs at the entrance to Junkers' Dessau factory. The upper sign reads: "Slogan of the week: Better to give up your life than to lose your honor." The lower sign reads: "This company stands man by man and united in the German Labor Front" (courtesy U.S. National Archives and Records Administration).

This new German dawn spread also to the Austrian aviation industry, which became part of the German aviation industry after the annexation of this country in March 1938. The most important aviation producer in Austria was the Wiener Neustädter Flugzeugwerke (WNF) firm. Soon after the German takeover this firm began to receive generous support from the RLM, which intended to use it as one of the main producers of the Me 109 fighter. The firm's main factory in the city of Wiener Neustadt was completely rebuilt and turned into an "exemplary national-socialist factory" (*Musterbetrieb*), largely along the lines set by the "Beauty Through Work" (*Schöneheit durch Arbeit*) organization.[54] Like its German equivalents, in its new form this factory also included a swimming pool and a sport stadium.[55]

It is clear that German aviation manufacturers sought not only to produce better aircraft during the Nazi time, but also to adjust themselves to the national-socialist ideology of the new regime. Returning to the good old lifestyle while modernizing was a striking feature of Heinkel and other firms and it represented Nazi *Volk* ideology at its peak of development.[56] Ernst Heinkel was an ardent supporter of National Socialism, as is evident from his speeches and dispatches to his factories "community." In December 1935, for instance, he openly praised Hitler, his new regime, their new values and their actions in his New Year's dispatch to his workers.[57]

Because the aviation industry had expanded so rapidly since the Nazis came to power, it is obvious that most firms — even those with less enthusiastic supporters of national-socialism at the top — sought to implement the new ideals concerning lifestyle, family values, health, hygiene and leisure in their new and expanding communities. "Community" was the keyword in those prewar years. By creating a so-called *Betriebsgemeinschaft*, each factory sought to structure the life of its employees in order to bring them in line with the new ideals. As a Junkers publication put it, under national-socialism it became the duty of the firm and the factory to take care of the workers' needs in all matters, any time and everywhere.[58]

As Nazi ideology preached a healthier lifestyle, aviation firms sought to construct "green" factories and countryside-like housing estates for their "communities." By putting most of their workers in one place the factories strengthened the sense of community prevailing among their workers and created a stronger bond not only among them, but between them and their workplace. After all, aviation firms were profit-oriented businesses, and in many respects Nazi community-building and social welfare schemes promoted their own interests — even during the war. Health care provided by the plants, for example, helped to control absenteeism on medical grounds. Historian Lutz Budrass viewed the establishment of health care and health insurance plans as one of the most important means used by the aviation companies to "discipline" (*disziplinieren*) their workers. Besides helping them fight absenteeism, it also enabled them to keep an eye on their workers and locate troublemakers and "strollers."[59]

Firm-based health care also reflected an ideal of social welfare and community function embedded in the aviation industry of the Third Reich. Junkers defines again the task at hand: "[The] individual is the central and ultimate object of all social work, the most important task of which is the care for his health. All our works enlisted medical staffs for the purpose of regularly monitoring the physical condition of their employees and workers."[60] In this context sport and state-organized vacations served the same ideals and the same purpose. Dr. Herbert Warning, chief doctor at Focke-Wulf's Bremen factory, clearly expressed in 1939 the main idea behind such schemes: "Our work tempo is very fast. The *Lebenskampf* (life struggle) of our nation is extremely hard. We are therefore *obliged* [emphasis in the original] to make our vacation a true source of relaxation."[61]

Nazi organizations acknowledged and praised several aviation factories for their contribution to the Nazi state and to the creation of the "national community." Focke-Wulf's Kamp-Siedelung housing complex in Bremen was declared by *Gau* Wasser-Ems in July 1941 as a "model workers' settlement."[62] Arado's Brandenburg factory was awarded a "*Gaudiplom* for outstanding performance" in 1938, and the Babelsberg, Warnemünde and Rathenow plants of the same firm soon won the same award. Local *Gauleitungen* also praised the Rathenow and Warnemünde factories for their excellence in professional education, healthy

The good life amidst ruins. The swimming pool and recreation center at Messerschmitt's Augsburg factory after bombing in 1943 (courtesy U.S. National Archives and Records Administration).

living environment, strong workers' community, and efforts to promote "strength through joy." Arado's Wittenberg plant was declared on May Day 1940 a "National-Socialist Model Undertaking" (*Nationalsozialistischen Musterbetrieb*) and Rathenow and Warnemünde were also declared as such on May Day of the following year.[63] Daimler-Benz's new aero-engine factory in Genshagen was awarded the prize of "National-Socialist Model Undertaking" in 1939. The award especially recognized the social achievements of this factory within the few years of its existence. The award mentioned specifically the luxurious workers' settlement financed by the firm and constructed around the factory.[64]

The booming aviation industry was therefore not only a trailblazer of modern technology and industrialism; it was also at the spearhead of the implementation of a new Nazi social order. The expanding aircraft factories sought to fully embody the National Socialist vision of changing industrial societies. They fully represented the work environment and conditions required for the transition to a modern industrial society. They also sought to transform the worker of the Third Reich from the miserable proletarian of the Weimar Republic to the self-conscious and proud industrial specialist of the "national community."[65] The close connection between aviation factories and National Socialism was well summed up in the closing remarks of an article published in the Junkers newspaper:

> After all, it is the spirit behind an organization that directs all its moves. True team spirit can develop only where the employer and employees form a community in pursuance of a common goal. It is this ethical foundation of the social work which was recognized by the award of the Golden Banner, a distinction which is bestowed by the Führer on National Socialist model undertakings.[66]

Teenager Josef Kellenberg from Sandbach started to work at Messerschmitt's Regensburg factory in 1939 and only shortly after finishing his apprenticeship. He recalled after the war:

> The hourly wage at the beginning was 48 Pfennig, which meant for me, in a 48-hour week, a good 23 RM per week. This was more than sevenfold what I was previously paid [as a machine builder with an agricultural machines manufacturer]. Not only that, I was able to work in a bright *Halle* engaged in final checks on the Me 109, and even in my search for a dwelling—I lived in a single room in the Ludwigstrasse—the plant rendered assistance. The social amenities were exemplary. In any case, I was very proud to be working for Messerschmitt. When I look back at it today, my activities at Messerschmitt in Regensburg and my later period in the Luftwaffe, seen from the specialist aspect, were the high points in my working career."[67]

These prewar visions, however, remained largely unfulfilled. The higher-than-average wages in the aviation industry, combined with its elite spirit and the high self-esteem prevailing in it, set it apart from the Volksgemeinschaft. The disparity between the workers in this industry and the workers in other areas of the economy was one indication that the Volksgemeinschaft was an illusion.[68] Furthermore, during World War II the manpower composition of the aviation industry changed rapidly and dramatically. Far from the supposedly exemplary element of the Volksgemeinschaft, it represented more and more a perverse sort of *Völkergemeinschaft* (a community of nationalities). The ideal of a highly motivated and healthy national-socialist society eroded rapidly in the face of wartime reality.

From Workbenches to Production Lines—Production Methods

During World War II the production process of aircraft gradually changed. This change should be viewed in a broader context of general changes influencing aircraft production methods. As the growing aviation industry received an increasing number of contracts following Hitler's rise to power, it looked for ways to expand its production capacity. Like the rest of the German armaments industry, the consequent increase of output was achieved not through the introduction of modern production technology, but through "internal streamlining," based on standardization, simplification of the products, bigger production facilities and longer working hours. The production process and production technology remained basically unchanged. The streamlining processes culminated in the attempt of state-owned Junkers to mass-produce the Ju 88 fast bomber on older-fashioned but massively expanded production lines.[69]

As historian Richard Overy pointed out, German prewar aircraft production was based largely on handwork preformed by a relatively small guild of expert artisans on static workbenches. The workers were trained technicians, well aware of their high status as expert craftsmen in an advanced branch of the industry. A professional production technician was trained for 4 years as an apprentice before gaining his status. Although aviation technology made swift progress in the prewar years, production methods remained largely static. Unlike the automobile industry of the 1930s, there was little mechanization of the production process in the aviation industry, and conveyor-belt-based production lines were nonexistent.[70]

This pattern of work restricted the expansion potential of the aviation industry. Since the production process was divided into a relatively small number of segments with several manufacturing tasks being preformed in each of them, the workers were required to possess detailed knowledge of their trade and high technical skills. Due to the long training period for technicians, it was difficult to provide the expanding industry with enough skilled manpower. Reluctance of the trained craftsmen to share their knowledge with trainee workers only aggravated the problems associated with prewar expansion. Workbench-based industry therefore meant a dead end when it came to expansion and mass production, as required under the circumstances of a total war.

Manufacturing a World War II aircraft was complicated. As renewed and experienced industrialists, like Henry Ford, failed to grasp, it was more complicated than car production and much different.[71] The normal workflow of a basic aircraft production was as follows[72]:

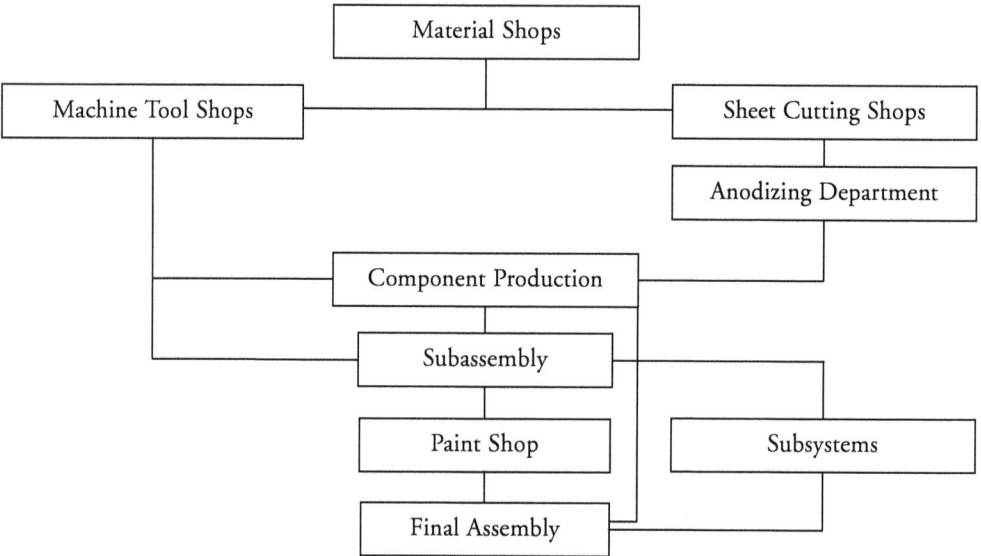

Subcontractors supplied subsystems, like engine, undercarriage, instruments and armament, that were added to the components or to the finished airframes. The plane was test flown, and upon the successful completion of the test flight the aircraft was delivered to the customer.

As part of Udet's rationalization program of 1941 and as part of worldwide trends in aviation production, the German aircraft industry started converting the old workbench system into modern conveyor-belt production lines (termed *Fliessband* or *Fliessstrasse*). This conversion of existing production facilities into mechanized production lines was a lengthy process. As late as February 1943, Milch still emphasized the need to establish conveyor-belt production lines in the aviation industry in order to match the high output of Soviet aviation production.[73] Conversion meant in most cases long production stoppages and delays, so factory managements tended to repeatedly postpone them. As a result, factories continued to produce aircraft using the old method well into the war. Even factories involved in several central production programs, like the production of the Me 109 and FW 190 fighters, converted to conveyor-belt production lines only during the second half of 1943.[74]

The Allies were quicker in adapting modern mass-production techniques and implementing them in aviation production. Both Great Britain and the USA benefited from mobilizing large parts of their automobile industry for aircraft production — something that

The old system. Production of Ju 86 light bombers at Junkers' Dessau factory before World War II. Work on the fuselages is being performed in parallel and not in line (courtesy U.S. National Archives and Records Administration).

the Germans largely failed to do. Although several German car manufacturers, like Opel, produced aero-engines under license, the only car manufacturer to become seriously involved in aircraft production was Volkswagen. The western Allies surpassed the Germans not only in terms of participation of the car industry in aviation production, but also, and maybe more importantly, in terms of organizational influence. In Great Britain William Morris, Lord Nuffield, owner and director of Morris Motors Ltd., and pioneer of inexpensive mass-produced cars in Britain, played a central role in the prewar "Shadow Factories" scheme. This scheme, started in autumn 1936, sought to massively expand British aircraft production within several years in order to face the emerging fascist threat. In May 1938, with the gathering war clouds on the horizon, the Air Ministry asked Morris to apply his car-manufac-

Old-fashioned production line for Ju 88 wings at Volkswagen's Fallersleben factory, probably 1942 (courtesy U.S. National Archives and Records Administration).

turing expertise to aircraft production. He established the first of six large "shadow factories" at Castle Bromwich, near Birmingham, taking advantage of the presence of several car manufacturers in this area and therefore of the availability of a relatively large skilled workforce. The factory used production lines similar to those used by the car industry. By 1940 Morris's factories became the main producers of British fighters and their engines.[75]

On the other side of the Atlantic, application of automotive methods to aircraft manufacture was carried even further, but less successfully. Car producer and mass-production pioneer Henry Ford offered the U.S. government in May 1940 the production of one thousand aircraft of standard design a day.[76] Both the American and British governments, shocked by the quick fall of France, soon started to consult him. Ford enjoyed enormous prestige as a car manufacturer and he also possessed some experience in aircraft production. His greatest rival, General Motors, also offered to produce aircraft before the United States entered World War II. Both firms soon led a comprehensive conversion of the American aviation industry while converting some of their own plants to aviation production. The conversion of the car industry into aviation production was far from smooth. Aircraft and aero-engines were much more complicated machines than cars and automakers encountered numerous problems when they started to mass-produce aircraft with their existing machinery and production lines. Furthermore, by definition the car industry was less flexible, since it was geared to manufacture standard civilian articles and was unable to incorporate the frequent changes of design that typify military production.[77]

Nevertheless, the influence of Ford and Nuffield was crucial in pushing the aviation industry to adopt the more efficient production methods of the car industry.[78] The American car industry reached its peak wartime efficiency with Ford's one-mile-long and 40,000-worker-strong Willow Run Bomber Plant in Ypsilanti, near Detroit. In March 1944 this factory produced fourteen B-24 heavy bombers, each made from 1.25 million parts, per day. Ford achieved this rate by redesigning the bomber for ease of manufacture and by creating a larger number of production breaks. Ford divided the aircraft into 73 subassemblies, while its original manufacturer used only 20.[79] Willow Run became a legend, but in contrast to what was portrayed at that time, it took it around two years to reach reasonable output. The plant suffered from severe problems during the critical years of 1942–43, leading to the nickname "Will it run?" In early 1943 it even came under congressional scrutiny due to its failures to reach satisfactory production rates. Ultimately it exemplified what could be achieved by using modern production lines, but it also proved that "building a bomber was not at all like building a car."[80]

Eventually, Willow Run was by no means representative of American aircraft production during World War II. More than 90 percent of all airframes, nearly half of all engines, and a substantial proportion of components were manufactured by the established aviation industry.[81]

The U.S. and international press widely publicized Willow Run after its opening in 1942 in a mostly glorifying way. Unaware of the fact that the factory was at that time in deep trouble, German leaders and managers viewed it as an example of what can be achieved using modern production technologies. Milch was the most energetic protagonist of converting the German aviation industry into the so-called *amerikanische Produktionsweise* (American production method) and viewed Ford as a role model for industrial efficiency. In Germany, Volkswagen was the only car manufacturer that produced aircraft, but it went into this business only because of the failure of the prewar *KdF-Wagen* (the Volkswagen car) scheme and because of its failure to offer a military version of this car before the war. Eventually, Volkswagen produced only subassemblies for the Ju 88 light bomber, munitions, wooden drop tanks, and a unique type of aircraft, namely the V-1 flying bomb.[82] It is quite obvious that Germany's modern car industry had little or no influence on aviation production. It was Willow Run, and not Volkswagen's Fallersleben factory, that served in 1942 as a model for a scheme of giant modern aviation factories. The scheme was born from the RLM's notion that massive increase of output would be possible only in newly constructed factories. Milch foresaw a quadruple output increase in both airframe and engine production with these state-financed factories, which the RLM planned to allocate eventually to different firms.[83] The large-factories plan reached its climax with the fantastic "thousand bomber factory," code named "Ultra." Plans for such a factory were first submitted in mid-1942. Junkers won the contract to construct this plant. With a floor space of 600,000 square meters it was supposed to be almost twice the size of Willow Run. Initial surveys were made in early 1943 in a proposed location in Oels, in Lower Silesia. The plant was supposed to use the most modern "American" production methods and bring to new heights the rationalization sought by Milch since replacing Udet. Construction of the plant was constantly delayed due to shifts in production priorities and it was finally canceled at the end of 1943.[84]

Efforts to convert existing aviation factories to modern mass-production methods continued independently of "Ultra." The main difference between the old production methods and the conveyor-belt production lines was the workflow. The production process was

A model of modern production. Final assembly of B-24 bombers at Henry Ford's "Willow Run" factory. Although a model of streamlining and efficiency, "Willow Run" also proved that producing aircraft was much more difficult than producing cars (courtesy National Air and Space Museum, Smithsonian Institution, SI 90-13145).

sectionalized into a large number of stations called *Takte*. Assembly of the main component began on one end, and parts and components were added as it moved along the line on a conveyor belt or cradles, until a complete product appeared on the other end of the line. The line moved at a predetermined pace, thereby dictating the time it took to accomplish the work at each station.[85] During the later part of the war the Germans constantly increased the number of stations along the line and therefore decreased the amount of work done in each of them. On 7 April 1945, for example, Messerschmitt's main office in Regensburg sent a new production plan to the "Bergkristall" underground factory, which operated a 13-station Me 262 fuselage assembly line. It stated the intention to increase the number of stations to 22 in the near future in order to bring this factory closer to the new standard of a 30-station production line.[86] By that time between one and five workers worked on each station of the fuselage assembly line. Most of the stations were operated by two workers.[87]

Perhaps the most important benefit offered by this fine division of the manufacturing process was the lower skill required from each worker. The workers on each station were required to know only the specific manufacturing task done on their station. Since only a

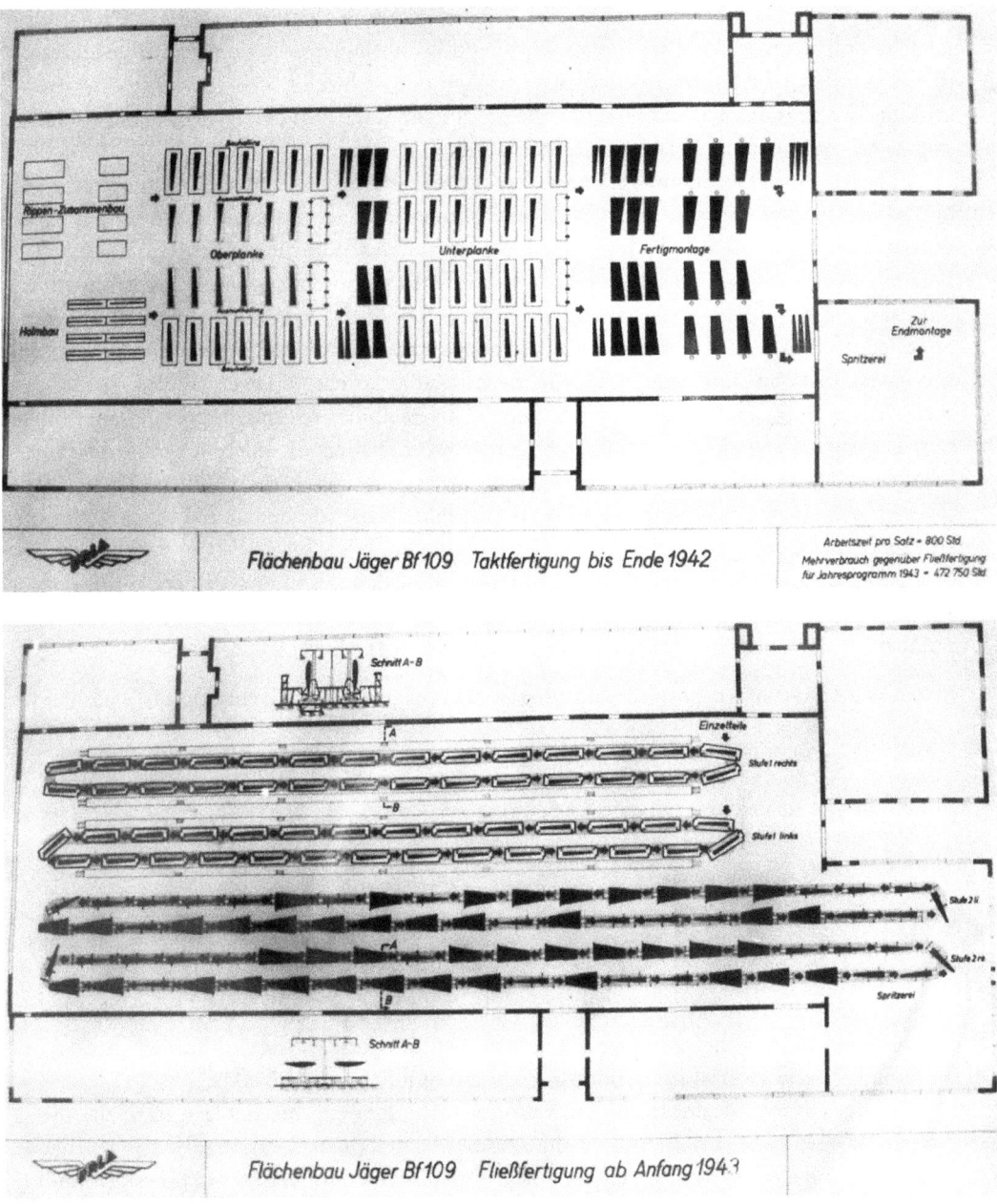

Top: Workbenches. An Erla diagram portraying the workflow of an old-fashioned production line for Me 109 wings at the end of 1942. The workflow begins to the left and goes through the following stations: rib and longeron construction, upper surface, lower surface, final assembly, paint shop and out to airframe final assembly (courtesy U.S. National Archives and Records Administration). *Bottom:* Conveyor belts. An Erla diagram portraying the same production line after conversion in early 1943. Four conveyor belts were installed in the same production hall (courtesy U.S. National Archives and Records Administration).

small amount of the whole manufacturing process was done in each station, only a narrow range of skill was required from the workers working on those production lines. The increase in the number of stations enabled the employment of briefly trained workers or even unskilled workers on each station along the production line. This change would have a deep impact on the character of late-war German aviation industry.

The leaders of Germany's aerial armament became fully aware of the advantage offered by the new production methods. Göring mentioned these techniques in November 1941 during a briefing at the RLM while referring to the increased output of Soviet industry. He noted that this increase was made possible by the introduction of conveyor-belt production lines and the use of automatic production machines operated by a relatively small number of skilled workers.[88] At least one German factory director learned later from his own practical experience that this development also made it possible to employ every skilled worker in every station in the factory according to need. This flexibility became important late in the war, as skilled German workers became scarce and foreigners of different skill levels were employed in growing numbers under German professional supervision.[89] By that time only foremen positions or positions requiring specific knowledge still demanded the presence of an expert craftsman.

Milch also acknowledged this new situation while describing to members of the Technical Office the way mechanized production lines enabled the Germans to use overwhelming numbers of unskilled foreigners and slave workers by March 1944:

> The manufacture of a high-quality double-row radial engine by unskilled foreigners is possible only if it is being done pattern-wise on a conveyor belt, where each man operates only a specific knob. This sort of training is possible with foreigners; we cannot make skilled workers out of these people. Skilled workers cannot be created within 4 weeks or even a year. A skilled worker is a man who studied his profession for 3 — mostly 4 — years, and then performed this work for 3 to 4 years. Someone who didn't do it is not a skilled worker. A semi-skilled worker can perform only specific work on a machine or on an object. You can see how difficult it is to perform the manufacturing tasks of so many types of aircraft.[90]

By the end of the war most factories used modern production lines. American intelligence officers, who surveyed several factories shortly after the end of the war, noted in their report: "As a general procedure, all final assembly, major sub-assembly, and sub-assembly was done on continuously moving conveyor lines, whether the work was accomplished in a large building, in underground tunnels, or in dispersal building in the woods."[91]

Another aspect of global modernization was the growing mechanization of the production process and the reduction of the amount of direct handwork involved in it. Parts, accessories, and engines were largely produced with different production machines: hydraulic presses, stretchers, punch presses, power brakes, metal shapers, shears, routers, etc. However, as the USSBS report about the German aircraft industry pointed out, not many machine tools were needed for machining forgings and castings. It was also found that German large-press capacity was larger than the Americans', and that some of these production machines worked for the aero industry outside the airframe plants.[92] German production machines were generally of good quality and easy to operate. These qualities enabled relatively high-quality and high-rate production by relatively unskilled workers with only little and specific training.[93]

Surprisingly, during this age of growing mechanization, riveting, a crucial production task to airframe manufacture and skinning, was still largely done with hand-held tools. Automatic riveting machines were designed for the production of several later types of aircraft, but saw relatively little use. Furthermore, the standard German production method demanded grounding the rivet heads (in order to make the aerodynamic surfaces smoother) using hand grinders, a process which cost time and was considered to yield results of lesser quality.[94]

The manufacture of airframes was done with the aid of jigs and holding fixtures. They held together, plan-wise, all the structural parts, ribs, longerons, stiffeners and skin in order to create a fuselage or a wing or any other large component of the aircraft. The jigs formed a sensitive part of the production process for two main reasons:

1. While it was possible to produce most parts on adjustable machines outside the main assembly plant, the set of jigs used to produce specific types of aircraft were unique and could not be simply "outsourced."

2. When aircraft production was dispersed in the later phases of the war, allocating the same sets of jigs to different and sometimes widely spread production centers was crucial in order to ensure the correct fit of all the components on the final assembly line. In order to achieve that, a master set of jigs was manufactured in several copies and each plant was allocated one copy as well as a normal set of jigs.[95]

Therefore, manufacture of tooling, and especially of jigs, was a crucial and sometimes lengthy process before initiating serial production of specific aircraft types.

While the above-mentioned type of jigs can be considered as a bottleneck, or a nonstandard part of the manufacturing machinery, other German assembly tooling was almost entirely of the universal and standard type. These universal jigs could be adjusted within certain limits and enabled the assembly of a fairly large number of designs. In most cases they were not fixed to the factory floor and could be moved easily to new locations.[96]

In contrast to the American methodology, different segments of German production lines were not concentrated in the same building, or even in the same complex. In Germany there was nothing comparable to the huge American factories, like Ford's "Willow Run," Consolidated's bomber factory at Fort Worth, Texas, or like North American's fighter factory at Inglewood, California, where most production of sub-assemblies as well as final assembly took place inside a single large building. The main reason for this German plant design characteristic was to make plants less vulnerable to bombing. Heinkel's Oranienburg-Germendorf factory is a good example of such an "internally dispersed" plant. It was composed of 7 main production halls. Six of them manufactured single components of the aircraft, while final assembly was done in the last one. The halls were placed far apart from each other and were connected by a network of roads.[97] In many cases production of certain components was even duplicated in other places. This initial dispersal of production meant that sub-assemblies and components had to be brought into the main assembly hall from different places.[98] Until 1942–1943 aircraft production took place mainly in large plants, where production of large components as well as final assembly took place in production hall complexes.

Understanding this system is crucial to understanding the changes that took place later on. For instance, the manufacture of the FW 190 single-engine fighter was supposed to be concentrated initially in Focke-Wulf's main factory in Bremen. Dispersal of its production began already in 1940–1941, because the Bremen factory was thought to be too vulnerable to air attacks. It was decided to move production eastwards and to disperse it among 4 different factories; each was supposed to produce 125 fighters annually. These plants were constructed in Marienburg (in East Prussia), Cottbus, Sorau and Posen (in Poland).[99] The four new factories, whose administrative center was in Marienburg, produced 652 of the 1,878 FW 190 fighters produced in 1942 — therefore above their projected output.[100] Although dispersed, these factories produced either completed aircraft, or one or another main subassembly for the final assembly factories. Dispersal of parts production and other facilities became much more intensified in 1944 and the original number of production centers multiplied towards the war's end to almost 30 times of its original 27. However, the pattern was set already in 1940 — long before Allied bombings became a serious threat.

Besides automation and mechanization there were a couple of other technical means to improve production rates: making aircraft simpler to produce in terms of machine tooling, simplified construction and standardization of parts. These refinements were included in existing types and were integrated into newer types during the design phase. The intended successor of the Ju 88 bomber was the Ju 288, which was considered very modern at the time it was designed in 1939–1940. It was not only technically more advanced than the Ju 88; it was also designed with ease of production in mind. While the Ju 88 used over 4,000 different types of bolts and screws, the Ju 288 used only 200 and was supposed to be largely produced with automatic riveting machines rather than by hand.[101] Simplification of design reduced the time it took to produce most late-war aircraft and lowered their price. In 1940 it took around 24,000 man-hours to build a single Ju 88 light bomber. In May 1944 it took the same amount of time to build the much larger and more complicated He 177 heavy bomber.[102] The tricycle landing gear of the Me 262 jet fighter cost only one-third of the price of similar earlier landing gear and was also lighter and stronger.[103] Generally the Me 262 was well adapted to late-war production facilities and manufacturing methods. As American intelligence officers determined after the war:

> It is obvious that the design criteria for the Me 262 included "ease of manufacture." The design breakdown is in almost perfect harmony with the most logical production breakdown.... Final assembly time is held to a minimum by the fact that each component assembly is a complete and self-contained unit requiring only a few bolts and connections when joined to the mating component. The components are broken down into minor assemblies in such a manner that most of the riveting can be done by machine. In fact 85 percent of the rivets on the Me 262 were machine driven according to Me production personnel.[104]

Perhaps the most striking indicator of the streamlining of the production process was the constant decrease in the unit price of the Me 109 fighter. In 1941 the price of a Me 109F was between 50,100 and 62,200 RM, depending on the size of the order. In 1945 unit price of the heavier and more powerful G and K models of the same aircraft decreased to only 43,700 RM regardless of the size of the order.[105] Another indicator of the rationalization was the increased efficiency of Heinkel when comparing its productivity in 1941 and in 1942. Although the firm still produced on the old production lines, at the end of 1942 the firm reported a 29 percent reduction in the time it took to produce a plane, while at the same time the output increased by 20 percent and the available workforce engaged in production decreased by 15 percent.[106]

These figures may be exaggerated, but they point to an important trend that made the German aviation industry more efficient in the post–Udet era.

Although modern aircraft were made largely of different metals, mainly aluminum, during the last two years of World War II wood had a renaissance in Germany as an aviation construction material. The resurrection of wood was mainly due to its cheapness and availability, especially as in 1942 Germany started to experience an aluminum shortage. Some aviation engineers also thought that it would be easier to work with wood under certain circumstances. Initially wood was used only in the construction of specific components. When incorporated into components like tailplanes and control surfaces, it was usually an efficient low-weight material, even on high-performance aircraft. Later, developers conceived aircraft with airframes made largely of wood, mainly in order to save valuable aluminum. This tendency started in the construction of transport planes. Junkers, a firm that pioneered the use of stressed metal skins in aircraft design, developed in 1940–1941 the Ju 252 as a new transport aircraft made of metal. It was supposed to replace the old Ju 52 airliner and transporter, which was also made of metal. As a consequence of the aluminum shortage the RLM ordered Junkers to redesign this aircraft and use wood in its construction. The redesign resulted in the Ju 352, which was produced in limited numbers in 1943–1944.

There was also an industrial advantage in the increased use of wood. The cabinetmaking industry offered convenient additional production capacity to the traditional aviation industry. Some of it was left unused by the war effort and in March 1944 the *Jägerstab* demanded immediate mobilization of idle cabinetmaking capacity for aircraft production.[107]

The use of wood caused some troubles, however, especially when used in high-performance aircraft. As wooden control surfaces and tailplanes were incorporated into the design of the Me 109 in 1944, failures of these parts caused a series of accidents. An investigation conducted by Messerschmitt and the Rechlin flight-test center revealed that defective gluing of the wooden parts and their rapid deterioration in field conditions were the most common causes.[108] At least two important late-war combat aircraft were conceived from the beginning as wooden planes and their production scheme relied heavily on the German cabinetmaking industry. We will deal with them later.

A mostly overlooked aspect of aircraft manufacture was production of different pieces of equipment, which were added to the basic airframe. As aircraft became more and more complicated the number of systems and subsystems built into them increased steadily. These included electric and hydraulic systems, electronic equipment, instruments, life-support systems, armament and other items. Their production composed a meaningful part of total production costs and time required for the production of a complete aircraft. In March 1944 Milch estimated that 52 percent of the total man-hours required to produce an aircraft were spent on equipment manufacture.[109] It meant that much manufacture took place not in the main production halls, but in smaller workshops within the factory's compound or completely outside it. Most items were provided by a large network of contractors specializing in the production of one or more items. For instance, in 1940 Messerschmitt's Regensburg factory worked with 38 subcontractors spread all over the Reich. They provided different parts: electrical equipment, propellers, engines, landing gear, raw materials, paints, ball-bearings, etc. Later a couple of larger engineering subcontractors were licensed to manufacture some of the firm's aircraft.[110] The growing complication of integrating so many pieces of equipment from so many places added to the complexity of aircraft production.

One of the most important and complicated pieces of equipment integrated into the airframe was the engine. Aero-engine production required extreme precision. Some researchers compared aero-engine production to the work of a watchmaker, while aircraft production was compared to the work of a plumber.[111] As another researcher put it, "An airframe plant can be regarded as a sheet-metal fabrication and assembly shop, whereas an engine plant was principally a precision machine tool shop."[112] An aero-engine is a machine made out of thousands of parts that must work together in great precision under powerful mechanical stresses and high temperatures. Therefore, the quality of the raw materials used to make them and the quality of their manufacture must be extremely high.

The aero-engine industry also moved during World War II from workbenches to modern mechanized production lines as part of its wartime streamlining.[113] Just like the airframe industry, this transformation progressed slowly. By the end of October 1942 not much had been done, and the chairman of the Aero Engine Main Committee, Dr. Wilhelm Werner, urged to hasten the construction of conveyor-belt production lines in all aero-engine factories. He viewed this technical change as the main way to bolster the productivity of the aero-engine industry.[114] Only in late 1942 did firms like Daimler-Benz started to convert their production to flow production with the aim of increasing monthly output to 1,000 engines. Conversion of Daimler-Benz's factories wasn't completed, however, until autumn 1943.[115] BMW's conversion started in earnest only in early 1943 under the leadership of Erich Zipprich, chairman of the BMW engines Special Committee, and was viewed as one of the main ways to enable the straggling firm to finally fulfill its output quotas. The different streamlining measures BMW had introduced since 1942 contributed to its increased productivity during later phases of World War II. While in 1941 it took 3,260 man-hours to produce a single BMW 801 radial engine, in early 1943 the figure dropped to 1,860 and in December 1944 to only 865 hours.[116]

As in the airframe industry, the use of modern production lines enabled the aero-engine industry to employ an increasing number of unskilled workers. This became especially obvious when series production of jet engines — an extremely complicated and advanced type of engine — started in summer 1944, using mainly slave labor.

Quality assurance along the production process, however, remained stricter in the aero-engine sector than in the airframe industry in order to ensure that the end product would function correctly after leaving the factory. These checks included intensive testing of the completed engines in special testing rigs available in every engine factory.[117] Constant technical developments made aero-engines more and more complicated and the reliance of most manufacturers on subcontractors and supply of parts and components from the outside increased steadily. According to Armaments Minister Albert Speer, this dependence caused a major bottleneck in aircraft production in 1943. Due to faulty subcontractors' management the aero-engine industry suffered a shortage of crankshafts. As a result many engines were left uncompleted in the factories. The problem was with the manufacture of different standard forged parts of the crankshafts, which required special machinery. There were only two or three factories producing these parts and they were unable to deliver the quotas required for the increased production programs. Even after this bottleneck was identified it took time to increase the production of these parts because it took time to produce the necessary machine tooling. The crankshaft shortage delayed the delivery of engines to the airframe industry, which in turn left completed airframes waiting at the factories for their engines.[118]

Production line of the Jumo 004 jet engine, probably at the Zittau factory. The rails, on which the engine trolleys traveled along the production line, are visible on the floor to the right (courtesy U.S. National Archives and Records Administration).

The crankshaft crisis demonstrates one of the main problems affecting late-war aircraft production in Germany. The increasing interdependence of different parts of the industry on supply from outside sources was a symptom of the production of modern and complicated machines. A functioning and well-coordinated network of manufacturers was necessary in order to perform production of such machines. Production of a complete aircraft from beginning to end could not have been taken place in one single factory. Final assembly was done in a main factory, but components and parts had to be fed into this main factory in order to enable it to assemble a complete aircraft. Modern industrial societies were able to accomplish this networking with ease, but as the crankshaft shortage demonstrated, it was a delicate network, which malfunctioned even if a single part of it failed.

Outsourcing—Aviation Production in Occupied and Axis Countries

A unique aspect of the wartime German aviation industry, which has hardly been dealt with by researchers, is aviation production for the Germans in the occupied and Axis countries. Different local firms outside Germany carried out a surprisingly large number of production and development programs for several leading German aviation firms and under the supervision of the RLM. Outsourcing of aviation production reached its peak rather late, even though it started early. Through outsourcing the Germans were not only able to increase total output and save their own capacity for priority projects; they also used outsourcing as an alternative business model in order to solve some of the structural problems of their aviation industry.

The quick succession of victories in Western Europe in 1940 improved Germany's strategic situation among other nations by allowing it access to a larger supply of raw materials (particularly tungsten and iron ores) and by offering additional industrial capacity. New manpower also became available in the form of occupied populations and hundreds of thousands of POWs. Overconfidence in *Blitzkrieg* strategy, however, influenced large parts of the Reich's political and military leadership, which generally thought that Germany would shortly win the war with current production rates. The failure to fully exploit the new gains was aggravated by the fact that the Germans generally chose to loot captured armament factories instead of reopening them under their control. This policy initially affected in particular the large and modern French aviation industry. Although a detailed survey of the French aviation industry was conducted as early as late June 1940 by the RLM, the Germans failed to take advantage of this largely intact industry.[119] The Germans closed most of the French factories and transferred to Germany a large proportion of their machine tools, where they were mostly stored because the Germans lacked the manpower needed to operate them.[120]

This inefficient policy changed, however, in late 1940–early 1941 when the RLM recognized the potential offered by the French aviation industry. In January 1941 Milch visited Paris and attended a conference of representatives and executives of the local aviation industry. In his keynote speech he called on them to accept German contracts and to cooperate with the Germans for their own and for their workers' benefit. He assured them that since the Vichy government approved such cooperation there should be no legal or moral obstacles for such mutually beneficial cooperation.[121] In order to promote and administer cooperation with foreign firms the RLM established liasion offices (*Verbindungsstelle*) within the GL (*Generalluftzeugmeister*). Each of them supervised the aviation industry of a single country. An RLM engineer directed each liaison office and was responsible for establishing contact with local producers and connecting them with relevant German firms or with the RLM.[122] Besides dealing with outsourcing, the liaison offices also negotiated the production of German aircraft under license by some Axis countries for their own use.

Generally, three models of industrial cooperation developed over the next few years. First and less common was that of German firms opening new branches in the occupied countries using existing factory complexes and local manpower. The second and the most widespread model was to outsource production of items to foreign firms under restrictive contracts. Such items ranged from minor parts to complete aircraft. The third model was outsourcing complete development and production projects to foreign firms. This type of outsourcing was more widespread than can be imagined and French firms were even contracted to develop several large and complicated aircraft.

German firms usually refrained from opening new branches in the occupied countries, although a cheap and readily available local workforce made such enterprises a logical option.

Foreseen difficulties in operating branches far from the well-established centers in Germany were probably the main reason for this neglect. However, business opportunities motivated a few aero-engine manufacturers to open factories in the occupied countries. One of them was Jumo — the aero-engine division of Junkers. Only a few months after the conquest of France, Jumo opened an engine factory near Strasbourg in Alsace. Junkers established the plant with its own equipment and machinery in a deserted Matford Ford car factory. The new factory soon reached a monthly output of 500 overhauled engines and 250 newly built engines. This factory relied entirely on local Alsatian manpower, which was regarded as friendly towards the Germans. Later, Soviet POWs were also employed in the factory under German and Alsatian supervision.[123] Jumo and BMW also opened several factories in the Czechoslovak Protectorate, taking advantage of the highly developed industrial infrastructure and skilled workforce in this country.[124]

Airframe firms used mostly the second business model in the occupied countries: contracting local businesses and firms to produce specific items. The focal point of aviation production for Germany was France, with its large aviation industry. Milch's speech in Paris in January 1941 signified a shift in Germany's policy towards the French aviation industry. From lukewarm interest the Germans now turned to full industrial cooperation — of course under German dominance. In spring 1941 the German and the Vichy government generally agreed to renew aviation production in France. France was allowed to keep one-sixth of the output for its own use, although this portion was composed entirely of French designs.[125] Large parts of the still existing French aviation industry were thus reactivated by the Germans in spring 1941 and orders from the RLM and from German firms started flowing to different French firms. By 30 June 1941, French aviation firms received from the Germans contracts worth 765,000,000 RM. These contracts provided work for some 62,800 Frenchmen.[126] Some factories continued for a while to produce French aircraft and engines, mainly using parts and components manufactured before the German occupation and still available in stores and depots. The Germans took over most of these aircraft and used some of them in second-line duties. They delivered other planes to their Axis allies, especially to Vichy France. French engines, especially the Gnôme et Rhône types, were used in some German designs, like the Me 323 six-engine transporter and the Hs 129 two-engine ground attack aircraft.

German firms mostly contracted French firms as parts and component manufacturers. French firms were initially contracted to produce only specific components. The German firms provided them the required blueprints and raw materials. The French were given only the minimum needed for the manufacture of their allocated product, therefore blocking further independent development by the contractor.

This type of cooperation was not restricted only to manufacture. Several German firms chose to hire French design offices and their staffs in order to expand their own design capacity. By November 1941 Focke-Wulf, Junkers, Messerschmitt, Fieseler and Heinkel opened design offices in France. Most of the employees working in these offices were French designers and draftsmen. There was more than enough work for them. As Focke-Wulf's representative in Paris reported at the end of November 1941, "According to our own observations, there are no more unemployed aircraft designers in France." Soon a competition developed between the German firms, as each sought to attract designers by offering higher wages and better contracts.[127]

In fewer cases French firms were licensed to produce complete German aircraft or large components for them. Messerschmitt was one of the first firms to contract a French firm for the production of complete planes. In early 1941 it licensed the firm SNCAN in Les Mureaux to produce its Me 108 liaison plane. Following the deal, Messerschmitt terminated the production of this aircraft at Regensburg, thus freeing the plant for increased production of the Me 109 and for the production of the new Me 210 heavy fighter.[128]

The German firm that used to the utmost the opportunity to use French capacity in order to expand its production of complete aircraft was probably Focke-Wulf. As early as late 1940 it helped the French aviation company Technique de Chatillon, based in Chatillon sur Bagneux, a suburb of Paris, to refurbish its facilities. After a short while Focke-Wulf fully acknowledged the capabilities of this firm and counted it as equivalent in any respect to similar German firms. It soon began transferring to the French firm the design work and production of some of its projects, and was fully satisfied with the results. Among others, Focke-Wulf planned to subcontract to Technique de Chatillon the design of its FW 206 mid-range airliner and of the FW 300 long-range airliner/maritime patrol aircraft. Focke-Wulf also planned to contract another French firm, Société Nationale de Constructions Aéronautiques du Sud Ouest (SNCASO), to produce them. It is important to note that this industrial cooperation was meant foremost to fulfill the German firm's postwar plans. Focke-Wulf developed these aircraft in cooperation with national airliner Lufthansa for commercial use in the postwar world. Because Focke-Wulf's highest priority during the war was military aircraft production, it was convenient to let French firms work on civilian projects for the postwar era. Focke-Wulf's long-term planning was typical of large portions of the German armaments industry. During the war, firms engaged in military production tended to look over the horizon and take into account long-term profitability in their corporate strategic planning.[129]

In order to enable SNCASO to produce the FW 206, Focke-Wulf arranged the reinstitution of its Paris factory, which was confiscated by the German military administration after the occupation of France.[130] When Udet ordered in early August 1941 to stop working on this civilian aircraft, he also approved a contract Focke-Wulf gave Technique de Chatillon concerning the production of parts for its new FW 190 fighter.[131] The French firm's initial production of these parts fully satisfied the German customer. This positive experience led Focke-Wulf's chief executive Kurt Tank to warmly recommend at the end of 1941 further contracting with this and other French firms in order to fully exploit the extra capacity they offered.[132] In the meanwhile Focke-Wulf tightened its cooperation with SNCASO in the field of military production. In February 1941 a German delegation settled down in SNCASO's main factory in Bordeaux. Soon afterwards SNCASO received an order for 419 (later reduced to 334) FW 189 tactical reconnaissance aircraft. The delivery of these aircraft was completed in April 1943. SNCASO's plants in Bordeaux and in Paris also produced some parts for the FW 200 maritime reconnaissance aircraft.

Focke-Wulf contracted SNCASO again in 1943 to produce more parts for the FW 190 fighter and to perform some of the development and design work of its further development, the Ta 152 high-altitude fighter.[133] Production in SNCASO's factories proceeded slowly, however. In May 1943 the USAAF heavily bombed its Bordeaux plant, which had just delivered the last FW 189 it produced and was gearing up for the FW 190 production. Following this setback the Germans and the French decided to move most of the production

General director, designer and test pilot Kurt Tank is congratulated by coworkers upon completion of a successful test flight in one of the aircraft he designed for Focke-Wulf (courtesy National Air and Space Museum, Smithsonian Institution, SI 80-2397).

facilities of the bombed factory to a nearby underground factory at St. Astier. This move proved to be a time-consuming operation and it dragged on well into 1944.[134] In the meanwhile SNCASO's Chatillon factory suffered from other difficulties typical to occupied France. On 13 December 1943, for example, workers of the factory went on strike and demanded increased salaries. The French management was able to resume work after only 25 minutes, but in the meantime a German SS and police detachment appeared at the factory and detained 18 workers. Although Focke-Wulf's representative in Paris thought that this was a successful action which improved discipline in the factory, overall French output indicates otherwise.[135]

Messerschmitt's experience with its partner SNCAN was generally similar. As revealed in the correspondence of Messerschmitt's Paris office and the French firm, inefficiency, passive resistance and sabotage caused difficulties and delays in the production of the quite simple Me 108 in France.[136]

Despite the difficulties encountered in France, production of complete German aircraft was increased after several German factories were damaged or destroyed by Allied air raids in the second half of 1943 and in early 1944. Focke-Wulf, for example, outsourced final assembly of FW 190 fighters to the French firm Morane-Saulnier in early 1944. Production was slow and largely ineffective, so practically no aircraft were completed by this firm before the liberation of France.[137]

In contrast to Focke-Wulf and other firms, Milch and Karl Otto Saur, head of the

Jägerstab, were not impressed by the French industry. Milch remarked on 1 June 1944 that once the Allies invaded France the French would cease working. He said that when that happened, French workers should be forcefully brought to work in Germany. Plans were made to evacuate as many as possible production machines and workers from France after the beginning of the invasion.[138] Massive Allied air interdiction of the French transportation network before and after the invasion was the main reason that this plan was never fully implemented.

Focke-Wulf, however, remained fixated on French production, especially when it was involved in long-term projects, like the FW 300 and its derivatives. Its story demonstrates the ups and downs of industrial cooperation with French firms. The FW 300 project began in 1941 and was initially proposed as an improved and enlarged civilian and military version of the FW 200 Condor maritime patrol aircraft.[139] In 1943 the original idea was abandoned and the project evolved into a much larger and completely new aircraft. It was renamed FW 300A or Ta 400, in honor of its designer Kurt Tank. Focke-Wulf and the RLM hoped to fly a prototype of the aircraft in early 1945 and to commence operational service in 1946. At least until April 1942 Focke-Wulf hoped to develop in parallel a large civilian airliner from the FW 300A for use in postwar long-range civilian aviation. SNCASO was supposed to begin development of the airliner in August 1942.[140]

Focke-Wulf planned from the outset to use French design capacity to design the FW 300 and to release German designers for work on main core projects, especially the FW 190 fighter and its derivatives. As the aircraft evolved into the more complicated FW 300A, some 80 percent of the work involved in its design and development was contracted to Technique de Chatillon. Later SNCASO was hired as the main FW 300A contractor. In this framework Focke-Wulf conceived in early 1943, in concert with SNCASO, a scheme to bring some of the French design staff working on the project to Germany and let them work there independently in their own design bureaus. The RLM approved this scheme on 14 August 1943 after negotiations with the French Ministry of Production and the Reich Plenipotentiary for Labor Mobilization (GBA) regarding the employment arrangements and conditions for the French workers.[141] French designers first settled in Focke-Wulf's main design center in Bad Eilsen, but later the German firm allocated them separate offices in Lage.[142] The significant involvement of foreigners in the design work raised security concerns and Focke-Wulf directed the Germans involved in the project to keep a watchful eye on their French colleagues.[143]

Work on the FW 300A/Ta 400 and other projects dragged on due to Germany's worsening war situation. In July 1943 the Technical Office noted that due to difficulties with the design work in France and due to lack of capacity, the program had suffered delays, and further recommended its termination.[144] The RLM decided, however, to allow Focke-Wulf to continue the development of the aircraft.

By the time the Allies landed in Normandy there was not even a prototype at the works. However, even during these desperate times Focke-Wulf kept working on the project. Its representatives even traveled to France on 10 June 1944 to discuss this and other projects with their French partners. The representatives met SNCASO engineers in Paris to discuss the progress of the FW 300A/Ta 400 project. The main topic of the meeting was the design and manufacture of the fuel tanks for the aircraft.[145] At the same time, and not very far from Paris, Allied troops had already established a beachhead in Normandy. Not a single FW 300A/Ta 400 was completed.

Further indication of the unrealistic approach of considering the prospect of further work in France after the invasion is the fact that on 20 June 1944 Focke-Wulf submitted a detailed plan for the production of wings for its new Ta 152 fighter by SNCASO's factory in Chatillon. The German firm decided to subcontract Chatillon for this purpose in May 1944, but going ahead with this scheme under the new situation seems to reflect a bad flow of information at the higher levels, or a loss of touch with reality by Focke-Wulf's executives.[146] This sort of over-optimism and detachment from the urgency of the military situation in the West is also strongly reflected by a remark Saur made during a Jägerstab meeting on 3 July. Answering a question regarding the prospects of producing the Ju 388 medium bomber in Hungary, he replied that he would like to move its production "partially to France and to produce there substantial numbers" of this plane.[147]

From 1942 the Germans outsourced other minor aviation research and development projects to French firms. Among them was the development of several light aircraft and trainers, but also of larger and complicated aircraft, like the He 274 and Ju 488 heavy bombers. None of these aircraft was ever completed, but after the liberation the French assembled two He 274 prototypes and used them for research and development of pressure cabins. There were other far-fetched programs outsourced to French firms. In August 1942 Blohm & Voss contracted the Breguet firm to produce two prototypes of its variable incidence wing BV 144 civilian transporter—another aircraft designed for the postwar

Outsourcing. A poor quality but unique photograph of the Ju 488 V401 heavy bomber prototype during assembly at the Latécoère factory in Toulouse (courtesy U.S. National Archives and Records Administration).

aviation market. The Germans supplied some of the jigs and production tools, and Breguet manufactured the rest. This aircraft also never went into series production as the RLM gradually forced firms to halt all civilian projects. Only one prototype of the plane was completed in France, but never took to the air and was taken over by the French after the liberation.[148] Like the BV 144 project, most of the other development projects handed over to the French were never finished before the end of the war.

The aero-engine industry also contracted several French firms to produce German engines under license. Prominent among them was BMW, whose chairman Popp had suggested to Göring as early as November 1940 the formation of business cooperation with several equivalent French firms. BMW was interested foremost in working together with Gnôme et Rhône, the largest French aero-engine producer. BMW's interest in this cooperation was rather narrow. It basically sought to free capacity for the production of the new BMW 801 radial engine by outsourcing the production of the older BMW 132 and of some key parts to Gnôme et Rhône. This business collaboration suffered from the same difficulties that affected the enterprises of the airframe industry, and plans to produce the BMW 801 engine in France never materialized. A total of 5,000 laborers worked in France for BMW, but the concern's management was never really happy with this enterprise.

BMW employed another outsourcing model by taking over a couple of Alsatian firms and using their factories to manufacture machinery and tooling for its main production centers.[149]

In sum, although the Germans made a real effort to mobilize the French industry for aviation production, final results were disappointing. The most important aircraft types produced in France were 334 FW 189 tactical reconnaissance aircraft, 784 Fi 156 Storch light planes, 516 Ju 52 transport aircraft, 175 Si 204 liaison aircraft,[150] and 107 Me 108 liaison aircraft.[151] The most productive year was 1943, during which a total of 1,285 aircraft were produced for the Germans. In 1944 production dropped sharply to only 502 aircraft.[152] The French also delivered several thousand components and engines, but the figures here are also negligible when compared to the German production figures.[153] The main reasons for this minor contribution were Germany's relatively late and slow mobilization of the French aviation industry, general inefficiency of French factories, low motivation of their workers, and different forms of resistance by French workers.

Although ineffective, French production for the Germans lived on after the war in different forms. When the Germans were driven out of France they left behind them not only a large number of aircraft production facilities, but also know-how and blueprints. As a result, several French contractors continued to produce German aircraft for several years after the liberation. Among them were 64 French-produced FW 190 fighters, which served with the French air force for a short time; Ju 52 transporters, which served with the French air force and with national airline Air France well into the fifties under the name Toucan; and Fi 156 light planes. The French even finished the development of several minor development projects handed over to them by the Germans. The SIPA firm near Paris, which developed the Ar 296 trainer for Arado, finished its development after the war and produced it for the French air force. SNCAN finished the development of the Me 208 light plane and produced it postwar alongside its Me 108 predecessor.[154]

The next important production center outside the Reich's original borders was occupied Czechoslovakia—the Protectorate. The German policy in Czechoslovakia evolved in a somewhat different way than in France. In 1939 the Germans captured Czechoslovakia's modern aviation industry practically intact and decided to activate it for their own purposes along with the rest of the relatively modern Czech armaments industry. By the end of 1939 Udet placed orders for 1,797 Czech aircraft, which were mainly produced by the two leading Czech aviation firms: Avia and Aero.[155] These aircraft were largely outdated by the time they were delivered, and the Germans either allocated them to second-line duties or handed them over to their allies, especially to fascist Slovakia.

Once the existing contracts to produce local planes ended, Czech firms were contracted to produce several German aircraft — mostly second-line types. It seems that training aircraft formed the most important contribution of the Czech aviation industry to Germany's aircraft production. Among the firms producing trainers for the Luftwaffe were Aero (Si 204 — 515 aircraft), Zlin (Bü 181 — 746 aircraft; Klemm 35 — 275 aircraft), Avia (Ar 96 — 1,744 aircraft) and Letov (Ar 96 — 512 aircraft, all of them produced in 1944, in addition to a single Ar 396 prototype).[156]

Czech firms were also contracted to manufacture several types of German aero-engines, particularly the AS 10, used mostly on second-line types, and the DB 601/605, used mainly on late models of the Me 109 fighter. The Germans also contracted several Czech general machinery firms to produce engine parts in late 1940.[157]

Focke-Wulf was active here also. By mid–1942 two Czech firms produced components for its FW 189 tactical reconnaissance aircraft and overhauled its FW 44 trainer.[158] The Czech industry also proved to be disappointing when it came to output and productivity. It took it some time to reach satisfactory production rates. In 1941 its output was still rather low, with only 819 aircraft. The highest output was registered in 1944 with 1,955 aircraft. Although just a small fraction of total German production in this year, it was still above the output of the larger French aviation industry.[159] However, Czech production can also be considered as a sideshow when compared to the German output, but by contributing second-line planes it enabled German firms to concentrate on the production of important combat types.

The Czechs also continued to produce German aircraft after World War II. Among others, Avia and Aero produced 394 Ar 96 trainers up to 1949. Aero produced between 1945 and 1949 some 179 Si 204 liaison planes for military and civilian use. Czech firms even produced after the war several German types, which were never manufactured on its territory during the war. Most numerous among them was the S-199 fighter, which was a re-engined version of the old Me 109G. Avia and Letov produced 450 such fighters for the Czech air force between 1947 and 1951. Ironically, 25 of these originally German fighters were delivered in 1948 to the embryonic Israeli Air Force and served during Israel's War of Independence. Avia even assembled several Me 262 jet fighters by cannibalizing aircraft and components the Germans left on Czech territory.[160]

The Germans also captured the Dutch aviation industry largely intact in spring 1940. In contrast to neighboring France, the Germans quickly reactivated the Dutch aviation industry. As Dutch officials reported after the war, as early as 22 May 1940, RLM officials appeared at the De Schelde factory at Flushing and ordered the management to resume production of the Dornier Do 24 flying boat it produced under license. Low-rate production of this aircraft continued until the Flushing factory was heavily damaged in a daytime air raid on 20 August 1943. Production of this factory was subsequently dispersed to several smaller locations. Dutch companies delivered to the Luftwaffe a total of 175 Do 24 flying boats during the German occupation.[161] Although this was a relatively small contribution, it helped to free Dornier's capacity to more important production programs and fulfilled most of the Luftwaffe's need for flying boats.

Beside this unique type of arrangement, based on prewar business cooperation, some German firms outsourced some of their production — again, mainly of second-line types — to Dutch firms. Junkers was the most enterprising German firm in Holland. By 1944 Dutch firms, especially Fokker, produced parts and main components for the outdated Ju 87 dive-bomber, for the even older Ju 52 transporter, and for the Ju 388 — Junkers' newest multi-role aircraft.[162] Fokker carried out at that time almost exclusively the production of Arado's Ar 196 seaplane, thereby freeing Arado to produce other and more important types. Surprisingly, Focke-Wulf was slow to realize the potential of the Dutch aviation industry. In 1944 Focke-Wulf followed the example of ATG, one of its FW 190 licensees, which outsourced in June 1944 the production of some parts to Fokker. In August 1944 Focke-Wulf sent the director of its Special Production Programs, Schuchardt, to tour the Dutch aviation industry. Upon his return he submitted a favorable report, recommending the immediate commencement of the production of complete FW 190 and Ta 152 fighters in Holland. According to his report, Fokker's top executives were especially eager to win important German contracts and asked for them explicitly. They were even ready to move some of their production facilities to caves in order to better protect them.

It must be stressed that although by that time aircraft production was centrally controlled, Focke-Wulf's contacts with foreign firms were done directly and higher authorities were largely left out of the loop. Schuchardt even urged his directors not to disclose Fokker's underground plant scheme to higher authorities, probably in order not to lose it to other firms.[163] This initiative came, however, much too late, and the Dutch industry never produced German fighters. The main reason Focke-Wulf turned to Holland was probably the loss of its French enterprises, which sent it looking for business opportunities elsewhere and in safer places.[164]

One of those places was Italy. Later in the war, after Italy changed sides and Germany occupied northern Italy, some German firms turned to Italian subcontractors based in the north. Among them were well-established firms like Breda, Fiat, Piaggio and SIAI. Focke-Wulf was the biggest contractor of Italian firms. While Messerschmitt chose in early 1944 to contract Italian firms only to overhaul its fighters, Focke-Wulf went far beyond that. Focke-Wulf's general director Kurt Tank expressed some ideas to use Italian production capacity in late 1943. The cause of these thoughts was almost certainly the problems Focke-Wulf experienced with its French partners, as indicated by Tank's decision to move large parts of the planned FW 300A production from France to Italy. In mid–December 1943 Focke-Wulf contacted the Armaments Ministry's representative in north Italy, Mr. Haberstolz. Haberstolz was a former Focke-Wulf executive and soon doubled as Focke-Wulf's representative in north Italy. The firm provided him a team of workers. On 22 December 1943, Tank and his production manager Willi Kaether met Haberholz in Italy and discussed with him the prospects of producing the fuselage of the FW 300A in Italy.[165] At the beginning of February 1944, Tank traveled again to Como and discussed with Haberstolz the prospects of outsourcing more work to Italian firms. They discussed among others the production of rigs and jigs for the Ta 152, overhaul of FW 190s and production of the FW 300A. Those present at the meeting even considered moving the entire FW 300A production from France to Italy.[166] Soon afterwards Focke-Wulf contracted several Italian firms to produce parts for its FW 190 fighter. In April 1944 Tank decided to outsource the production of the fuselage of his FW 300A to the Italian firms Fiat, Breda and Piaggio, which were supposed to form an industrial group dedicated to this project.[167] In this way the production of this large aircraft was supposed to take place in Italy, France and Germany — a scheme somewhat reminiscent of the current Airbus consortium production scheme. This scheme, however, was extremely pretentious and unrealistic under contemporary circumstances. Although the participating firms made some preparatory work, not much progress was made until the RLM terminated the project on 6 September 1944.

As was mentioned above, Focke-Wulf also hoped to contract Italian firms to manufacture parts for the new Ta 152 fighter. Furthermore, it planned to produce each month at least 250 aircraft of this type in Italy, mainly in two new underground or bunker factories. The Jägerstab approved this plan in early June 1944 and sought the best ways to implement it.[168] Fiat and Savoia-Marchetti — two of the most important Italian aircraft producers — were chosen as the main contractors of this ambitious scheme. Fiat was expected to establish its production line in an unused water tunnel in the Po Valley and in a new bunker plant in Savoia.[169] Focke-Wulf made detailed plans for the production run, but since the projected final assembly plants were never constructed, complete Focke-Wulf fighters were never produced in Italy and Italian firms delivered only parts and jigs to Germany. Their work continued until quite late in the war. As late as January 1945 Focke-Wulf received different parts worth 36,800,000 lire from Italian firms for its FW 190 and Ta 152 fighters.

The main problem Focke-Wulf encountered with Italian production was shipment. The products had to make a long journey to Focke-Wulf's factories in north and northeast Germany. At that time rail transportation was badly disrupted both in north Italy and in Germany, so most of the parts were sent to Germany on trucks.[170] Partisan attacks on rail and road transportation in northern Italy made an already difficult situation even worse. The commissioner for aircraft production in Italy reported in July 1944 that partisan activity in the area of Turin, where the Breda and Fiat factories were located, made production there practically impossible. Furthermore, the partisans confiscated almost all the vehicles used by the factories, scared off workers and even attacked and destroyed some workshops. The partisans enjoyed popular support in these areas, partially because of the German practice of forcing local people to work for them.[171]

No complete German aircraft was ever produced in Italy, but the Germans allowed Italian firms to continue producing some of their own fighters and allocated them either to the Luftwaffe or to the Italian National Republican Air Force. Fiat continued, for example, to produce its G-55 fighter, but its production run suffered from raw material shortage in northern Italy and therefore never reached the projected output.[172] The Italian outsourcing enterprises failed to provide either a meaningful workforce or production capacity. In early July 1944 Hitler ordered to cease all aviation production activities in Italy and to move all production machines and workforce available to factories in the Reich.[173]

Some production for the Luftwaffe was carried out also in Hungary. Most prominent was local production of the Me 109 fighter and of the Me 210 heavy fighter under license. However, most of the Hungarian output went to the Hungarian air force. Generally the Germans considered Hungarian productivity rather poor, although German pilots considered the Hungarian-built Me 210 as better than its German counterpart.[174] Junkers explored some possible Hungarian partnerships, but its representatives were not impressed. While discussing where to produce the Ju 388 outside Germany in 1944, director Richard Thiedemann of Junkers told Saur: "It was easier to get in Paris 50 complete planes with 9 German engineers and otherwise only Frenchmen, than 5 planes in friendly Hungary."[175]

Although widespread, aviation production for the Luftwaffe outside Germany proved to be disappointing in terms of output and efficiency. Apart from France, the Germans realized relatively quickly the extra production capacity offered by captured factories. Although the RLM provided little guidance and strategic planning in this regard, some firms quickly grasped the business opportunities offered by Nazi conquests. Partnership with foreign firms served the long-term plans of individual firms, especially regarding production for the civilian aviation market. Individual firms chose in most cases to contract foreign firms to produce parts and second-line aircraft types, thus allowing them to concentrate their own precious capacity in the production of high-priority types. Initially the development work of some minor projects was handed over to some foreign design bureaus, but later in the war, as those bureaus proved their efficiency and German firms became overloaded, foreign firms became deeply involved in some advanced projects. Their involvement was seemingly in disregard of security risks involved in such outsourcing. Allied intelligence indeed received in several instances useful information about German aircraft and their technologies from people working in aviation factories in the occupied countries.[176] Production in the occupied countries formed only a fraction of total German production, although it relieved the German aviation industry from several minor production programs. Antiquated facilities, raw materials shortages, low motivation and general low efficiency contributed to this low yield.

In the broader context it is important to emphasize that by contracting firms outside Germany, German firms not only sought new business opportunities, but also sought solutions to some of the structural problems of the German aviation industry. Most important among them was the manpower shortage. Outsourcing failed to solve this and other problems, therefore forcing the RLM and some firms to seek other solutions to the problems they faced while trying to increase output.

2

The Aviation Industry and the Air War

Developments in the air struggle over Europe in 1939–1944 influenced the German aviation industry in direct and indirect ways. On the one hand the aviation industry supported the war effort of its own air force by providing it the hardware it needed. Therefore, developments on the industrial sector determined to a large extent the Luftwaffe's successes and failures, its needs and its actions. On the other hand, the actions of the Luftwaffe also affected the industry. Most importantly, German air superiority meant security from enemy air attacks. Loss of air superiority enabled the enemies' air forces to attack the aviation industry. The outcome of the aerial struggle largely determined the developments and changes of the German aviation industry, and its emerging new character. Narrating the history of air power in World War II or of the Luftwaffe is not the aim of this research,[1] but a short description of the major trends in the air war up to early 1944 is mandatory here in order to understand some of the most important changes in Germany's late-war aviation industry.

Towards the Abyss—The War of the Luftwaffe

Germany entered World War II with the best tactical air force in the world. By the beginning of 1944 this air force was in shambles. In fact, when looking into its operational history, it is obvious that after several early brilliant campaigns, the Luftwaffe's efficiency deteriorated rapidly. Although it was able to support German ground troops well into World War II (especially on the Eastern front), it lost air control over Germany and western occupied Europe quite early, therefore allowing Allied aircraft to operate over them with increasing freedom.

There were several reasons for this failure. One of them was Germany's reliance on tactical offensive air force and negligence of strategic and purely defensive capabilities. Limitations of the national economy and limited industrial capacity played an important role in the prewar decision to create a tactical Luftwaffe rather than a strategic one. One of the main determining factors in this crucial decision was Germany's inability to build and support an expensive strategic bomber force due to lack of resources.[2] Initially this decision proved to be a lucky one, because superior tactical air power was a decisive factor in the success of early German blitzkrieg campaigns. The technical and tactical superiority of the Luftwaffe were fully exploited during the campaign in the West in 1940, where it fought mostly enemies right across the Reich's border. The Luftwaffe even succeeded in supporting the *Wehrmacht* on a strategic level during the Norwegian campaign, but this was largely due to weak opposition and its ability to establish air bases in Norway soon after the invasion of this country.

The first serious problems appeared when Germany tried to project its air power across the English Channel against Great Britain. The use of aircraft designed mainly for short- or intermediate-range tactical use proved problematic when engaged in a strategic air campaign. Furthermore, the Luftwaffe encountered over England what is termed today an integrated air defense system, composed of early-warning radar stations, an efficient communication network, modern defensive fighters and extensive ground defenses. This system was conceived precisely to counter the kind of threat posed by the Luftwaffe in 1940. The Luftwaffe lost the Battle of Britain and as a result the Germans were forced to abandon Operation "Sea Lion"—their planned invasion to England.

At the same time the Luftwaffe also failed to prevent British bombers from attacking targets in western Germany. Early British daylight attacks proved to be too costly and as a result the Royal Air Force (RAF) started to operate its bombers only at night. Night bombing was largely inefficient due to the inability of most bombers to find and hit their assigned targets. Furthermore, it was almost impossible to hit pinpoint strategic targets at night. Nevertheless British bombers were able to regularly drop bombs on Germany without the Luftwaffe being able to stop them. It was a bad omen, which was only a prelude to the British night onslaught on German cities in 1942–43.

In the aftermath of the Battle of Britain the Luftwaffe was still a powerful and effective force, as was proved by its crucial support of the Wehrmacht's operations in 1941. These included the conquests of Yugoslavia and Greece, operations over the Mediterranean, the airborne invasion of Crete, and the early successes against the Soviet Union. Despite these successes the Luftwaffe started losing its edge in 1941, when its theatres of operations expanded beyond its capabilities. The inability of the aviation industry under Udet to supply the Luftwaffe at that time with the aircraft it needed has already been discussed here. The industrial failure was made even worse by the heavy losses of flight crews and aircraft the Luftwaffe suffered in the East and elsewhere. By the end of 1941 German aircraft production and aircrew training could no longer keep up with the losses.

Now the Luftwaffe entered a vicious circle with disastrous implications for its combat effectiveness: in order to complement losses, the Luftwaffe's high command pulled out growing numbers of aircraft and experienced personnel from flight training centers and sent them to replenish combat units. As a consequence, flight-training programs were cut back and the level of training deteriorated continuously.[3] Because new flight crews received lesser training, the loss rate increased and the crisis escalated. At the same time, the Allies mobilized their resources more efficiently in order to keep up the strength of their air forces despite constant attrition. The British, for example, established the Empire Air Training Scheme, which expended enormously the prewar training establishment and enabled training of new flight crews in great numbers and to high standards. The United States' entry into the war, with its enormous production and training capacity, aggravated Germany's position also in this regard.[4]

Although the Luftwaffe continued operating on a massive scale in 1942 and 1943, the seeds of disaster were sown by the vicious cycle of attrition and overstretching commitments it had entered in 1941.

The crisis in the air war came to a head in 1943, after the Luftwaffe was heavily bruised in the East, and as western air power increased its operations in the West. Among others, in 1943 the U.S. Army Air Force (USAAF) started targeting directly the German aviation industry in order to eliminate the Luftwaffe's combat effectiveness by cutting off its supply artery. The focal point of the air struggle shifted increasingly to a battle for air superiority over the Reich. The outcome of this struggle determined to a large extent the fate of the aviation industry and other key elements of the German armaments industry.

From the Battle of Britain to "Big Week"—Allied Attacks on the Luftwaffe and the Aviation Industry

While the grinding air campaign on the Eastern Front continued, the failures of the Luftwaffe enabled the western Allies to strengthen their strategic air campaign against Germany. In 1943 Allied air power started to target aviation production facilities directly.

The RAF bombed some facilities connected to the aviation industry in 1940–1942, before the arrival of the USAAF. These attacks were mostly inaccurate, because of the RAF's inability to attack precision targets at night. Therefore, the RAF carpet-bombed cities related to the aviation industry and not the factories. Furthermore, these sporadic attacks were not part of a general bombing policy targeting the aviation industry.[5] Another mode of operation used by the RAF from 1941 in an effort to weaken the Luftwaffe was to offensively engage German fighters over western France by commencing fighter sweeps. These operations failed largely because most German fighter units had been transferred by that time to the East. Additionally, the RAF suffered heavy losses in the process.

The turning point came in 1942. While British Bomber Command increased its carpet-bombing of German cities at night, the USAAF's Eighth Air Force, operating from England, started its own campaign of daylight precision bombing, using bombers heavily armed with defensive armament and equipped with the modern Norden bombsight. As it built up its force throughout 1942, the Eighth Air Force restricted its attacks to targets in Western Europe and refrained from engaging the heavy defenses of the Reich. The American bombers usually flew their missions without fighter escort and relied on their own defensive armament in order to deal with enemy fighters. Although early attacks concentrated mainly on submarine-related targets in France, from August 1942 the Americans started bombing aviation factories producing for the Germans in France, the Netherlands and Belgium. Among the attacked factories was Fokker's factory near Amsterdam, the Gnôme et Rhône aero-engine factory at Le Mans (attacked twice), and a couple of Erla repair shops in Belgium.[6] These attacks were widely dispersed and formed no serious threat to German aircraft production. Furthermore, they formed no part of a strategy aimed at disabling Germany's aircraft production.

As the Eighth Air Force gathered strength and prepared to attack Germany, the need for a clearly defined bombing strategy came up. The basic assumption (and a correct one) was that German fighters posed the greatest threat to the Combined Bomber Offensive

Tranquility before the storm. Messerschmitt's Augsburg factory, probably in mid–1943. Visible is a sole Me 410 heavy fighter. Note the camouflaging of the buildings and tree-like camouflage devices in the background (courtesy U.S. National Archives and Records Administration).

(CBO), which had been decided upon in the Casablanca Conference in January 1943. Therefore American plans for the CBO viewed the reduction of German fighter strength as an intermediate objective in order to obtain air supremacy over Europe. As a result the German aviation industry was listed as the second highest priority target for the CBO after submarine construction yards. At that time the Battle of the Atlantic was raging and submarines posed a serious threat to the Allied war effort. The menace of the submarines was therefore reflected in the top priority allocated to their construction facilities and their bases.

The aviation industry became priority target number one of the CBO following a new directive the Allies' Combined Chiefs of Staff issued on 14 June 1943, code named "Pointblank." This change of priorities reflected, on the one hand, the turn of the tide in the Battle of the Atlantic in the previous month, and on the other hand the intensification of the bombing campaign against Germany. Pointblank divided the aviation industry into 4 main target systems: airframes, engines, propellers and accessories. It was assumed that the complete disruption of any one of these systems would bring aircraft production to a halt.[7] The Allies decided to concentrate their attacks on the airframe factories. This decision was part of the widely accepted notion that achieving air superiority was the basic requirement for the conduct of successful strategic bombing campaign. It is noteworthy that a year later another aviation industry was placed on top of the target list when the USAAF prepared to begin its strategic bombing campaign against Japan. This decision reflected lessons learned over Europe. Second priority target complexes in the Pointblank directive were German submarine yards and bases — an objective that reflected the still-burning needs of the Battle of the Atlantic, even after it finally tipped in favor of the Allies. The third objective was the rest of the aviation industry, the ball-bearings industry, and oil production. These last two items drew directly from prewar researches that had attempted to identify "bottleneck" industries, the destruction of which would cause the collapse of the enemy's entire war economy.[8] Pointblank was a central strategic directive of World War II, because it served as a guideline for the strategic bombing of Germany until 17 April 1944. As we shall see, these intervening 10 months were most crucial in the history of the German aviation industry.

Once deciding to target the aviation industry the Allies started to collect detailed intelligence about factories involved in aviation production and about the workflow of aircraft production in Germany. Different branches of Allied intelligence prepared elaborate organizational charts and flow charts in order to find out exactly how aircraft were produced in Germany, by whom and where. By studying and analyzing this material Allied intelligence officers tried to locate bottlenecks and particularly vulnerable points along the production process.[9] It was a long and tedious job that involved the use of different intelligence sources, from interpretation of thousands of aerial images through agents' reports and highly secret deciphering of German coded radio traffic.

The quality of the intelligence was excellent and it enabled the USAAF to concentrate its attacks on some of the most important fighter-producing factories of the Third Reich right from the beginning.

One of the first aircraft factories to be attacked by the Americans was the WNF Wiener Neustadt factory in Austria, license-producing Me 109 fighters. It was attacked on 13 March 1943 by bombers of the Fifteenth Air Force operating from North Africa.[10] The larger Eighth Air Force, operating from England, started its offensive on 17 April 1943 by dropping 262 tons of bombs from 105 bombers on Focke-Wulf's main plant in Bremen. This attack destroyed or heavily damaged 50 percent of the factory, as well as destroying 10 FW 190 fighters and damaging another 12.[11] The main offensive, however, began only in June 1943, as weather conditions over Europe improved. A total of 6 airframe factories producing Me 109 and FW 190 fighters were bombed. In August two more factories were bombed, including Messerschmitt's important Regensburg plant.[12] No aircraft production factory or related target was bombed in September, but in October four aircraft factories and one propeller factory (VDM in Frankfurt/M) were bombed. Among those targets bombed in October was Focke-Wulf's Marienburg factory, which was the most distant target attacked so far by the Eighth Air Force.[13]

In addition to these pinpoint attacks, Hamburg suffered in July the worst carpet-bombing so far. A rare combined effort by the RAF and the USAAF killed around 40,000 and destroyed large parts of the city in a series of raids. Furthermore, the RAF dropped for the first time during its Hamburg raids stripes of chaff—stripes of aluminum foil used to deceive radar. These stripes, code named "Window," practically paralyzed German defenses and enabled the RAF to heavily bomb the city with a relatively small number of losses. The Hamburg disaster was just the kind of wake-up call the German leadership needed in order to shift priorities. Within days after the end of the raids Göring ordered to switch the focal point of German airpower to defense of the Reich. Next to the Luftwaffe, the RLM and the aviation industry were also ordered to adjust to the new priorities.[14] From now on aircraft production switched more and more to fighters and fighters' equipment.

Throughout the second half of 1943 German defenses caused increasing losses to the attacking bombers — both by day and by night. Initially U.S. bombers flew without friendly fighter escort all the way to the target, relying on their heavy defensive armament. As a consequence defending German fighters took an increasingly heavy toll of the bombers. In July, U.S. fighters began to escort the bombers part of the way to their targets, but the Germans learned quickly to attack the bomber formations only after U.S. fighters were forced to depart due to their limited range. The struggle in the skies over Germany now became bloodier. On 17 August 1943 the Eighth Air Force launched a large synchronized attack against a ball-bearing factory in Schweinfurt and against Messerschmitt's plant at Regensburg, which was thought to be the center of Me 109 production. Although the targets were heavily damaged and it took four months to regain pre-attack output at Regensburg,[15] the attackers suffered grave losses. Sixty B-17 bombers were shot down—10.3 percent of the inventory of these planes available to the Eighth Air Force at that time. A second raid on Schweinfurt on 14 October 1943 also cost 60 bombers. Most of the losses were attributed to German fighters. It became clear that without achieving air supremacy over Europe, the Eighth Air Force was going to bleed to death. Providing fighter escort all the way to the target and back was the obvious solution, but posed a huge technical challenge. Mounting losses of B-17s to defending fighters was one of the main reasons why most deep penetration daylight bombing raids were suspended during the winter of 1943–1944. At the same time German night defenses caused heavy losses to the British Bomber Command. After Hamburg, Bomber Command sought to destroy Berlin in a series of raids, known as the Battle of Berlin. The bombing campaign failed to achieve its aims and Bomber Command lost 1,047 bombers in the process.

In the meanwhile the American aviation industry succeeded in developing what was thought to be technically impossible: a long-range escort fighter, not only with sufficient range to escort bombers all the way from England to Berlin and back, but also equivalent in performance to all contemporary operational German fighters. The North American P-51 Mustang entered large-scale operational service at the end of 1943, enabling the Eighth Air Force a forceful return to the skies of Germany. On the eve of the renewed bombing campaign enemy air power was more than ever a top-priority target. In his Christmas dispatch to the USAAF, its commander General Arnold emphasized the need to destroy Germany's air power: "Destroy the enemy air force wherever you find them, in the air, on the ground, and in the factories."[16] At the end of 1943 it was clear that although hard hit, the German aviation industry was only slightly damaged. The effort against the ball-bearing industry also largely failed, although all the main ball-bearing factories were bombed and heavily damaged. The main reason for this failure was the fact that both the industry and the armed forces possessed large stores of ball bearings and these stores sufficed for 3 months, during which time production was partially restored.[17]

In November 1943 the USAAF began planning a massive strike against the German aviation industry, but during most of the winter of 1943–1944 weather conditions over Europe prevented the execution of this offensive. Nevertheless, during January 1944, the skies cleared somewhat and several aircraft factories were bombed. One of them was the AGO factory in Oschersleben, producing FW 190 fighters. Losses were heavy again; of the 174 bombers taking part in the raid, 34 were shot down. However, in the same month German losses were also high: 30.3 percent of the Luftwaffe's single-engine fighter force was destroyed, mostly due to air combat and accidents.[18] This figure was an encouraging sign for the Allies.

In February the USAAF and the RAF carried out the most concentrated attack on Germany so far. This offensive was code named "Operation Argument" but became better known as "Big Week." The focal point of these attacks was the German aviation industry and its supporting infrastructure. Fighter planes' final assembly and component plants were the main targets. The Allies hoped to eliminate almost the entire German aircraft industry in a ten-day all-out attack, but due to bad weather the offensive was called off after only 5 days.

Nevertheless, during these five days attacks were highly concentrated and mostly effective. It was easy to locate the targets against the snowy ground and American bombardiers took full advantage of this situation. Although final assembly and component plants were the main targets, aero-engine and ball-bearing factories were also targeted. The targeting system was elaborate and was based on comprehensive intelligence the Allies had gathered up to this point. Thus, for example, bombing of the Erla final assembly plant for Me 109 fighters in Leipzig-Möckau was supplemented by an attack on an Me 109 components plant in the same area. Junkers' Bernburg factory, producing Ju 88 night fighters, shared a raid with the fuselage factory in Oschersleben and the wing factory in Halberstadt, on which it was dependent.[19]

The USAAF dropped a total of around 10,000 tons of bombs between 20 and 25 February 1944 on 23 airframe and 3 aero-engine factories; most of them were heavily damaged. The RAF dropped during the same period an additional 9,200 tons on Leipzig, Stuttgart,

Big Week. An Me 109G-6 fighter, still bearing factory registration, destroyed in the bombing of the Augsburg factory on 25 February 1944 (courtesy U.S. National Archives and Records Administration).

Steyr, Schweinfurt and Augsburg — cities that hosted some important factories related to the aviation industry.[20]

Besides direct damage caused to factories, American air power also struck a heavy blow at the Luftwaffe in the air. During "Big Week," U.S. fighters escorted the bombers all the way to their targets. At the beginning of February escorting U.S. fighter pilots were allowed for the first time to go after German fighters wherever they were encountered and not just stick to the bomber formations. During "Big Week" this change of tactics contributed to the massive losses suffered by German fighters. This policy contributed significantly in the following months to the heavy attrition of the Luftwaffe in the air. Prowling American fighters in these operations shot down not only German fighters. Bombers, transporters and trainers flying in what was considered a relatively safe airspace deep inside the Reich were also destroyed in the air and during strafing attacks on their bases.[21]

The attackers also suffered heavy cumulative losses. On the last day alone the USAAF lost 70 bombers and 7 fighters, but unlike the Germans, the Allies could bear these losses and easily replace them.

Crucial to the success of "Argument" was the damage caused to some of the most important production centers of the German aviation industry. The largest attack was mounted on the first day of the operation. In its course 941 American bombers attacked 12 factories, mainly in the Braunschweig-Leipzig area. Thanks to the strong fighter escort only 21 bombers were lost. Among the factories attacked on the first day were Focke-Wulf's plant in Posen, Arado's plant in Tutow, AGO's plant in Oschersleben, and different Erla plants in and around Leipzig.[22] All these plants produced Me 109 and FW 190 fighters. During "Big Week" a total of 12 factories producing these fighters were attacked. These attacks perfectly served the strategy of the USAAF by contributing directly to the destruction of the German fighter force.[23] Additional important plants were attacked in the following days. Messerschmitt's factories in Regensburg-Prüfening and at Obertraubling were attacked and heavily damaged on 22 and 25 February 1944. On 25 February 1944 Messerschmitt's second main factory in Augsburg was also heavily bombed. Production capacity of this factory was reduced by about 35 percent and 30 of its structures were gutted by fire.[24] These attacks and their consequent damage caused a heavy disruption of Messerschmitt's production programs. Among others, the production workshops of the nose section of the Me 262 were heavily damaged and the entire production program of the future fighter of the Luftwaffe had to be relocated to other places. The Germans estimated that this strike caused a three-month delay in the supply of nose components. Furthermore, a bomb hit on a shelter trench killed or injured the entire instrumentation production specialist team, which again caused long delays until replacements were found.[25]

"Big Week" marked a turning point for the Luftwaffe and for the German aviation industry. The attacks caused unprecedented damage and interruption to a large number of important production centers. They signified a worrying trend. Now the Luftwaffe was not only unable to defend its own country; it also failed to defend the infrastructure and industry upon which it was dependent for its very existence. If losses on the scale of "Big Week" were to continue, Germany could lose the main core of its aviation industry in a matter of weeks. Since no immediate improvement in the air war situation was in sight, it became obvious that in order to survive further onslaught the aviation industry must resort to other measures. It was a traumatic moment and a grim realization for the German leadership.

In retrospect the Allies were unable to continue the momentum of their attack on the aviation industry, and there was no follow-up to "Big Week." Allied intelligence overestimated the success of "Big Week," and as a result the pressure on the aviation industry was eased up.[26] In the following months Allied air power and especially the American heavy bomber fleet was pulled out of the Reich's sky and concentrated on the aerial preparation for the Normandy invasion. The RAF's night bombing campaign also eased up in the first half of 1944. In March 1944, technically and tactically upgraded German defenses practically defeated British Bomber Command by inflicting upon it unprecedented losses. As a result Bomber Command also turned most of its attention to the invasion preparation task — at least until new tactical and technological solutions could be found.

After the war Allied intelligence discovered that the USAAF had missed the real bottleneck of the German aviation industry: aero-engines production.[27] As mentioned before, only 3 engine factories were attacked while the main force of the offensive was directed against the airframe industry. Furthermore, it was found out that the damage caused by the February raids was not as severe as was thought at that time. After the Germans thoroughly surveyed the damaged factories they found out that most of the production machines survived the destruction of the structures in which they were placed. It was discovered that, though bomb blasts damaged structures or collapsed them, when the roof and the walls fell in they simply buried the machinery without completely ruining it. The Germans soon started to clear the ruins, dig out the machines and place them in newly constructed or hastily repaired structures. This kind or repair work enabled the resumption of production in seemingly totally demolished factories within a relatively short time. The raid of 25 February 1944 on the Augsburg plant, for example, damaged only 30 percent of all production machines. Due to a quick salvage and the implementation of lessons learned from the 1943 raids, the Augsburg factory resumed its pre-raid output within a single month.[28] At the end of February 1944, however, these developments were unknown to the Allies, and the immediate German response to "Big Week" was a massive reorganization of the entire aviation industry and the production processes. The extreme measures taken within this reorganization were expected to save German aircraft production from what was perceived as a complete annihilation.

New and Old Technologies

The deep changes caused by the "Big Week" trauma happened at a time when German aviation technology stood on a historical watershed. In 1944 several German scientific and technological breakthroughs matured and became ready for series production. Transforming such projects into series-produced hardware posed a major challenge to industrialists and administrators. The task became much more difficult due to the special circumstances caused by "Big Week" and its subsequent reshuffle.

It is important to look at the ways the Germans prepared for the production of modern aerial weapons based on revolutionary technologies and to what extent, if any, they formed a marked departure from "traditional" production programs. This look should not distract from the fact that after all, most aircraft and aero-engines produced in Germany during the last phase of the war were of a conventional design. Thus, while the industry geared up for producing the fruits of the research and development establishment, conventional technologies remained dominant. It should be also remembered that the application of most revolutionary technologies was restricted to a narrow line of products. New propulsion systems, for example, were intended for use only in combat aircraft, at least until after the war.

The problem of developing new technologies in a time of war and using them on operational platforms was an important aspect of late-war aviation industry and one that kept decision-makers rather busy. The main question facing the decision-makers was which path to choose while planning the future composition of the Luftwaffe. An early possibility that was not seriously considered was to give up production of advanced aircraft for the duration of the war and concentrate all efforts on mass-producing existing designs. Another option, which became dominant only in late 1944 and 1945, was to fully revolutionize the composition of the Luftwaffe and to equip it almost exclusively with jet- and rocket-propelled combat aircraft before the end of the war. The third option was to continue conventional aircraft production while moving forward in parallel with the development of new technologies. In practice this is exactly what happened, and although plans were laid out for a Luftwaffe equipped almost exclusively with modern combat aircraft, these were too farfetched, and so production of conventional fighters continued in full tempo until the end of the war.

The two most important technological developments influencing German aviation production were the jet engine and the rocket engine. Their development began in academic research during the 1920s and 1930s. One of the main reasons for Germany's lead in the development of these engines was the personal involvement of powerful aviation industrialist Ernst Heinkel. He became interested in their development in the mid–1930s because of his own dreams of high-speed flight.[29] Other leading aero-engine firms also initiated studies of different forms of jet propulsion at the end of the 1930s. Furthermore, the RLM and the Army actively supported practical research and development of these engines.[30] It is true that initial research activities of revolutionary propulsion enjoyed only lukewarm support by the RLM or the Luftwaffe, but once they realized the military potential of these technologies they increasingly supported them. After the outbreak of World War II the RLM forced Heinkel to transfer his technological know-how to other firms working on jet propulsion in order to enlarge the research and industrial basis of these technologies. As a result the company that pioneered the development of modern propulsion technologies was sidelined and lost its edge, although it continued its own development work using the firm's own funds and facilities.

Wartime development of revolutionary aviation technologies can be divided into four main areas:

Jet and Rocket Engine Technology

Heinkel manufactured the first practical jet engines, but after he was forced to give away his research, Junkers' aero-engine division (Jumo) and BMW became the main developers and manufacturers of jet engines. The jet engine was a major technological breakthrough. It was developed to a similar extent only in Great Britain, but not on the same order of magnitude. While British developers concentrated their efforts on the development of centrifugal flow engines, which were easier to develop and produce, most German developers worked on axial flow engines, which were more advanced and offered better performance.

The development of such a cutting-edge technology was a protracted process, which was aggravated by shortages of exotic metals required for the manufacture of some parts operating under extremely high temperatures and strong mechanical forces. German metallurgists managed to reduce their use to a minimum by developing special alloys, in which only small proportions of the rare metals were used. This technical compromise solved some problems but caused other technical problems, because these alloys were not as strong as they were supposed to be.[31] Early jet engines thus suffered from low reliability. In autumn 1944 the average running life of the Jumo 004 engine was only 10 hours[32] and pilots complained repeatedly about their poor reliability.[33] The equivalent BMW 003 engine was even more problematic: in early November 1944 one of Heinkel's test pilots reported an average of 7.4 failures per running hour with this engine after a visit to Arado's flight test base. Most of the failures were due to fires and breaking turbine blades.[34] It is interesting to note that besides offering dramatic increases in aircraft performance, early jet engines did not require special fuels and could operate on almost any type of fuel. It was reported that successful experiments were even carried out in operating the Jumo 004 jet engine with Rumanian crude oil.[35] The main type of fuel used for their propulsion was J2, a diesel-type fuel. It was a low-octane fuel that required less refining than the fuel used in normal piston engines. Therefore, jet propulsion also offered simplified fuel logistics.

The Jumo 004 axial-flow jet engine (foreground) represented the future of aviation. It suffered, though, from a long gestation and poor reliability until the end of World War II. The engine behind is the Ishikawajima NA-20, a Japanese copy of the BMW 003 engine.

The rocket engine was another type of revolutionary propulsion. Several German firms and research facilities developed different forms of rocket engines before and during World War II. In this field the Germans were even further ahead of the Allies. The Americans estimated after the war that the Germans were at least four years ahead of the USA, particularly in the development of liquid-fuel rocket engines.[36] The most important implementation of liquid rocket propulsion was the Army's V-2 ballistic missile, but it was also a central component of several Luftwaffe projects. Unlike the jet engine, rocket engines required special fuels, which were both difficult to manufacture and sometimes difficult and even dangerous to handle.

Jet Aircraft

While early jet aircraft were basically propeller aircraft designs with a jet engine, it was clear that the enhanced performances offered by jet engines required new approaches to aircraft design. This realization led, for example, to completely new wing designs, including swept wings.[37] The German leadership was quick to realize the military potential of jet aircraft and initiated, in parallel to the development of jet engine, the design of jet aircraft. The Messerschmitt's Me 262 twin-engine fighter became Germany's leading jet aircraft project but others soon followed. Most German designers proposed their own jet aircraft designs after the potential of this propulsion became clear, but only a small number of them ever took to the air.

Rocket Aircraft

The extremely higher speeds offered by this engine also required completely new aerodynamic design. Messerschmitt led the way again with the operational Me 163 rocket interceptor, but other rocket-propelled aircraft flew experimentally by the end of the war. The main disadvantage of rocket-propelled aircraft was their short endurance. They carried enough fuel for only several minutes of powered flight and were forced to glide back to base after fuel supply was exhausted. Although rocket planes were a spectacular technological achievement, this deficiency severely limited their military value, and thus they were never produced in significant numbers.

Missile and Guided Weapons Technology

The German Army led the way in the mid–1930s with its A-series surface-to-surface missiles developed at Peenemünde and elsewhere by a group of scientists led by Wernher von Braun. This work culminated in the V-2/A-4 ballistic missile, which was produced for the army and later for the SS by firms outside the aviation industry. The Luftwaffe as well as the aviation industry launched during the war a multitude of missile and guided weapons development programs. Research and development activity in this field proliferated immensely during the war, as evident from several bulky reports prepared by Allied intelligence after the war.[38] Among these weapons were air-to-surface missiles, air-to-air missiles, surface-to-air missiles and all sorts of precision guided munitions.

As Rolf Schabel pointed out, politics and faulty decision-making regarding some of the advanced technology projects, attributed primarily to Hitler in postwar German memoirs literature, were overshadowed by technological difficulties in delaying their introduction to operational service. These difficulties were inherent to such advanced technological developments

Waiting for engines. The second prototype of the Me 262, constructed in April 1942, sits on the factory airfield tarmac at Augsburg waiting for its engines, probably in mid–1942 (courtesy U.S. National Archives and Records Administration).

and simply could not have been solved overnight. The protracted development work required in order to make these new technologies ripe for operational use was the main reason jet and rocket aircraft entered operational service only in 1944.[39] The main problem with jet aircraft was the low reliability of their engines. Only in summer 1944 did German jet engines become reliable enough for operational use, even though their development had begun in late 1939. The same problem plagued rocket fighters. Prototypes of the Me 163 rocket interceptor began powerless flight testing already in 1941, but entered limited operational service only at the end of May 1944. This long delay was caused by the difficulties involved in making a rocket-propelled aircraft ready for operational service. These problems were never fully solved and rocket-propelled aircraft were somewhat sidelined after initial enthusiasm.

Guided missiles and "smart" weapons were an even tougher technological nut to crack with contemporary technologies. The revolutionary rocket engines used on most missiles were problematic enough, but complicated guidance systems posed an even bigger technological challenge, which the Germans were unable to solve before the end of the war. Some guidance systems the Germans experimented with, like TV guidance and anti-radiation systems (homing in on enemy radar emitters)[40] matured only in the mid–1960s. It is worth noting that American developers encountered the same difficulties with similar guided weapons they designed and tested during the war. Thus the few guided weapons that entered

Great hopes. An Me 262 A-1a fighter of operational conversion squadron III/EJG2 is being prepared for flight. This specific aircraft served previously with *Kommando Nowotny*. It survived the unit's short and mostly unsuccessful combat debut with the type in October–November 1944 (courtesy National Air and Space Museum, Smithsonian Institution, SI 79-4069).

operational service were imperfect weapon systems, which suffered from many "childhood diseases." All these exotic weapons, including the V-missiles, were of limited military value due to their technological immaturity.

Some of the new jets figured prominently in production plans submitted from mid-1943. The RLM planned a massive procurement of some of the more developed types and the aviation industry geared up to produce them *en masse*. They subsequently appeared in increasing numbers on the production lines in 1944–45. These modern aircraft entered series production just as the aviation industry went through the greatest transformation in its history. Here is a short description of the most important advanced aerial weapons that entered series production before war's end.

Me 262

This fighter was arguably the most important jet aircraft of World War II and the one on which the Germans hung their highest hopes. It was a twin-engine, single-seat, heavily armed fighter. The basic airframe was ready before its projected BMW 003 and Jumo 004 engines. After its first jet-propelled flight on 25 March 1942 (during which both BMW P.3302 engines failed), it took more than 2 years before the revolutionary jet engines became ready for operational service. Even as series production of the Me 262 started, its engines were still plagued with problems. The Germans viewed the Me 262 as a preliminary basic design, which could have been constantly upgraded with new technical developments. Projected future developments of the basic airframe included strongly swept-back wings, more refined aerodynamic shapes, mixed propulsion (jet/rocket), radar and air-to-air missiles.

The production plans of the Me 262 became a hotly debated topic in top-level discussions. Principally, the RLM decided in 1943 to produce the Me 262 as a replacement of all other fighter types. Hitler intervened and demanded to produce it as a fast bomber to combat the expected Allied invasion of western Europe. The debate concerning the preferred role of the aircraft raged for several months and caused some spectacular conflicts within the German leadership. Messerschmitt was expected to be the main producer of the Me 262 and deliver thousands of these aircraft to the Luftwaffe. Its Regensburg plant was to be the main production center.[41] Large parts of the story of the German aviation industry in World War II are related in one way or another to the production of this aircraft.

Ar 234

A twin-engine, single-seat jet bomber and reconnaissance aircraft. Its design was first proposed in late 1941. It flew for the first time on 30 July 1943, but its development was protracted, mainly by a major redesign, which replaced its initial droppable takeoff trolley and landing skid with a conventional landing gear. Just as the Me 262, it entered series production only when its engine, the BMW 003, became ready for operational use in summer 1944. The RLM gave Arado an initial production contract for 100 aircraft in November 1943. Further orders increased the number to 200 aircraft scheduled for delivery by the end of 1944 and Arado planned to deliver 1,930 aircraft by September 1945. The main production centers of the aircraft were Arado's plants in Brandenburg, Wittenberg and Freiberg.[42] Like the Me 262, the Ar 234 was a basic design, which could be further developed in various ways and incorporate emerging new technologies as they became available.

Me 163

An aerodynamically advanced tiny rocket fighter, designed as a point defense interceptor. It was the only World War II aircraft that regularly reached speeds close to the speed of sound, but its military value was limited. Its greatest tactical disadvantage was the limited endurance of its engine. Unlike the previous projects described here, due to its limitations the Germans never intended to produce it in large numbers. Therefore, only limited procurement of this aircraft was planned.[43] By the end of 1944 the limited production of the Me 163 was further reduced due to difficulties in the supply of its unique fuels.[44]

V-1

A jet-propelled cruise missile, which was developed from summer 1942 by the Luftwaffe in parallel to the army's A-4 (later known as the V-2) missile and as its competitor. The V-1 was developed by Fieseler and its initial production run was carried out in Fieseler's Kassel main factory. Later another production line was established in a separate and highly secret part of Volkswagen's giant factory complex at Fallersleben. Other V-1 production lines were constructed in the Mittelwerk underground factory, originally used for the V-2 production, and in several smaller underground facilities. The V-1 was a cheap and simple weapon system using a primitive form of jet engine called a pulse-jet, which was suitable only for a one-way flight. The engine was developed by the Argus aero-engine firm from 1939 and was ready for production at the end of 1941.[45] It took only 250 work hours to produce each missile and it was largely constructed from cheap metal and not from expensive aviation-grade aluminum. Similarly to other development programs supervised by the RLM, over-

The Me 163 *Komet* rocket fighter was also viewed as a promising concept, but it failed because of short endurance, protracted development and the difficult nature of its novel power plant (courtesy National Air and Space Museum, Smithsonian Institution, SI 93-4859).

optimism was also a dominant feature in the V-1 program. Full production plans were drawn up and then had to be canceled and redrawn as development of the weapon dragged on due to different technical issues. The projected entry into operational service at the end of 1943 proved to be completely illusory, and the first operational launch against England took place only a week after the D-Day landings. In 1944, as bomber production declined and then completely stopped, the V-1 became the Luftwaffe's principal offensive weapon and its only weapon system capable of regularly penetrating Great Britain's air defenses. Although it was very inaccurate, it caused heavy damage and loss of life when fired indiscriminately at big cities.[46]

He 162

A last-resort jet fighter, conceived, developed and first flown by Heinkel within 3 months. It was a single-engine, single-seat compact fighter of mixed metal-wood construction, which made it suitable for mass production under relatively primitive conditions. Its unique production history will be dealt with in detail later on.

All these advanced weapons can be considered as fully developed weapon systems that entered series production in different rates before the end of World War II. However, their introduction into operational service was far from smooth, and the elimination of their teething technical problems took time and demanded continues development work. As a result, jets never even came close to fully replacing piston engine aircraft in production or in operational service. Even in 1945 conventional designs continued to form the main bulk of German aircraft production. A short description of the main conventional types produced at that time is the following.[47]

Me 109[48]

This single-seat, single-engine fighter was designed by Willy Messerschmitt and entered operational service with the Luftwaffe before the outbreak of the war. It was even used in the Spanish Civil War. It was the most produced fighter ever, with around 35,000 units built before and during World War II. Some 13,942 Me 109s were produced in 1944 alone.[49] It was supposed to be replaced as the main Luftwaffe fighter in 1942, but delays with the introduction of a replacement, and subsequently the decision to produce the Me 262 as its replacement, meant that it was never replaced. Mass production of its modified versions continued right until the end of the war, even though by 1944 it was inferior to both Western and Eastern fighters in many respects. It was considered difficult to fly and the masses of young and inexperienced late-war German fighter pilots mostly lacked the skills needed in order to fly it effectively in combat.

FW 190

Another single-seat, single-engine fighter that was somewhat outdated by 1944. It was developed and purchased as a more advanced complement to the Me 109. It entered operational service with the Luftwaffe in autumn 1941, but suffered continuous problems with its BMW engine. Advanced and much improved versions of this fighter were supposed to enter series production in mid–1944. These were the re-engined and redesigned FW 190D and the Ta 152. The latter was designed primarily as a high-altitude fighter. These fighters were considered to be more than equal to modern Allied fighters. Severe delays in their production meant that the older models soldiered on and remained the most dominant types.

The old warhorse. A Me 109G-5 of 7./JG27 in flight over the Adriatic, January 1944. Although somewhat outdated at that time, constant updates kept this fighter in mass production until the end of the war (courtesy National Air and Space Museum, Smithsonian Institution, SI 73-1989).

The next generation. FW 190V29/U1—a development prototype of the Ta 152 high-altitude fighter, in which the BMW 801 engine was replaced with a more powerful Daimler-Benz 603 equipped with a turbocharger (courtesy National Air and Space Museum, Smithsonian Institution, SI 82-11848).

The FW 190 was also used as a fighter-bomber and as ground support aircraft.

Ju 88, Ju 188, Ju 388

The initial model was the Junkers 88, which was developed before World War II as the Luftwaffe's "wonder bomber." The production of this aircraft formed one of the largest contracts in the history of the German aviation history and turned state-owned Junkers into one of the giants of German industry. This twin-engine aircraft was later converted into an efficient night fighter, and in this role it stayed in series production almost until the end of the war. Its wartime development was the improved Ju 188, which entered service in early 1943. The completely different Ju 388 bomber-reconnaissance aircraft came too late to enter service in meaningful numbers. The production of these three Junkers aircraft was terminated in February 1945.[50]

Other less important types were produced in ever-decreasing numbers during 1944, but all of them were stricken from the production plans before the end of that year. In any case, of all the 20 types produced in significant numbers in 1944, the piston-engine types described above constituted 74 percent of total aircraft production for the year.[51] Taking into account that other planes included in the 20 types were also propelled by piston engine, the dominance of "traditional" aircraft over jets on the production lines is obvious.

Paradoxically, at the time Germany commenced production of some of the most advanced aerial weapon systems of their time, there was a marked drop in the production quality of German aviation products. Hermann Göring generally confirmed the deterioration of quality after the war and attributed it to the dispersal of the aviation industry in 1944.[52] Adolf Galland, former influential General of the Fighters, gave the following explanation when asked by American interrogators why dispersal caused deterioration of quality: "Because the assembly lines were interrupted. The planes no longer were in assembly halls, but somewhere the control surfaces were manufactured, in another plant fuselages were made, construction took place in destroyed halls and construction took place in the open air instead of under roofs."[53]

The explanations offered by these two wartime leaders provide only a partial explanation for this phenomenon. It appears that particularly the high-priority modern jets and other new aircraft types suffered from these declining production standards.[54] Following a comparison flight test of an Me 262 against an Ar 234 prototype in June 1944, Messerschmitt's main development office at Oberammergau complained about the quality of production Me 262s: "**Workmanship**— The workmanship carried out on production Me 262s leaves much to be desired. Armament hatch covers, sheet steel cockpit, engine cowlings, etc., were all poor as were those of most production machines. The surface finish is also coarse."[55] Problems with series Me 262 continued and became even worse. In February 1945 Messerschmitt's chief test pilot Fritz Wendel reported severe technical problems originating from poor quality control after visiting operational units flying the plane.[56]

Conventional types also suffered from this deterioration. Poor quality of the initial production runs of the Ta 152 fighters led to a brief production halt in order to enable the engineers to locate all the defects and to fix them. The main problem appeared to be faulty welding of the aileron pushrods. Further technical problems and other sticky production difficulties finally led to the cancellation of this aircraft at the end of March 1945, when its production capacity was allocated to the production of older proven types.[57]

Unit commanders and pilots receiving the new aircraft experienced severe problems with their new mounts, caused by poor workmanship and slack quality control. Erich Sommer, flying Ar 234 reconnaissance aircraft since the summer of 1944, reported that "spare engines and accessories had to be stripped before use as quality control in the factory became less and less reliable."[58]

It seems that production standards of series Ar 234s was particularly low. Captain Dieter Lukesch, who commanded the first operational Ar 234 squadron, remarked about the production quality of the brand-new aircraft his unit received in July 1944: "Hardly any aircraft arrived without defects which covered all systems and were caused by hasty completion and shortage of skilled labor at the factories. The same applied to the Arado's equipment."[59]

These problems were well-known and higher authorities tried to solve them in different ways. Among others, the Germans tried to use Allied standards in order to restore their own standards of quality. In this framework the Airframe Main Committee arranged to have wing sections of shot-down American aircraft sent around to aircraft factories to show the relative superiority of American workmanship in this regard, and serve as an incentive to do as well.[60] At the end of September 1944 Ernst Heinkel pointed at the excellent polish of the wings of the American Mustang fighters and at the way such finish could improve the performance of German planes. He blamed the inferior German finish on a poorly trained workforce and demanded better training in order to achieve better workmanship.[61]

The main solution was to improve quality control. Quality control was normally preformed on the final products upon leaving the production line. From 1944 it became generally slacker and less efficient.[62] The dispersal of most factories in 1944 made it much more difficult to perform proper quality control because quality control departments usually stayed at the main factory and could not satisfactorily perform their tasks in the dispersed facilities. Furthermore, while efforts were invested in expanding production, much less effort was invested in expanding quality control departments to cope with the new situation.[63]

The countermeasures failed in most cases to improve the quality of the end products and the production standards of German aviation products kept deteriorating until the end of the war. Arguably, the main reason for this failure, and for the worsening quality, was the composition of the workforce the Germans used at that time for aviation production. This huge army of foreigners, POWs, and concentration camp inmates lacked the motivation that prevailed in the German aviation industry before the war. With such manpower it was extremely difficult to preserve the prewar standards.

3

Reorganization of Aircraft Production

"Big Week" turned out to be the most traumatic event experienced by the German aviation industry in its history. The concentrated attacks caused great alarm in the German leadership and made it turn its full attention to the problems of aircraft production under strategic bombing.[1] The concentrated attacks caused grave damage to some highly developed production facilities and severely disrupted important production programs. Karl Frydag, chairman of the Airframe Main Committee, claimed after the war that the Luftwaffe lost around 4,000 aircraft due to "Big Week" bombings.[2] The impact of "Big Week" was deep enough to set in motion a general reorganization of the entire aviation industry and its management. It should be noted, however, that some of the dramatic measures taken after the February attacks were deeply rooted in earlier developments and trends. The main trend apparent after "Big Week" was to reinforce "two central premises of modern industrial management: control and standardization."[3] Udet tried to assert these two elements, so central to military related production, but failed miserably. Milch tried again, and although he succeeded in some areas, in early 1944 the German aviation industry was still badly in need of more efficient control and a higher degree of standardization.

The problems the Germans faced in early 1944 were not only organizational in nature. "Big Week" proved the vulnerability of the German industry to concentrated bombing efforts. The Germans therefore urgently needed a solution to the protection of the aviation industry. Since the Luftwaffe was not in a position to challenge Allied air power in the air, other solutions had to be found on the ground.

The reorganization of 1944 helped first to restore and then to increase aircraft production. Nonetheless, although output reached record figures in 1944, it did not prevent Germany from losing the war in the air. The aircraft produced in 1944 could not turn the tide and prevent Allied air power from roaming freely over the Third Reich. Hitler tacitly admitted the failure of German air crews and their aircraft in August 1944 when he ordered to dramatically increase the output of antiaircraft artillery and its equipment.[4] This order signaled that from this point on Germany's air defense would be primarily ground-based. However, the Germans never fully gave up the hope to regain air superiority over the Reich or at least to inflict heavy enough damage on Allied air power to force it to ease its aerial onslaught. Therefore, the Reich's leadership viewed the reorganization of the aviation industry as the main means to this end.

New Bosses

As already mentioned, much of the German aviation industry was nationalized before the outbreak of World War II. The need to tightly manage the industry in order to supply the war machine enough aircraft demanded increased centralization, which failed under Udet. Even before Udet's suicide, attempts were made to create a unified and more effective central control mechanism, combining executives from the industry and RLM officials. On 14 May 1941, Göring ordered the establishment of the *Industrierat des Reichsmarshalls für die Fertigung von Luftwaffengerät* (Reich Marshall's Industrial Committee for the Production of Air Force Equipment). Its nominated chairman was Udet (later Milch), and most of its members were top executives from leading aviation firms. Interestingly, non-aviation-related firms were also represented in this committee — an indication of an effort to bind other industrial sectors to the aviation production effort. This move was possible because Göring was also the Reich's Commissioner of the Four-Year Plan, theoretically having Germany's entire economy under his control.[5] The Reich Marshall's Industrial Committee was the first attempt to integrate the aviation firms into a general policy-making and planning forum.

Several months later more important reforms were made in the general management of war production. The most important change was the appointment of a new Minister of Armaments. After Minister Fritz Todt died in a plane crash in February 1942, Hitler appointed young and energetic architect Albert Speer as his successor. Speer believed that the only way to improve Germany's war production was through a higher degree of centralization and tighter control. He strove, among other reforms, to take the designers and chief executives of most firms out of the loop, because he viewed the double role of most of them — especially of prominent aircraft designers — as counterproductive.[6] His first initiative to improve aviation production was through the establishment of the *Zentrale Planung im Vierjahresplan* (Central Planning in the Four-Year Plan), established in early 1942 in cooperation with the RLM. This supreme committee was based on the earlier *Industrierat,* which was reorganized into 3 main committees: Airframes, Engines and Equipment. These main committees were composed of several subcommittees, which were appointed to supervise the production run of a specific firm or of a specific item. They were responsible for the allocation of plants for the production of specific types of aircraft, and subsequently for supervising the efficient allocation of resources within the specific factories.[7] Heinkel's supervisory committee, for example, was called F3 and supervised the production of the He 111, 177 and 219 aircraft.

These supervisory committees were also expected to create additional production capacity within the factories they supervised in order to enable future expansion.[8] By placing industrialists as supervisors of other industrialists, the RLM tried to compel the industry to assume broader responsibilities in fulfilling production plans and streamlining their work.[9]

Additional supervisory committees were later established in every military district in order to supervise regional armaments production. This maze of committees and subcommittees was designed to supervise aircraft production at almost every level and in every region. The committees supervised the activities of the factories and firms and rang the alarm bell if they found something wrong.

Such an alarm rang over Focke-Wulf in early 1942, when its supervisory F4 committee sent unfavorable reports to the RLM about the firm's failure to fulfill its assigned output. The firm's top executives were summoned to the RLM and were reprimanded for their mismanagement.[10]

3. Reorganization of Aircraft Production

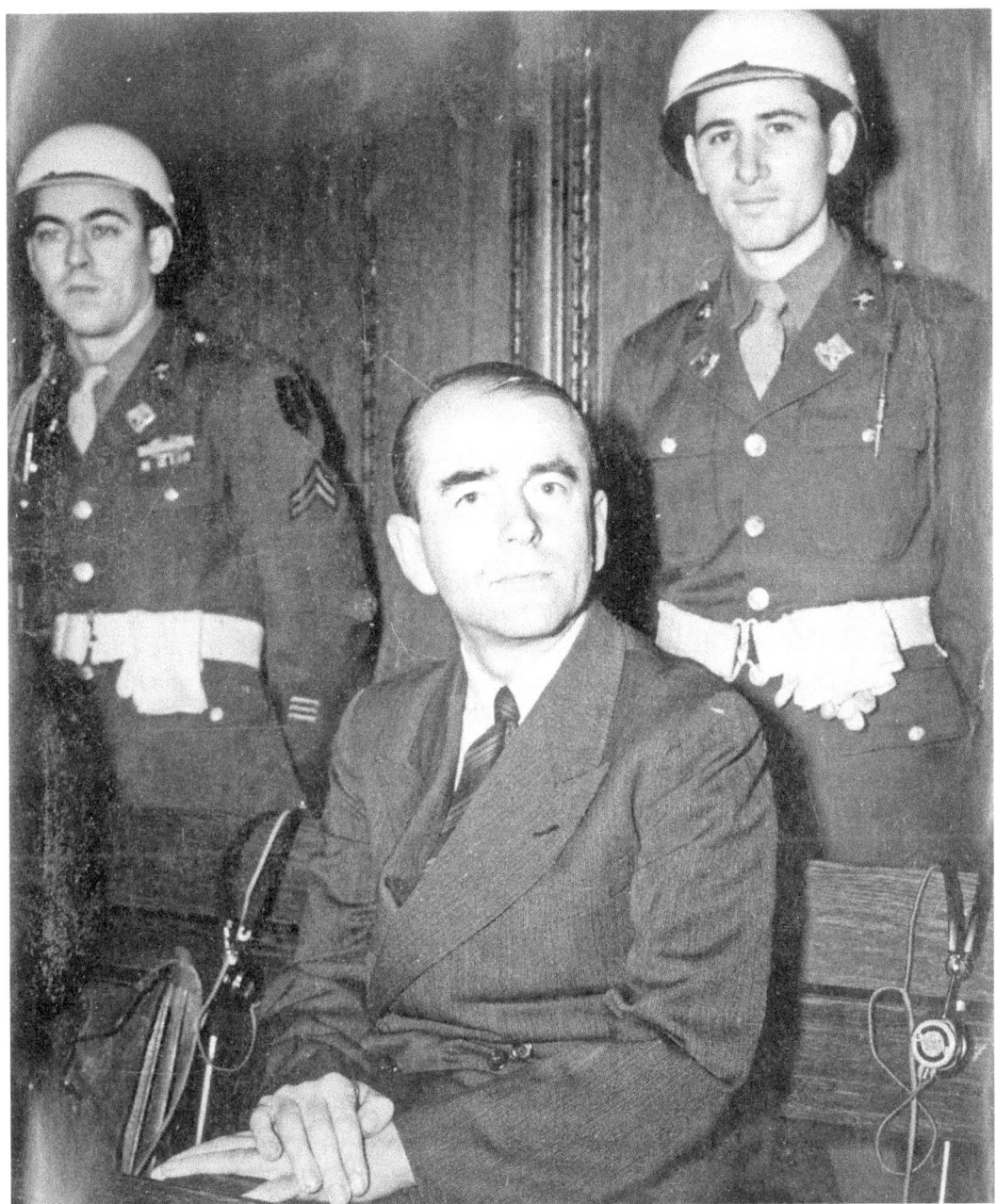

Albert Speer during the Nürnberg Trials (courtesy Yad Vashem Archives).

The committees of the *Zentrale Planung* and the Technical Office of the RLM became the two main pillars of aircraft production management under Milch. Under the new leadership of Speer and Milch this system brought immediate improvement and increase of productivity from 1942, particularly through better allocation of resources and making the industry use them more efficiently.

Another method used by Milch since 1942 in order to tighten his control over the industry was to call in special commissioners from the outside and give them broad powers to reshuffle and reorganize specific firms. These appointments were usually accompanied by the dismissal of top executives. Such special commissioners replaced, for example, several executives of the VDM propeller manufacturing firm after its output fell 20 percent behind the projected figures.

In other cases Milch simply fired executives for failing to deliver the goods. Heinrich Koppenberg, the powerful general director of Junkers, was sacked because his firm failed in the development and production of the Ju 288 medium bomber — another costly failure of the German aviation industry during World War II.[11] Dr. Hans Klemm, the owner and chairman of the Klemm firm, was forced to resign on 23 May 1943 after he objected the RLM's initiative to use the free capacity of his firm for the production of the Me 163 rocket fighter. Until then the Klemm firm produced in its Böblingen factory in southern Germany only light wooden aircraft and some components for other firms. Klemm was therefore of the opinion that his firm was not suitable for the production of a revolutionary and complicated rocket fighter. The RLM overruled his objection, forced him to resign and replaced him with one of its own men.[12] Speer reported after the war that at a certain point Göring wanted to court-martial industry executives who failed to fulfill production plans or to report such failures, and that Milch supported this idea, but nothing came of it.[13]

All the organizational reforms of 1942 helped to largely eliminate the division between state and private interests in the aviation industry. On the one hand it was done by inclusion — making the heads of the industry permanent partners in the planning and supervising organizations. On the other hand it was done through disciplining industrialists and coercing them to fall in line with the policies dictated by the RLM and the Armaments Ministry.

Although the reforms corrected to large extent the disarray of the Udet era, they proved insufficient to solve the problems of a country involved in a multi-front war and suffering from a limited industrial and economic base. All through 1943 pressure had increased to bring all the armament effort under one central direction. The RLM's powerful position enabled it to keep aerial armament independent and separated from the rest. While the army and the navy handed over their armament programs to the Armaments Ministry in 1943, Milch kept the aviation industry under his direct authority. This independence was possible largely through Göring's high status in the Nazi hierarchy, which enabled the RLM, and therefore its state secretary Milch, to enjoy a special status.

This status withered rapidly towards the end of 1943 following the dramatic increase in the bombing campaign against the Reich and the failures of the Luftwaffe on the different fronts. Göring's fall from grace and the increasing competition between the armament programs of the different services caused deterioration in the relations between the RLM and the Armaments Ministry. At the beginning of 1944 the Armaments Ministry tended to support the army's tank production and the navy's submarine construction programs at the expense of the Luftwaffe. This policy caused growing tension in meetings between RLM officials and their counterparts from the Armaments Ministry.[14]

An additional important trend, which had begun in 1942, was the growing involvement of the SS in armaments projects. This trend should be viewed within the general context of the increasing powers of the SS and its organizational expansion. By 1942 this organization, which began in the 1920s as a small bodyguard squad, became an extraordinarily powerful and central organization. Besides holding the Reich's entire police structure under its authority, it also owned numerous economic and industrial enterprises, deployed several crack military formations, controlled certain occupied territories, executed the "Final Solution," and operated a bewildering range of "think tanks," planning staffs and research organizations. The inclination of the SS towards military matters and modern industrial enterprises, combined with its control over thousands of prisoners and other resources, put it in a favorable position to support the armaments industry.

The SS possessed one specific resource that was extremely valuable to the military and the industry: a cheap and readily available workforce. By early 1942 the SS viewed its expanding network of concentration camps not only as a means to protect the state and to "educate" criminals, but foremost as an economical enterprise, a source of inmates for use in war production and in other places related to the war economy.[15] This attitude was the main reason for the genesis of cooperation between the RLM and the aviation industry with the SS. Two main characteristics integral to the ideology and self-consciousness or self-identity of the SS played an important role in this trend. First was the general fascination of the SS with everything modern (at least along its own perception of modernism): data processing, mass production, hygiene, healthier environment, modern policing methods and new military technologies. Its leadership was especially fascinated by new technologies—a tendency Himmler exhibited during his early business enterprises in the 1930s.[16] The second element was the full commitment of the SS as a whole to the total war, and specifically the commitment of its organizations dealing with business enterprises, industry and construction. The unconditional submission of the industry to the national cause under the circumstances of total war fitted perfectly with SS ideology. As Michel T. Allen wrote, "At a time when quotas and raw yield replaced prices as the quantifiable measures in account, and when every capacity was mobilized at all costs, many SS executives found what they had wished for all their professional lives."[17]

Motivated by these factors—and by some business failures—the SS started to shift the focus of its business enterprises.

Already in 1941 the SS started shifting away from expanding its own industrial enterprises and started to seek cooperation with different firms.[18] This development increased SS involvement in key armaments production programs.

The best example of what can be termed a "total commitment" high-tech project was the cooperation of the army and the SS in the production of the V-2 missile.[19] Although the army's ballistic missile was not a traditional aviation industry project and was not related to the Luftwaffe, it showed how far the SS and its partners were ready to go in order to mass-produce a fascinating high-tech and high-priority weapon system. The V-2 project also perfectly served the desire of the SS to get rid of certain groups of prisoners by using "extermination through labor." As historian Michael J. Neufeld pointed out, the V-2 was probably the first modern weapon that killed more people in its production phase than in combat.[20] The SS also supported aviation production projects, even though they proved to be less deadly to the prisoners than the V-2.

Due to its interest in modern technology the SS even initiated its own aviation research and development program. It employed several jet engine developers in the so-called *Aussenstelle Alfred der Kraftfahrttechnische Lehranstalt (KTL) der Waffen SS* (Alfred Station of the Waffen SS School of Motorization) at St. Aegyd, Austria. Max Adolf Müller, a jet propulsion pioneer, who had worked on jet engine designs since 1937 while working for Junkers, led this group. In mid-1939 he moved to Heinkel, where he worked on the axial flow HeS 30 jet engine. He competed with Von Ohain, the inventor of the jet engine, who concentrated on centrifugal flow jet engine. Müller, who was well connected to the Nazi Party, apparently fell into dispute with Heinkel after blaming him for ruining jet engine development. He left the firm in May 1942.[21] After several months he approached the SS with his design proposal. Initially the SS wanted to adapt his engine as a gas turbine for use in tanks, but while working for the SS Müller also developed a new jet aero-engine, the TL-300, which never proceeded beyond the design stage. In 1945 the KTL was evacuated to Sulzhayn in the Herz Mountains and then to Schlatt in the Austrian Tyrol.[22]

This activity was a mere sideshow compared to the growing commitment of the SS to the production of military equipment. As Himmler proudly announced, by the beginning of 1944 the SS was shifting large portions of its workforce from low-tech industrial enterprises — mainly in stone quarries and earthmoving — into one of the most technologically advanced branches of the German industry: the aviation industry.[23]

While the SS started providing manpower and construction capacity to the armaments industry, the air war crisis escalated rapidly. After the air battles of late 1943 it became increasingly obvious to the leaders of Germany's armament that the lost air superiority over the Reich must be restored in order to avert defeat.

Although generally good cooperation existed earlier between the RLM and the Armaments Ministry, towards the end of 1943 the Armaments Ministry began to block some of the RLM's priority requests. Speer's staff kept arguing that there were not enough resources to attain the goals of production program 224/1, which formed the core of the RLM's aircraft production plan for 1944. Milch's requests to give aviation production a special status regarding the allocation of resources were also denied by the Armaments Ministry. It seems that at that time Speer tried to "snatch" aircraft production from Milch by assuming blocking tactics.[24] However, some of Speer's problems were real. In 1943 Germany suffered from the worst drought in 90 years, which heavily affected hydroelectric power production. This in turn caused considerable loss of production in several sectors of the heavy industry — including the aluminum industry. By mid-November 1943 it was realized that less aluminum would be available in 1944 than ever before. The aviation industry could expect therefore a supply of only 22,000 tons in 1944. The supply of aluminum became so desperate that in 1944 the Germans began to systematically recover shot-down Allied aircraft and melt them down in order to reuse their aluminum. Since 10 to 12 tons of aluminum were used in the construction of each Allied heavy bomber, shot-down bombers were viewed as a valuable resource for aluminum.[25] The aluminum shortage appeared to rule out even the revised production program 225/1 submitted in December 1943.[26]

"Big Week" practically ended the conflict between the RLM and Speer's ministry. By that time Speer fully realized the consequences of lost air control over the Reich and started pushing hard for increased fighter production. The shock "Big Week" caused convinced him and others to fully turn their attention to air armament. At the same time, Milch also became convinced that under the new circumstances the RLM could not continue aircraft production on its own. On 23 February 1944 Milch started a tour of inspection of the damaged factories. He returned to Berlin on the last day of "Big Week" — on 25 February. On the following day he went to visit Speer, who became very ill at around that time and spent most of his time in bed. Milch approached Speer with the idea to create a unified directorate under his own and Speer's authority and to empower it to take every measure required in order to attain the goals set by the aircraft production programs. Milch declared that he saw no other way to keep aircraft output on reasonable level.[27] It was supposed to be a temporary solution for a maximum duration of six months, aimed at solving the immediate crisis in fighter production. In reality, Milch's suggestion signified the bankruptcy of the RLM. Speer viewed it correctly as a clear admittance that Milch and his ministry could not longer manage aircraft production.[28] The two were able to convince Göring to follow the example of the Army and the Navy and to hand over the management of the Luftwaffe's production programs to a new organization.

On 1 March 1944 Göring ordered all the committees existing at that time to be put under a unified control organ named Jägerstab (Fighter Staff) — a supreme committee for the management of fighter production in the next six months. According to its establishment order, the Jägerstab was supposed to "carry out the repair and relocation of the damaged factories with a direct authority and without any bureaucratic scruples. It should also contribute to the restoration and increase of fighters' production."[29] Hitler soon approved all the executive orders regarding the establishment of the staff, but raised concerns regarding the title Jägerstab because it hinted at its tasks, and because it could create the impression that it had something to do with hunting, which became unpopular during the war.[30] Despite these concerns the title was kept as was initially decided.

The new staff was supposed to manage the production of specific types of aircraft: day and night fighters, short-range reconnaissance aircraft, destroyers (twin-engine day fighters), and close-support aircraft. These types became collectively called *Jägerstabsbetreute Maschinen* (aircraft supervised by the Fighter Staff).[31] With the establishment of Jägerstab, aviation production finally received the highest priority in Germany's war production. Consequently the Jägerstab received particularly generous budgets for its construction, reconstruction and relocation programs. At the beginning of April 1944 the Armaments Ministry approved a budget of 550 million RM to the Jägerstab. The RLM received only 200 million RM.[32] Its special status was confirmed by an official list published in mid–April 1944 in order to clarify the prioritization of manpower allocation. This list prioritized first specific weapon systems and components, and then general industrial programs. The Luftwaffe's Me 262, Me 163, Do 335 and Ar 234 aircraft, including their armament and equipment, were on top of the weapon systems list. Next were assault guns and then the V-2, V-1, guided torpedoes, air-defense searchlights, electronic capacitors and the High Pressure Pump — a code name for a long-range cannon later known as the V-3. The top industrial programs were fighter production, their engines, armament, ammunition and other related equipment. Next on the list were submarine production, armored vehicles, motor vehicles and tires.[33]

Although Milch and Speer shared overall supervision of this new organization, its daily management was given to chairman Karl-Otto Saur, 42 years old at that time. Hitler personally approved Saur's appointment and he started to work immediately. This trained engineer directed the Technical Office in the Armaments Ministry and possessed no knowledge or experience whatsoever with aircraft design or production. However, in the past he had been involved in streamlining several important war production programs and won much respect among his bosses in the process. He and his staff significantly reduced, for example, the amount of certain raw materials used in the production of submarines and locomotives.[34] Besides the fact that he had had nothing to do with aircraft before, he was known for his hostile attitude towards the aviation industry and for favoring the army in the allocation of resources. Since Speer became ill and was largely absent for the next couple of months, Saur replaced him and therefore became the real boss of the new aircraft production scheme.[35]

The Jägerstab conferred daily at 10 o'clock. Present were its permanent members, among them the chairmen of the 12 main committees dealing with different aspects of aviation production and its related administration. Each member reported the relevant statistics of his committee and the recently encountered problems. Afterwards individual matters were discussed and once each member finished his part he was sent away to continue his work.[36] Saur's management style was strict and made full use of his powers. He followed closely the production figures and the problems associated with them. If something went wrong he reacted immediately and sought immediate solutions. It is clear from the protocols of the staff meetings that even during the first couple of months, when Milch was still present, Saur was the real boss. He ruled his new post with an iron hand. Experts were often sent to bombed factories in order to supervise their repairs. If staff members reported unsatisfactory production figures, "he would tell them [the chairmen] to go there and push them up, or that we must go there, and we would go around to the different companies and speak to the men. The whip — we called that."[37]

One of the whips used by Saur was the use of special SS commissioners. It was similar to old trick used earlier by Milch, but the use of SS men made it even more powerful. If a firm or a factory failed to produce the projected output it was threatened with an SS commissioner. In November 1944 Saur received unfavorable reports about the situation in Messerschmitt from the Armaments Ministry's representative in Augsburg, Prof. Overlach. The directorate of the firm under chairman Friedrich Seiler was especially criticized. Messerschmitt's Augsburg factory had already attracted Overlach's attention in late August 1944, as a group of employees sent to Overlach an anonymous complaint letter, in which they reported the scandalous misuse of cars and precious gasoline for private use by some of the firm's executives.[38] As a result Overlach conducted his own investigation and found more problems.

Since this state of affairs obviously put the Me 262 program at risk, Saur took immediate action and ordered to send *SS-Standartenführer* (SS Colonel) Albert Kloth, the head of the SS Raw Materials Office,[39] to look into the matter and solve the problems by all available means.[40] Unfortunately we do not know if Kloth ever went to Bavaria, but almost two months later the commander of the Armaments Command in Augsburg reported that Messerschmitt still failed to deliver even half of the projected output. He reported that Gerhard Degenkolb, the newly appointed Armaments Ministry's special commissioner for Messerschmitt production, also visited the firm and remarked that it must be totally reformed. The *Gauleiter* of Swabia, Karl Wahl, was the next inspector to visit the plant and he promised to report the dire situation to the Führer. The inspectors generally recommended to sack Seiler and to reinstate Willy Messerschmitt as a chairman.[41]

The Jägerstab went to work immediately after its establishment. Within a week from its establishment, Saur and his staff toured the aviation industry all over Germany in order to get an idea of the damage caused by the recent wave of bombings and of the general situation. In order to save time, factory representatives entered the staff's train one station before the factory's station and briefed the staff about the current conditions in their factory. Once arriving at the factory, a 10-minute briefing took place and then the party broke up into individual discussion groups, dealing with specific issues like repairs, workforce issues, etc. The staff members heard what each department required and decided on the spot if the demand was justified or not. In case it was, the members sent telegrams even before the end of the discussion in order to set things in motion as quickly as possible. In some cases, such as workforce or repair teams allocation, these arrived sometimes within 24 hours after the discussion.[42]

During this tour, code named "Operation Hubertus" (subsequent tours also received this code name), the Jägerstab also prepared preliminary plans for the reorganization of the aviation industry.[43] The immediate tasks at hand were the dispersal of the aviation industry, the relocation of the most important production programs to underground facilities and repairs to the damaged factories. In order to accomplish the repairs, the Jägerstab initiated a general repair operation worth 606 million RM. Repair teams were quickly established and sent to the damaged factories with the required repair equipment. These teams worked under the supervision of repairs commissioner Roluf Lucht.

Further bombings forced the staff to react in real time. On the morning of 12 April 1944, for example, Saur opened the daily meeting with a report about the bombing of aviation factories in Stettin, Sorau, Cottbus, Bernburg, Aschersleben, Oschersleben and several airfields. This emergency forced the staff to immediately redirect repair teams to the damaged factories. Saur viewed this recent wave of attacks as a reminder of the urgency of the staff's other main tasks, namely dispersal and relocation to underground facilities.[44] While the repair work turned out to be relatively easy and was largely completed within several weeks, the above ground dispersal was expected to take until August 1944, with the relocation to underground facilities completed by the end of that year.[45]

The Jägerstab was one of Milch's last initiatives before losing his power and being sidelined. His downfall had begun already in 1943, when Göring often blamed him for continuing Udet's blunders.[46] This attitude originated from Göring's old fears that his talented protégé would become too powerful. These fears looked reasonable at the end of 1943, as Hitler showed great respect towards Milch, especially after he was told that the V-1 cruise missile would be ready for operational use in March 1944 and that the Me 262 could be used as a fast bomber. Once again over-optimism regarding the progress of major procurement programs proved to be disastrous — this time personally for Milch.[47] While Milch presented impressive increased production figures after "Big Week," his days as Göring's deputy were now numbered. Already in January 1944 Hitler was bitterly surprised to hear from Milch that in contrast to his earlier optimistic briefings, continuing problems with the V-1 would delay its serial production. Hitler was furious, because it meant a long delay in the beginning of the revenge bombardment of England he anxiously expected.[48] Even after the establishment of the Jägerstab, Göring continued to disfavor his former protégé and blamed him for different failures of the aviation industry. Among others, in early May he blamed him publicly for wasting construction capacity by rebuilding those bombed factories that were on top of the Allies' target list, only to see them being bombed again.[49]

Milch's downfall finally came on 23 May 1944 during a meeting with Hitler in his Obersalzberg residence. This meeting was part of a series of high-ranking discussions presided over by Göring in his own Obersalzberg villa, which aimed at solving some of the urgent problems associated with the development and purchase of new aircraft types. The Me 262, which was now far behind schedule, stood at the center of the discussions. At that time there were only 10 flying prototypes of the aircraft; three other prototypes had crashed and a small pre-production batch was just about to leave the factories. It was clear that something drastic had to be done in order to speed up this project. After several morning sessions with Göring the meeting moved to Hitler's place in the afternoon. After hearing a briefing about the current status of the Me 262 program, Hitler asked Milch how many bomb-carrying Me 262s were already produced. Milch's answer, that all Me 262s produced so far were pure fighters, enraged Hitler.[50] His outburst signaled the end of Milch's career, and a further delay in the Me 262 program.

Milch was left alone to bear the results of the misconduct of many others. This was a clear case of collective disobedience of a Führer order, in which both civilians and Luftwaffe officers simply ignored a direct order issued by Hitler to adapt the Me 262 to the role of a fast bomber. Although Messerschmitt developed a bomb-carrying capability for the plane, this adaptation was given low priority and was not incorporated into the initial production series. Hitler hoped to use this fast bomber to strike the expected Allied invasion of Europe. This disobedience therefore meant the collapse of an important pillar of his plans to fight off the invasion.

No one was punished directly for this insubordination, but 3 days after the meeting Speer attended a Jägerstab meeting after a long absence from office. He announced that from now on his ministry would fully manage aircraft production, therefore turning Milch's position into an empty shell.[51] Although Milch was not formally suspended from his office, he slipped out of active life in the following weeks and came to his office only rarely. The shift of power was symbolized by the relocation of the Jägerstab meetings. While early meetings took place at the RLM or in the Luftwaffe barracks at the Tempelhof airfield, later meetings took place in the Armaments Ministry.[52]

The establishment of the Jägerstab signified an increased involvement of the SS in aviation production. Only a week and a half after the staff was created, Himmler sent Göring a report about the involvement of the SS in the aerial armament programs of the RLM. One of the main points he made was the increased productivity wherever the SS became involved. He thus suggested to Göring to extend the responsibilities of the SS in order to increase production even further.[53] In light of these developments, Göring and Himmler dispatched on 4 March 1944 the chief engineer of the SS, Dr. Hans Kammler, to work with Saur and his team.[54] Kammler was a highly efficient and ruthless SS officer and engineer, who served before the war in the Luftwaffe and worked in the construction department of the RLM in Berlin. In late May 1941 he was released from the Luftwaffe and became a full-time SS officer. He rapidly moved up in the hierarchy, and in February 1942 Himmler appointed him to head the main construction agency of the SS, named Office Group C. He and his group already showed their organizational skills in managing the construction of the Mittelwerk underground V-2 plant, in which concentration camp inmates were brutally used. Speer highly praised his work in this project, and wrote to him a personal congratulatory letter in December 1943.[55]

Kammler's task as a member of the Jägerstab was to lead the underground relocation of the aviation industry and to direct all construction projects related to aircraft production in cooperation with Franz Xaver Dorsch, the head of *Organisation Todt* (OT). This paramilitary construction organization was established in 1938 to carry out the construction of the *Westwall* line of fortifications along the Franco–German border. Afterwards OT's construction activities were restricted to the occupied territories, where it constructed, among others, giant bombproof submarine bunkers along the French Atlantic coast, and a network of roads in the east. Now, for the first time since 1938, OT was tasked with construction projects within the Reich.[56] The division of work was clear: while Kammler and the SS were responsible for the conversion of existing underground spaces and for the creation of new tunnel systems,[57] OT was tasked with constructing six bombproof bunker factories. Kammler brought with him a small team of SS specialists named *Sonderstab Kammler* (Special Staff Kammler). It was to work alongside the Jägerstab and run the huge construction projects required in order to complete the underground relocation. The spirit prevailing at the Jägerstab and at Kammler's SS staff was quite identical and made the cooperation of the two teams simple and effective. Saur and his men approached their task in a similar manner to the way the SS engineers approached theirs: decisively and with total commitment, while at the same time bearing a special kind of elite team spirit. They also believed in the same kind of modernism as the SS. For the Jägerstab, modernization meant mechanized production lines, production multipliers, and the use of statistics and other scientific tools of management in order to achieve higher yields — the same values that appealed to SS industrial specialists.[58]

These means were already deeply rooted in the aviation industry. As aircraft became more and more complicated, firms meticulously planned their aircraft production using precise calculations, modern scheduling and resources allocation planning. Such methods were, however, no guarantee of the successful completion of a development and production program. The Ta 154 multi-role aircraft project formed a textbook example of a well-planned complicated project involving numerous contractors spread over large parts of the Reich, which went miserably wrong.[59]

The establishment of the Jägerstab weakened the RLM as the top aviation organization of the Third Reich after Milch left the stage. Saur emphasized the need for a close cooperation between his staff and the RLM in the organization of production and in the sphere of research and development.[60] In practice, however, the RLM rapidly lost power through the weakening of its two heads, Göring and Milch. This loss of power was obviously not Milch's original intention. He foresaw the Jägerstab only as a strong interim executive body to serve the interests of the Luftwaffe and the RLM in a time of emergency. Along with his own fall from grace and with the successes of the Jägerstab came important consequences. Gradually the Armaments Ministry, with a strong SS representation became the real manager of the aviation industry. In early June the Jägerstab became responsible also for bomber production, and soon afterwards Hitler principally approved Speer's formal request to completely hand over aircraft production to his ministry. Speer promised Hitler to convince Göring to approve this dramatic change, and Göring had no choice.[61] He formally acknowledged the handover on 20 June 1944. In a special decree he officially transferred the responsibility for the aviation industry from the RLM to Speer and his ministry.[62] This decree sealed a process which formed a major break from the way the aviation industry was run until 1944.

Although the RLM remained responsible for research and development, overall management of aviation production was now in the hands of "non-flyer" technocrats and industrial experts. These production experts were interested foremost in production rates and statistics rather than in technological finesse and the performances of the end products. Just like Speer, they viewed some of the chief aircraft designers who functioned as general directors with some suspicion.[63] For them the means were the ends; increasing the output of the aviation industry was viewed as the ultimate goal. For Saur and for many of his associates it did not really matter what happened to the masses of aircraft after they left the production lines. It did not matter that many aircraft were never delivered or that superior Allied fighters flown by better-trained pilots easily shot down those delivered.[64] Even later in the war, after the successful D-Day invasion, Saur regularly opened meetings of his staff with self-congratulatory announcements of the recent output achievements. He especially pointed out the difficult circumstances under which this output was achieved, and the fact that people in Germany and abroad closely followed this "struggle of the outputs."[65]

As Göring remarked somewhat bitterly after the war: "Saur was a man completely sold on figures. All he wanted was a pat on the shoulder when he managed to increase the number of aircraft from 2,000 to 2,500.... Saur lived only for his numbers, numbers and numbers."[66] Saur and his staff acknowledged only rarely and very late the futility of their efforts. In a Rüstungsstab meeting on 6 November 1944, several members remarked that there were not many pilots left to fly the aircraft they struggled so hard to produce, and that even these were undertrained. Colonel Gordon Gollob, at that time the leader of a special fighter command, defended his service by saying that training of new pilots suffered from severe fuel shortage—an explanation that only underlined the grave situation.[67]

It seems that at least Speer grasped the basic problem facing Germany's air power: although the output of fighters multiplied, the Luftwaffe had no pilots and no fuel to fly them. He acknowledged the first problem in a speech at the Luftwaffe's flight test center in Rechlin on 3 December 1944 by teasing Galland, the General of the Fighters, for his inability to find enough pilots to fly the fighters provided by the industry. He joyfully congratulated the industry for winning the "first round" against the Luftwaffe and expressed the hope that the Luftwaffe would win the next "round."[68] In any case, increased output was everything for the production experts. It is therefore not surprising that the most severe axing of aircraft types was done under the reign of late-war leadership. It is also hardly surprising that the brutal use of slave labor reached its climax under the leadership of these "non-flyers." It must be remembered, however, that the professional aviation masters of the RLM and the Luftwaffe had initiated these two developments much earlier. The "non-flyers" simply picked up their earlier work and refined it to new extremes using their rational management style.

The Reich's leadership viewed the Jägerstab as a great success by the time its original six-month term approached its end. It was clear when looking at its achievements that the partnership of the Armaments Ministry and the SS with the RLM was successful. Although it was initially intended to disband the Jägerstab at the end of the six-month term, already in late May Milch and Speer discussed with Göring the possibility of making the Jägerstab responsible for the entire Luftwaffe production.[69] This idea was discussed during the infamous Obersalzberg conference which resulted in Milch's downfall. By the time Göring approved the scheme in June, Speer was already looking beyond. In mid-July Speer decided to expand the responsibilities of the Jägerstab to control and reorganize the entire armaments production.[70] On 1 August 1944 he ordered the creation of the Rüstungsstab (Armament Staff), into which the old Jägerstab was fully incorporated. Speer's establishment order explicitly acknowledged Saur's successful management style: "In order to enable other important programs to take advantage of the proven work methods and organization form of the Jägerstab, I hereby order the establishment of a Rüstungsstab. Its task is the execution of the given programs without bureaucratic obstruction through direct control."[71]

Saur stayed as the powerful chairman, but overall supervision was now solely in Speer's hands, although Göring kept some authority in specific matters, like the development and purchase of new aircraft types. The maze of committees formerly under the Jägerstab was now placed under the authority of the new staff. At the beginning of October 1944 the organization of the committees was streamlined with the unification of the Airframes Main Committee and the Equipment Main Committee into an Aircraft Construction Main Committee (*Hauptausschuss Flugzeugbau*) chaired by Karl Frydag. This committee was composed of only 4 groups: single-piston engine fighters, jet fighters, bombers and special aircraft. The Special Committees (F2, F3, etc.) supervising the production at single firms and of specific items were placed within these groups.[72] This latest reorganization meant further weakening of the RLM. Several RLM departments that were formerly placed under Milch, including the once formidable Technical Office, were disbanded or transferred to the Rüstungsstab. All research and development work was taken out of the RLM and was entrusted to the new *Chef der Technischen Luftrüstung* (TLR — Chief of Technical Air Armament), headed by Major-General Ulrich Diesing and placed within the High Command of the Luftwaffe (OKL).[73]

TLR now conducted its own development and testing programs, in most cases completely independent of the Rüstungsstab. Sometimes its projects aroused suspicion and inquiries from different organizations that wondered under which authority it operated.

Such was the case of *Sonderkommando Nebel*— a small operational research and development unit that originally was established in July 1944 in order to test-fly Messerschmitt's *Amerika Bomber*, the Me 264. It was stationed from late 1944 in Offingen under the command of Captain Nebel and was tasked to develop a night-fighting capability for the Me 262 in cooperation with the Offingen Metallbau firm — a branch of Messerschmitt. The unit consisted of two flight crews, 29 ground personnel and around 100 firm workers. In addition, 300 inmates and POWs were employed in the construction of a production hall for the unit near the Me 262 forest factory at Burgau (code named either "Kuno I" or "Kiesweg I"). Captain Nebel reported in March 1945 that besides converting some Me 262s into night fighters, Offingen Metallbau was also involved in the production of nose sections for the *Enzian* air-defense missile and wings for the Do 635 long-range fighter (which was never built).[74] The existence of *Sonderkommando Nebel* became known to the Armaments Ministry only after a local bus company complained that the unit rented a bus and never paid the bill. After looking into the matter, the local office of the Armaments Ministry bitterly complained that by using unauthorized capacity and resources, this project disrupted other important and authorized production programs. It then asked to find out who authorized its operation and why it was not coordinated with the Armaments Ministry.[75] It was found out, among other things, that the construction work carried out by the *Kommando* near "Kuno I"/"Kiesweg I" disrupted Messerschmitt's own construction of this important forest factory.[76]

Although the new Rüstungsstab received much broader responsibilities than the narrowly tasked Jägerstab, fighter production continued to be Germany's top war priority. In May 1944 the Allies finally started to systematically strike Germany's real Achilles heel: the oil industry. The ease with which these strikes were carried out and their immediate catastrophic impact on Germany's oil supply underlined dramatically Germany's failure in the air. Speer drew the conclusion that the only solution to the escalating crisis was increased fighter production, as he outlined in a memo to Hitler at the end of June and then again at the end of July.[77] It was clear to Speer that the bombing offensive against the oil industry signaled the beginning of a vicious circle: without adequate fighter protection, the oil industry would be unable to produce the fuel required for this fighter protection. The grave importance of fighter production was clearly underlined in June 1944 in a ministerial decree issued by Speer.[78] He repeated this message in front of the Rüstungsstab on 21 August 1944:

> I want to say something about a matter, which is one of the most important tasks we need to fulfill. It is fighter production. I clearly emphasize that we must keep fighter production at the highest possible levels at any cost, and that the still existing bottlenecks, especially with engines, must be cleared by all available means. It is unthinkable that we will not be able to keep the word we gave the *Reichsmarshall* regarding the execution of the program by all means due to incompetence in the Rüstungsstab. The production figures for August, which were much higher than those of July, as Saur will later tell us, are a proof of what we can do.[79]

Three days later Speer repeated the same message in a conference of armament executives and factory managers.[80]

The establishment of the Rüstungsstab was the climax of the rationalization process of Germany's war production. Finally a single authority managed Germany's entire war production, as well as the most crucial elements of its economic infrastructure. It was good news for the aviation industry and for the Luftwaffe, because those in charge of national resources allocation clearly placed aircraft production on top of their priority list. Hitler also started to issue specific orders regarding the high priority of specific aviation programs, obliging every concerned state and Nazi Party organization to support them. On 12 October 1944, for example, he empowered several special commissioners as heads of important production programs, like the Me 262, the Kahla bombproof factory, jet engine production etc. These commissioners were empowered to solve any problem plaguing these projects by all available means. All the relevant organizations of the Nazi Party, the Wehrmacht and the state were ordered to fulfill all their requirements.[81]

Speer also repeatedly intervened in order to reinforce the support of aviation production. On 26 October 1944, for instance, Speer personally intervened in order to secure extra skilled manpower needed for the construction of production rigs for "high performance fighters." All sectors of the German industry as well as local offices of the Nazi Party were ordered to cooperate in locating and allocating the required workers. Since the measures taken afterwards failed to fulfill the demand, Speer's order was reissued in mid–December 1944.[82]

Hitler continued to use special commissioners well into 1945, especially for the Me 262 program. In early February 1945 he appointed Gerhard Degenkolb as a special commissioner for Messerschmitt's production in order to increase the Me 262 output.[83] Degenkolb won an excellent reputation as a formidable organizer and administrator. He first managed the increased production of war locomotives and was then assigned to organize the production of the V-2 as chairman of Special Committee A-4. Under the circumstances of February 1945, however, Degenkolb faced an impossible task, and he failed to achieve a meaningful increase of the Me 262 output.

The growing involvement of the SS also gave an extra push to the aviation industry at that time. Hans Kammler in particular reinforced his reputation as a man who could get things moving and overcome huge obstacles. Even the Luftwaffe fully appreciated his contribution, and in August and October 1944 he was recommended for decoration with the Knight's Cross of the War Service Cross with Swords — one of the highest military decorations for non-combat service. This recommendation was based on the brilliant work he had done in the underground relocation of fighter production and on his efforts to restore and increase aircraft production.[84]

Hitler further recognized Kammler's services shortly before the end of the war. On 27 March 1945, Hitler entrusted to the hands of this construction-site engineer overall control of jet aircraft production. In the twilight hours of the Third Reich and under the title *Generalbevöllmächtigter für Strahlflugzeuge* (General Commissioner for Jet Aircraft), Kammler became responsible for development, production and even operations of jets.[85] Although he was formally placed directly under the Armaments Ministry, his position was powerful enough to enable him to independently set priorities regarding resources allocation. His position was superior to the RLM and the Luftwaffe, and even General Kammhuber, Göring's own Commissioner for Jet Aircraft, was subordinated to him. With this appointment Kammler became the big boss of Germany's advanced aerial weapons programs: the V-weapons, the surface-to-air missiles, and jet and rocket aircraft. As Ralf Schabel has pointed out, now the SS took over the development, production and operations of most of Germany's advanced aerial weapons inventory.[86] This achievement was of course, completely futile and hollow, because by that time the Third Reich was on the verge of total defeat and there was not much left to do.

It is interesting to compare the way aircraft production was managed in Germany and in Great Britain in a time of crisis. On 14 May 1940, after the British leadership fully realized the country's dire situation, all aviation development and production activities were taken from the Air Ministry and were handed over to the newly formed Ministry of Aircraft Production. Its minister, Lord Beaverbrook, was a successful businessman with no former experience in aviation. He enjoyed full backing of the prime minister and was empowered to overrule any objection. His compact ministry combined production experts, representatives of important manufacturers, and RAF officials. Soon after entering office, Beaverbrook decided that fighter production should have priority over virtually all other types of weapon systems. He used an energetic and aggressive management style in order to tightly control the entire British aviation industry.[87] This leadership structure bore similarity to the one instituted in Germany after "Big Week."

It was not the only similarity. As British aviation historian Sebastian Cox has remarked: "In both instances [the British and the German], however, the foundations of the supposedly miraculous increases in fighter production that followed had been laid by their predecessors."[88] In Britain it was the prewar infrastructure and production plans developed by the Air Ministry before the war that contributed to the achievements of 1940. In Germany there were, among others, the huge investments in facilities and infrastructure, Milch's takeover after Udet's suicide and his rationalization measures that prepared the ground for the huge outputs of 1944. Therefore, Speer and Saur, although becoming the main players in 1944, were able to base their success on a foundation prepared earlier by the RLM and the aviation industry.[89]

Type Reduction

One of the problems the RLM failed to solve before the arrival of the Jägerstab was the lack of standardization and the multitude of aircraft types and subtypes in production. This problem was closely related to the deficiencies in the RLM's procurement policies.

Udet and his staff failed to streamline production programs to just a few standard types, and thus concentrate available industrial capacity in the production of the most important aircraft. As historian Richard Overy wrote, poor management, disproportionate influence of the Luftwaffe on design changes, and inter-firm rivalry were the main reasons for this failure.[90]

The general institutional chaos prevailing in the Nazi polycracy was also evident in the aviation industry at that time. It contributed to turning potentially mild failures such as the Me 210 and the He 177 into major procurement catastrophes. Designers, who were also in many cases the general directors or chairmen of their firms, had their own agenda and seem to have invested more in development of new types and their variants than in solving problems associated with serial production.[91] The Me 209, which was supposed to replace the veteran Me 109 fighter, was a good example for this state of affairs in the aviation industry. Although at the time it was proposed in early 1943, Messerschmitt was already contracted to produce the next-generation Me 262 jet fighter, the firm kept meddling around with the piston-engine Me 209. It tried to convince the RLM to purchase it as a backup to the Me 262. Willy Messerschmitt managed to influence Hitler in early August 1943 to order the purchase of this aircraft in spite of Milch's decision three months earlier to cancel the project altogether.[92] Messerschmitt's efforts went on for at least one year until the aircraft was finally stricken from the production plans. In the meantime, however, Messerschmitt wasted resources badly needed elsewhere in designing this aircraft and preparing it for production.[93]

Messerschmitt's campaigning for the Me 209 was part of his effort to restore his status after the Me 210 debacle. This campaigning formed part of an initiative to standardize the Luftwaffe's aircraft. He first advised Hitler in September 1943 to reduce the number of types produced at that time from 36 to 2 jet types and 4 or 5 conventional types—most of them of his design, of course. These aircraft were supposed to fulfill all tactical requirements through modular construction (called *Baukastenflugzeug*), which Messerschmitt had proposed since the end of 1942. Such design was supposed to enable interchangeability of components and easy incorporation of specific conversion packages (*Rüstsätze*). Modular construction was also supposed to make these planes easier to produce on modern production lines in smaller factories, therefore neatly fitting the overall rationalization drive of that time.[94]

Serious attempts to streamline production and to adapt it to Germany's deteriorating war situation were made in the production plans prepared under the impression of increased strategic bombing on 8 August 1943 (no. 223/1) and 1 October 1943 (no. 224/1). In these plans the focal point shifted heavily towards increased fighter production, while bomber production rate was kept at its current level. This shift was advocated by Milch, who was well aware of the status of Germany's battered air defenses and of the need to regain air superiority in order to avert military defeat. At that time Hitler also pressed for type reduction. In September 1943, probably following the aforementioned meeting with Messerschmitt, he asked Speer to convince Göring and Milch that fewer aircraft types should be produced.[95] Production plan 224/1, submitted after Hitler's meeting with Speer, projected a monthly output of 3,327 single-engine fighters and 577 twin-engine fighters by July 1944.[96] This plan set the goals for the 1944 production rate, but production plan 225/1, published in December 1943, was less ambitious and more balanced. Upon Hitler's request the production figures of fighters were somewhat reduced in order to enable production of the massive He 177 bomber.[97] Göring also pressed for increased bomber production in order to strengthen his bomber arm for the expected resumption of the night bombing campaign against England.[98] This campaign started on the night of 21–22 January 1944 under the codename "Steinbock," when 447 bombers attacked London. The attacks continued with an ever-decreasing force until May. Not much was achieved, because the attacks were not concentrated and the attackers suffered grave losses. This so-called Little Blitz practically finished off the Luftwaffe's long-range bomber force and thus wasted most of the increased bomber output demanded by Hitler.

Since production plan 225/1 (December 1943) was the last formal production program published until July 1944, it was supposed to represent aircraft production for most of 1944, including a fairly large number of bombers, which proved to be completely ineffective during the "Little Blitz." However, events unfolding in the following months dictated a sharp deviation from this plan towards vastly increased fighter production. This shift of policy happened in spite of Hitler's repeated demands to continue bomber production. Yielding somewhat to this pressure, the Jägerstab approved in early March 1944 reduced bomber production regardless of the decision to concentrate all efforts on fighter production. It also decided to equip 30 percent of the He 177 bombers with modern guided bombs and aerial torpedoes in order to improve their operational capabilities.[99] Even in April 1944, when Saur submitted a revised production plan that included no bombers, Hitler demanded to continue production of the same troubled He 177. The German navy also expressed interest in the aircraft as a long-range maritime reconnaissance platform to support the new submarine offensive it hoped to start in 1944–45 with its revolutionary new submarines.[100] In late May and early June, plans were made to produce the aircraft in a new forest factory in Eger, Czechoslovakia, in order to free German capacity for fighter production. All these meddling explains why quite large numbers of this costly and ineffective aircraft were produced in 1944.[101] Hitler finally declared the He 177 "*vollkommen uninteressant*" (completely uninteresting) in mid–June 1944, but its production continued at a low rate for several more weeks, mainly for the maritime reconnaissance role.[102] The ax finally fall on this bomber at the beginning of July 1944 after a key discussion chaired by Göring, aimed at terminating or limiting the production of less important aircraft types. It was decided, among other matters, to terminate all conventional bomber production. Even afterwards it took a couple of months until the production lines came to a complete stop.[103] After the He 177 was finally terminated in autumn 1944, the only strategic offensive weapon left to the Luftwaffe was the V-1 (which was never included in the aircraft production programs).

Not everything was done wrong in the streamlining process. Postwar Allied analysts noted that although a large number of aircraft types were mentioned in the production programs, at least until the last three programs, the overwhelming proportion of aircraft produced were of a relatively small number of types. It even stated that in this regard, Germany was probably in a better position than the United States, where a considerable number of second- and third-line types were produced in large numbers.[104] This conclusion appears true when looking at the proportions of aircraft produced in 1943 and 1944: Of the 25,527 aircraft produced in 1943, some 20,327 (79.6 percent) were combat types. Of the 39,807 aircraft produced in 1944, 28,926 (72.7 percent) were fighters.[105] The German industry was therefore able to use large portions of its capacity to produce the main combat types included in the production programs as early as 1943.

The end of the bombers. Gun camera photograph taken during a strafing run by an American fighter on grounded He 177 bombers in 1944. By the time the He 177 production was terminated, most of these aircraft were grounded due to fuel shortage and operational constraints (U.S. Air Force via National Air and Space Museum, USAF-51947AC).

Type	1939		1940		1941		1942		1943		1944	
	No.	%	No.	%	No.	%	No.	%	No.	%	No.	%
Fighters 1 eng.	1541	18.8	1870	17.3	2852	24.2	4542	29.2	9626	37.7	25860	65.0
Fighters 2 eng.	315	3.9	1236	11.4	880	7.5	671	4.3	2112	8.3	3066	7.7
Bombers 2 eng.	2314	28.2	3348	30.9	3816	32.4	5371	34.6	6254	24.5	5041	12.7
Bombers 1 eng.	557	6.8	611	5.6	476	4.0	917	5.9	1844	7.2	909	2.3
Bombers 4 eng.	6	0.1	38	0.4	58	0.5	251	1.6	491	1.9	518	1.3
Total combat	4733	57.8	7103	65.6	8087	68.6	11752	75.6	20327	79.6	35394	89.0
Others	3462	42.2	3723	34.4	3694	31.4	3804	24.4	5200	20.4	4413	11.0
Grand total	8195	100.0	10826	100.0	11776	100.0	15556	100.0	25527	100.0	39807	100.0

German Aircraft production by type and by year, 1939–1944[106]

Type	January	February
Fighters — all types	2,552	1,971
Bombers and long-range reconnaissance	82	36
Ground-support and seaplanes	369	372
Transporters	0	0
Liaison and trainers	182	100
Total	3,185	2,479

German aircraft production in the first two months of 1945[107]

Close examination of individual production figures clearly shows that less important types were indeed produced in negligible numbers from 1943. For instance, in 1943 only

104 Ar 196 seaplanes were produced, and in 1944 only 80 (and these were largely produced in French or Dutch factories). In 1943 only 76 FW 200 long-range maritime patrol planes were produced, and in 1944 only 7. These two aircraft were well-established types that had been produced since the outbreak of the war. Their replacements were produced in even smaller numbers, if at all.[108] Obviously, production lines that continued to produce even small numbers of second- and third-line types wasted much-needed capacity, but by outsourcing the production of most of these types to firms outside Germany, the Germans were able to effectively use the capacity of their own industry at time of crisis.

As the number of less important types constantly decreased, some minor production programs were completely stopped. Thus in October 1943 the RLM ordered to stop all helicopter production. As a consequence the main helicopter producer, Focke-Achgelis, was ordered to transfer a large proportion of its workforce to Messerschmitt's Leipheim factory, where the Me 163 rocket fighter was produced. The firm was allowed, however, to continue development work with a much reduced staff. The Jägerstab finally terminated development work of helicopters with Hitler's consent at the beginning of March 1944.[109]

Therefore, under Milch, by the end of 1943 aircraft production was already quite rationalized and streamlined. Although waste of time and resources existed in far-fetched development projects and in several unsuccessful production programs, generally the German aviation industry produced just what Germany needed at that time.

Germany's Wooden Wonder. A Ta 154 prototype roars over an FW 190A fighter in one of Focke-Wulf's factory airfields (courtesy National Air and Space Museum, Smithsonian Institution, SI 82-11774).

While production of conventional aircraft types and aircraft already in series production was indeed becoming more effective in 1943, there were several repeats of the Me 210/He 177 debacles. One of the most wasteful projects of that time of increasing rationalization was the Ta 154 twin-engine multi-role aircraft project. This aircraft, designed by Focke-Wulf's chief designer and general director Kurt Tank, was supposed to replace most twin-engine fighters in 1944 — especially in the night-fighting role. It was constructed largely of wood (54.2 percent), and therefore was supposed to be easier and cheaper to produce than metal aircraft. Most of its production was expected to take place outside the mainstream aviation industry — a feature viewed by Milch in February 1943 as a "most pleasant relief."[110] In May 1943, Special Committee F4, responsible for Focke-Wulf and therefore for the new plane, estimated that it would take only 7,376 man hours to produce a single plane and that 3,866 workers — 880 of them cabinetmakers — would be required to produce 100 units per month. Because of its wood construction the RLM planned to manufacture the plane in three separate and independent "production rings" (*Fertigungsringen*). Each consisted of a central final assembly factory and a host of smaller cabinetmaking workshops and factories in its region. These regionally clustered "production rings" were initially largely determined by the availability of a large wood industry and by being located in the relatively safe Eastern parts of the Reich. They included rings in Thuringia (around Erfurt), Silesia (around Freiburg and Breslau) and around Posen (capital of the annexed Polish *Wartheland* region). A fourth "ring" was later conceived around Detmold in western Germany, based on the capacity of the Teutoberger sawmills of this region. The "rings" were supposed to form a largely autarkic production network, independent of the existing aviation industry.[111] Only the Posen "ring" was connected to an existing Focke-Wulf factory. The production line at Posen was a model of modern mechanized production. Separate production lines manufactured main components and performed final assembly. Final assembly was done on two parallel production lines, which were divided into 17 stations. Four workers manned each station and the production line moved at a rate of 6.05 cm per minute. Focke-Wulf's management expected to operate these two final assembly lines on a two-daily-shifts basis with a combined output of 200 aircraft per month — or 1 aircraft every 4 hours.[112] Collectively the "rings" were supposed to produce a total of 500 aircraft monthly.

Focke-Wulf was the main contractor of the program, responsible for the design and for the supply of parts and raw materials, but other firms also took part in the program. The decision to pick Focke-Wulf as the main contractor was based, among others, on the firm's experience in operating widely dispersed production centers. Focke-Wulf was responsible for the Posen ring, but the other rings were placed under the firms Reperaturwerk Erfurt and ATG. Special Committee F4 supervised the entire operation on behalf of the RLM and also coordinated the supply of the Jumo engines for the aircraft. Plans were made with Jumo to disperse the production of the engine in accordance with the dispersal of the airframe production.[113] Special effort was made to prepare the wood industry to produce the wooden parts of the aircraft. The wood and furniture industry sent selected cabinetmakers to Focke-Wulf, where they were trained in the fine techniques of aircraft wooden parts manufacture. This retraining proved to be a difficult task that caused some delays.[114]

Despite the warning of some experts from the Technical Office, the RLM ignored the fact that the German aviation industry had little experience with high-performance aircraft made of wood, and that the traditional wood industry had no experience at all with aviation production.[115] Designer Kurt Tank successfully convinced the RLM using the argument that the British had been able to construct such an aircraft: the famous Mosquito multi-role plane.[116]

Although it was first flown in July 1943, troubles with the power plant, the fragility of the landing gear, and problems with the basic structure of the aircraft severely delayed the Ta 154 program. It was also found out that quality control of the prototype's manufacture was sloppy — an extraordinary deficiency in the manufacture of prototypes. These problems led to several accidents and to further delays. In the meantime Heinkel's competitor, the He 219, entered initial operational service and even became the Luftwaffe's preferred contender for the night fighting role. However, due to Milch's insistence, its production was restricted while the Ta 154 stayed the RLM's favorite candidate. The main reason for Milch's stance was the elaborate production plan of Focke-Wulf's contender. The situation became more complicated after Allied bombing in early 1944 destroyed the Goldmann factory in Wuppertal, where a special adhesive used to bond the aircraft's wooden parts was produced. The Dynamit AG firm produced replacement glue called Kaurit, which turned out to be too weak. Since it was insufficiently acid-neutralized it actually ate away at the wood it was supposed to join, weakening the structure and causing several crashes. Amazingly, Milch stuck to this unfortunate aircraft and refused to give it up as late as May 1944, although by that time he was aware of the fact that Allied air superiority over Germany limited it only to night operations.[117]

By that time a large amount of effort was invested in preparing for the series production of the aircraft. Two completely new factories dedicated to the Ta 154 production were constructed in Posen-Kreising by around 2,000 Jewish slave workers (who were deported to Auschwitz after the work was done in September 1943).[118] An important factor influencing the decision to construct the factories in Posen-Kreising was the availability of a Polish workforce in this area.[119]

In 1944 the production scheme was broadened and more firms became involved. Another production line was later constructed in an underground factory of the Gothaer Waggonfabrik firm, located in a salt mine. Volkswagen's Fallersleben factory also became involved in the project and was contracted to provide some components for the engine mounts. Volkswagen and the RLM even hoped to produce complete aircraft in some free space left at Fallersleben, and from May 1944 in free space available in a V-1 underground factory at Tiercelet.[120] Typically, Focke-Wulf also tried to outsource some of the design work as well as the production of some components to French firms. Some 57 French draftsmen and designers under French direction were employed in September 1943 in Focke-Wulf's main design office in Bad-Eilsen and many of them worked on the Ta 154.[121] At about the same time Focke-Wulf and Volkswagen, which by now had become an important subcontractor, sent a number of technicians and draftsmen to a Peugeot factory in Alsace in order to make the required preparations for the production of different parts and components there and to instruct French workers.[122] The cooperation with Volkswagen and Peugeot continued well into 1944.

As the development of the aircraft dragged on with no meaningful improvement in sight, Kurt Tank joined the line of German aircraft designers ordered to explain their failures to higher authority. He was summoned to an inquiry in front of the furious Göring in Nürnberg — reportedly after an overzealous worker reported he was sabotaging his own creation — and was asked to explain this costly failure.[123] The Ta 154 project continued until

The fourth *Muskito*, Ta 154V4, after a landing accident on 24 August 1944. It was one of several prototypes and development aircraft that crashed during the plane's unhappy development, which continued, as this photograph proves, even after the Ta 154 program was terminated on 27 June 1944 (courtesy National Air and Space Museum, Smithsonian Institution, SI 78-1716).

the cold-headed Saur and his technocrats finally axed it on 27 June 1944. It was estimated that at least 10,000 workers wasted an entire year on this project and that at least 1,027 suppliers wasted time and resources on it at one time or another.[124] Even in October and November 1944 the Rüstungsstab still reallocated to other places workers that were supposed to be engaged in the Ta 154 production.[125] All the factories engaged in the Ta 154 production either received other tasks or lost most of their skilled workers to other firms engaged in fighter production. Some of the wooden parts producers continued working for Focke-Wulf and started to produce parts for the Ta 152 conventional single-engine fighter.[126]

Another case of a severely protracted and faulty procurement program was the V-1 flying bomb/cruise missile. The weapon was an evolution of an earlier pilotless bomb-carrying aircraft design. An early suggestion to develop an unmanned flying bomb was submitted collectively by the firms Argus, Siemens/Askania and Fieseler in April 1942. In June 1942 the firms submitted concrete and detailed plans for the weapon, which the RLM approved on 19 June.[127] Since this project was viewed as the Luftwaffe's answer to the army's A-4/V-2 missile, development of the missile was rushed and corners were cut in an effort to save time. Ignoring the revolutionary nature of this weapon system, the RLM decided to skip individual testing of the different components and instruments used in the missile and to start right away testing fully equipped prototypes. The first V-1 took to the air on 24 December 1942 and soon more followed. The rushed development quickly resulted in a long series of malfunctions and crashes, whose causes were difficult to identify and correct. Particularly the novel Argus pulse-jet engine caused constant troubles that were solved only after intensive high-speed wind-tunnel testing.[128]

In the meantime, under pressure from Hitler and from the army's rival project, the RLM made schedules for series production and service entry, completely disregarding the severe development difficulties. In August 1943 the deadly bombing of Hamburg and a bombing of the V-2 development and production center att Peenemünde caused Milch to press for immediate high-rate production of the V-1. Earlier it has been decided to integrate Volkswagen as a second main contractor. Fieseler handed the required blueprints to Volkswagen in July 1943, although at that time no serial V-1 flew. A completely separated production hall for this highly secret weapon was constructed at Fallersleben and after some initial difficulties Volkswagen started producing the missile in August 1943. A monthly output of 2,000 missiles was expected by December 1943 and initial launches against England were planned for 15 December 1943.[129]

These elaborate plans soon proved to be completely illusory. The development of this modern and revolutionary weapon was highly complicated and it took almost a year to solve all the technical problems. Further delay was caused when an RAF night bombing of the city of Kassel on 22 October 1943 severely damaged Fieseler's main V-1 production center as it was gearing up for production. The firm was forced to relocate the production line to a facility outside Kassel and was largely out of business until completion of the move.

The RLM's decision to start series production in late summer 1943 resulted in spectacular waste of resources. Volkswagen, which after the bombing of Kassel became the main contractor of the program, was forced to completely stop production in November 1943. It was discovered that the welded structure of series missiles could not withstand the forces of ground launch, and as a result the missile disintegrated in midair soon afterwards. The general quality of the final product was also considered to be low. As Karl Frydag reported in a development meeting at the RLM on 3 December 1943, only one-third of the missiles delivered by Volkswagen to the flight-test center in Peenemünde were ready for launch. As a result of this deficiency the wings and the midsections of around 2,000 missiles produced up to this time were scrapped and the V-1 had to be redesigned and tested again.[130] The V-1 never really matured, and even as it finally became operational after the D-Day landings, it still suffered from high failure rates and from poor accuracy and reliability.

Although such failures caused much waste, in other places things improved. It was, however, a lengthy and complicated decision-making process. Under the reign of the Jägerstab and then of the Rüstungsstab the number of aircraft types was finally reduced. Saur claimed that in February 1944 there were 45 types in the production program. Some second-line types were terminated soon after the establishment of the Jägerstab and more type reduction was seriously discussed. In the three-day emergency conference that took place in Göring's villa in Obersalzberg on 23–25 May 1944, it was decided to reduce the 32 aircraft types produced at that time to only 16 by the end of 1945 and early 1946. It was an ambitious plan, which in retrospect included several largely unrealistic items. Among others, the RLM/Jägerstab leadership planned to completely terminate the Me 109 and replace it and the older FW 190 fighter variants with the Ta 152 piston-engine fighter and moreover with the jet propelled Me 262. The Ju 388 was to become a multi-role aircraft and replace several other types. The troublesome Ta 154 was still expected to enter series production, also as a multi-role aircraft. Due to Hitler's explicit wish, the He 177 was to stay in production. The production of Heinkel's old war-horse (and moneymaker) the He 111 bomber was left intact, but the plane was relegated to the transport role.[131] A new heavy jet bomber, the Ju 287, was also expected to enter series production by June–July 1945 and reach a monthly output of 100 aircraft by the end of 1945.[132]

The new production plan, which never received the customary plan number, replaced several aircraft designed to fulfill the same role with one or two types and terminated a number of older types. As we saw, it was not always efficient in task allocation and still foresaw mass production of pricey bombers. Even shortly before the Normandy invasion, Hitler — still obsessed with offensive thinking — demanded increased bomber production.[133] Only after the invasion was the number of types reduced again, to 20. Type reduction was now viewed as a sure means to boost output. Even the Technical Office of the RLM pressed for more reductions. In late June 1944, for example, the Technical Office estimated that further reduction of types would lead to an increase of monthly output by 1,100 aircraft.[134] The Jägerstab made another step in this direction during a discussion with Göring on 1 July 1944. In accordance with Hitler's newly issued order, all heavy and medium bombers were finally terminated and their pilots were retrained to fly fighters. The aim was set at a monthly output of 5,000 fighters and a total strength of 10,000 operational fighters at any time. Production of several second-line types was terminated and production of those left was limited.[135] By September, after the Rüstungsstab took over, the number of types was reduced to only 9 by eliminating all the second-line types that were left. This number increased to 10 following the approval of the new He 162 light fighter soon afterwards.[136] Almost all the twin-engine types disappeared, and generally the only twin-engine piston aircraft still produced from now on was a small number of Junkers night fighters and reconnaissance planes.[137]

At the end of 1944 further prioritization was done by dividing remaining types into special categories. At the top were the *Hochleistungsflugzeuge* (high-performance aircraft): the different jet and rocket aircraft. Below them were the so-called *Leistungsgesteigerte Flugzeuge* (improved performance aircraft), which included a couple of piston engine fighters with more powerful engines. Then there were different bombers, night fighters, liaison aircraft, trainers and one helicopter type (although development and production of helicopters was terminated).[138] By concentrating almost all the capacity of the aviation industry on fighter production, the Germans hoped to equal Allied output — especially American. The problem was that this effort was done partially through playing with numbers and through largely flawed logic.

During 1944 the Allies attributed the declining power of the Luftwaffe to their concentrated strikes against the German aviation industry. Only after the war did they discover that at the same time aircraft output figures actually soared to new records, so that on face value the Allied effort was obviously unsuccessful. A quick look at the German production figures for 1944 supposedly confirms this assumption.

Month	Jan	Feb	Mar	Apr	May	June	July	Aug	Sep	Oct.	Nov	Dec
Output	2445	2015	2672	3034	3248	3626	4219	4007	4103	3586	3697	3155

Monthly aircraft production in 1944[139]

The main point, however, is that this outstanding increase of output throughout most of 1944 was meaningless, because no practical use was made of the end products. In other words, while the industry and its workers (including the slave workers) excelled and increased output despite all the odds, failures elsewhere and Allied excellence in airpower doomed this industrial achievement. Although July 1944 was the most productive month ever in the history of the German aviation industry, on 10 July 1944 only 178 operational single-engine fighters were available for defensive operations over Germany.[140] Several factors caused this amazing failure: shortage of trained pilots, lack of ground crewmen, high attrition rate (both combat and non-combat), general low serviceability due to crumbling logistical support, fuel shortage, and difficulties in delivering finished aircraft from the factories to operational units.

Besides the real increase in output, there was a mathematical-economic factor that partially helped this increase look more impressive than it really was. It was largely the disappearance of most heavy and multi-engine aircraft from the production lines in 1944 that enabled the increased output. The production of single-engine fighters, which predominate the production lines in 1944, demanded less raw material, fewer man-hours and fewer machine tools. In theory, every canceled heavy bomber or transporter meant three additional fighters. It was a simple mathematical fact, which enabled a massive increase of output once the Germans gave up bomber and transporter production.

As Speer saw it and expressed it in a speech at the Luftwaffe's flight-test center at Rechlin, although the Americans produced a staggering number of 69,600 aircraft in 1944, only 26,800 of them were fighters, while Germany produced at the same time 30,500 fighters. Therefore, as he viewed it, even in December 1944 Germany's position was not so bad because it was able to produce more fighters than the Americans.[141] This logic was, of course, self-deceiving and highly flawed in many respects. It failed to acknowledge Soviet and British outputs, the superior quality of most Allied late-war aircraft, and the fact that bombers and ground-attack aircraft could also destroy other aircraft, especially on the ground. This way of thinking only underlined the loss of touch with reality which became widespread among German leaders towards the end of the war.

The ultimate streamlining and type reduction came just few weeks before the end of the war, as Hans Kammler became the General Commissioner for Jet Aircraft. He immediately threw overboard the last official production program, 228/2, submitted on 16 March 1945, and ordered to stop the production of all piston-engine aircraft. From now on all efforts were to be concentrated on the production of the Me 262 — the final "wonder weapon."[142] It was, however, too late for such a drastic change and production of all other types produced prior to this decision continued until the end of the war.

The concentration of available production capacity on several main types was done at the expense of spare parts production. In early 1944 the German industry received orders for around 4.2 billion spare parts for aircraft. The RLM and then the Jägerstab sought continuously to cut this number. Milch reflected on his own experiences from World War I, when as technical officer he was responsible for the spare parts of his squadron. Like quartermasters in every army in history, he kept far more parts in his stores than he actually needed. As he commented at a meeting with high-ranking Luftwaffe technical and supply officers in March 1944: "I believe that such idiots, as I used to be, still exist today. The war can be lost through such idiocy if we don't stop it, if we do not put an immediate end to this exaggerated stocking."[143] Milch regarded this issue so important that he threatened to shoot supply officers found in possession of excessive numbers of spares in their stores.[144] Based on the logic that there were too many spare parts around anyway, it was decided to reduce dramatically the production of spare parts. Plainly said, almost all the parts and components produced after the Jägerstab took the helm were used in the production of new aircraft. Only minimal numbers of spares were sent to depots and operational units. This cut soon reduced serviceability in the operational units of the Luftwaffe due to lack of spare parts. In early 1945, for example, most units flying the Me 262 jet fighter suffered from acute shortage of spare parts for the airframe and the engines. The stores of the supply centers of these units were simply empty.[145] Spare parts shortage led to widespread cannibalization of used parts from grounded and crashed aircraft.[146]

Although cutting the production of spare parts freed some production capacity, there was another phenomenon that continued to cause a lot of wastage. While main types were indeed reduced, many subtypes and conversions were still regularly developed and introduced into operational service. As we saw, most of the aircraft produced by the Germans late in the war were new variants of older types. The effort to continuously update older types indicated a serious problem: the lack of a stopgap generation of combat aircraft between the prewar designs and the new jets. The main technique to try to fill this void was to repeatedly introduce new variants of the same design. It was done in order to keep some standard types updated and to adapt them to changing tactical situations. New variants were also introduced in order to adapt existing aircraft for special tasks different from their original. The demands and initiatives for new versions usually came from the Luftwaffe, but the RLM and the manufacturers proved to be too eager to fulfill them. Lack of a structured screening process and a clear decision-making system on behalf of the RLM — even under Milch — contributed immensely to this state of affairs.[147]

In many cases the Luftwaffe demanded special additions that were not part of the original design. As Peter Schmoll noted, this ill-advised procedure was a leitmotif in almost every German World War II aircraft.[148] It was one of the main factors contributing to the He 177 debacle. One engineer involved in this program graphically described this situation: "From the end of 1942 an officer from the front appeared every four weeks or so with full powers of the *Reichsmarschall* and laid down changes that were based on the most recent combat experience. In principle previously made decisions were revoked in such cases. The firm could not comply with the changes and, understandably, after a few months became timid."[149] Milch also described this system in a somewhat exaggerated way: "I think especially about the demands put forward by some generals. They are quite creative in this field. Some generals easily think that with a monthly output of 3,000 aircraft we could have also got 3,000 variants. In fact, we were not so far from that."[150] Introducing many frequent changes in existing types proved to be counterproductive in times when productivity became most important. The problem of multiple variants became so acute that soon after the establishment of the Jägerstab, Göring discussed its solution with Hitler on 18 or 19 March 1944.[151]

At the end of March 1944, in a high-ranking meeting at the RLM, Milch begged the Luftwaffe generals to hold back their regular demands for different improvements and upgrades for the next 3 to 5 months in order to make it easier for the Jägerstab. He also ordered a stop to modifying fighters for multi-role use by installing bomb-carrying equipment on them, and instead to employ them only in the pure fighter role against Allied air power.[152]

These measures failed to put a complete stop to the practice. In some cases Hitler himself intervened and ordered further complicated developments of basic types. As the Me 262 finally appeared in meaningful numbers in early 1945, Hitler was pleased by reports

The FW 190G, a long-range ground attack version of the FW 190 fighter, was just one of a bewildering array of versions evolved from the basic FW 190 design (courtesy National Air and Space Museum, Smithsonian Institution, SI 82-11858).

about its successes against heavy Allied bombers. Consequently he ordered the installation of different and heavier guns on the aircraft in order to increase its firepower. As late as 8 March 1945 he ordered the development of Me 262 variants carrying six 30mm cannons (instead of the normal four), two 37mm cannons and even a heavy 50mm cannon, originally mounted on armored cars.[153] All these modifications demanded extra development and testing, which were hardly realistic at this phase of the war.

The FW 190 fighter was another good example of this wasteful practice. It was produced in a huge array of subtypes and special variants. These included different day fighters, night fighters, two-seat trainers, ground-attack planes (finally replacing the outdated Ju 87 *Stuka* dive-bomber in the close-support role) and even torpedo bombers. While generally production was streamlined, the frequent introduction of new subtypes continuously burdened series production.[154] Focke-Wulf performed so many design changes and modifications to the basic design that by the end of the war there were still between 500 and 600 draftsmen working on blueprints for FW 190 modifications — a situation that disrupted other design tasks. Among the affected projects was Kurt Tank's advanced jet fighter design, the Ta 183, for which only 200 draftsmen were available.[155]

Amazingly, designers continuously submitted designs for new subtypes of aircraft that never even reached full production. At the time the ill-fated Ta 154 was canceled, there were 12 subtypes and conversions of the basic design in different design stages. On the drawing board there was even a further development — the Ta 254, which also included 2 subtypes — before a prototype was even constructed.[156]

Large portions of the variants were created by installing specific modifications on existing variants. In accordance with the German practice, some changes were introduced as field modification kits (*Rüstsätze*), which were usually installed at service depots by Luftwaffe technical teams. More complicated updates were retrofitted to older models. This procedure required pulling aircraft out of front-line service and shipping them back to a factory for the required conversion. Furthermore, some of the field modification kits became standard upgrades, which were incorporated into new aircraft on the production lines and required redesign and retooling. The FW 190 fighter was again one of the most retrofitted aircraft. In March 1944 the technical director of Focke-Wulf's main design office complained that his department was unable to support the numerous variants and changes made on the basic airframe.[157] This memo failed to stop the ordering of further variants and retrofits. On 27 October 1944, for example, the Luftwaffe's High Command ordered from the Klemm firm the following monthly conversions of standard FW 190 ground-attack planes:

- 20 planes for night attack.
- 20 planes for carrying the BT 700 torpedo bomb.
- 5 planes for carrying the BT 1400 torpedo bomb.
- 20 planes for conventional torpedo bombing.[158]

Ordering such conversions at that time of the war was senseless. Although the Germans desperately needed ground-attack aircraft on the Eastern front, equipping them with anti-shipping weapons was not quite useful at that phase of the war. To underline this point, the Rüstungsstab had stopped all production of aerial torpedoes already in September 1944. This case obviously points to the fact that the Luftwaffe and the Rüstungsstab were not always fully coordinated.[159] In any case, this conversion order meant that each month 65 aircraft were supposed to be pulled out from operational service or diverted from final assembly plants to Klemm's workshops.[160]

By November 1944 series production of the different FW 190 versions became so complicated that Focke-Wulf's main design and production supervision office was forced to reorganize in order to keep track of the plane's production. The problems were aggravated by the fact that large numbers of firms were now involved in the production of this aircraft. Besides Focke-Wulf, also Arado, AGO, NDW, Weser, Fieseler, Mimetall, Erla, Gotha, Gustloff and Langenwerke license-produced complete aircraft or main components.[161] This proliferation of manufacturers created further local variants of the basic series aircraft. The special *Serienbetreuung* department, first established in July 1944 within Focke-Wulf, was reorganized and its representatives were given broader authority in order to solve the problems arising from the large number of variants and to streamline production in the different firms by reducing the number of variants.[162]

The other main German piston engine fighter, the Me 109, also "suffered" from a multitude of subtypes and modifications. This lack of standardization contributed to the steadily deteriorating serviceability in operational units equipped with this aircraft. On top of the general short supply of spare parts, it became difficult to supply these units with the spare parts required for the specific subtypes they were using. In contrast to the FW 190, Messerschmitt sought to solve this problem by initiating a standardized version of the Me 109 in the summer of 1944. This version, the Me 109K, incorporated all the features of the newest subtypes of the G variant produced at that time. It started to roll off the production lines in September 1944, but only 754 such aircraft were produced by the end of the year. The initiative failed, however, because production of former versions never stopped and the Me 109K itself soon also evolved into several subtypes.[163]

The process of constant evolution was well established within the German aviation industry. One should only follow the evolution of the Me 109 fighter (it first flew in September 1935) to see how far a basic design could go. It must be said, however, that such an evolution process was also common on the Allied side. Famous Allied fighters, such as the American P-51 Mustang, the British Spitfire and the Soviet Yak fighters, proliferated into many variants during Word War II, but in contrast to the German practice, only a few main types of these fighters were mass-produced. The evolution of these aircraft was quite straightforward, and although sub-variants existed, each nation formed its own system of incorporating them in the most efficient way. The Soviets simply restricted the number of subvariants approved for series production. In the USA frequent changes and updates demanded by the armed services were initially added to newly completed aircraft in special dedicated depots, but this system proved to be highly inefficient and costly.[164] Later, changes were usually accumulated into batches and were incorporated into series-produced variants as "blocks." This system allowed fewer disruptions of running production programs. The Aircraft Modifications Committee of the British Ministry of Aircraft Production carefully screened any suggested modification and then discussed with the industry the best way to integrate it. The British solution was of incremental improvements and continuous modifications that could be "spliced in" more quickly and with less disruption to production schedules. This practice minimized frequent retooling and delays needed when implementing immediately any new change. As a result of this policy, the Spitfire fighter between 1938 and 1944 went through more than 20 significant revisions, none of which required more than a fraction of the man-hours spent in developing and tooling up the original design.[165] In contrast to a common belief,[166] thanks to this policy the Allies not only produced larger numbers of aircraft, but also high-quality ones. The latest models of the Spitfire and the Mustang were equal or superior to all of Germany's late-war piston engine fighters and were even able to successfully engage enemy jets on many occasions.

While the new German centralized production leadership tried to streamline the production programs, it also prepared the production of completely new types. Prominent among them were the Me 262 and the Ar 234.[167] While the Ar 234 was a later development, the Me 262 had been on the scene since 1941 and its status was higher. Although this research does not deal with their history, several central issues regarding these planes should be highlighted here. Although the RLM decided to go ahead with the Me 262 in early 1943 as the leading future aircraft of the Luftwaffe, starting its serial production was a painfully slow process. Two main problems plagued it at this stage:

1. Technical problems with the jet engine, which were solved only in the summer of 1944.
2. Lack of skilled workforce for its production. This problem is less well-known and was part of a chronic problem affecting the entire aviation industry since the beginning of World War II.

In early December 1943, Hitler ordered to make every effort to start series production of the Me 262 and the Ar 234 as soon as possible. He ordered Göring to report to him about their production status every 14 days. Göring and Milch were therefore under growing pressure to push these projects forward.[168] The main factories of each firm were foreseen as the main production centers of these aircraft. Messerschmitt's Regensburg and Augsburg factories were supposed to produce the Me 262, while Arado's new factory and flight test center at Alt Lönnewitz was supposed to produce the Ar 234. It was logical to allocate the production of these advanced aircraft to well-established and modern factories.

Although they were allocated the highest priority, the new jets seemingly competed with the V-weapon programs. While the Luftwaffe was able to largely save its aircraft-producing capacity by handing over the production of the cheap V-1 "wonder weapon" to car manufacturer Volkswagen (which already manufactured components for Junkers bombers), the army's A-4/V-2 was a much more serious rival in terms of resources allocation. The question of giving the Me 262 higher priority than the V-2 was raised several times by the RLM and the Armaments Ministry, but the two programs basically received the same high-priority level. In June 1944 the supporters of Luftwaffe production apparently achieved a major success by influencing Hitler to restrict monthly production of the V-2 to only 150 pieces. According to an order Hitler issued, "manpower and materials made free [by this restriction] should be used foremost for the high-rate production of 'Kirschkern' [code name of the V-1 program] and then — as far as possible — in the increased production of jet bombers."[169]

This decision was largely meaningless, though, because the V-2 remained a high-priority program, and as an army project fully backed by the SS it could rely on the resources of these two organizations.[170]

Therefore, the V-2 posed no real threat to jet aircraft programs and their production delays were caused mostly by the previously mentioned difficulties.

Flight of Fantasy — Late War Research and Development

While preparing these aircraft for production, German designers started working on the next generation of military aircraft. Among them were — again — numerous variants of the Me 262 and the Ar 234, but also a host of futuristic designs. One of the most popular topics of the "what if" school of aviation and military history are those secret advanced aviation projects carried out by aviation firms, research organizations and the Luftwaffe towards the end of World War II.[171] The number of advanced aviation projects carried out while Germany rapidly lost the war is indeed striking, and the topic cries for a closer look and some analysis.

Historian Lutz Budrass saw in this "projects inflation during the last weeks of the war the best sign of the fact that the armaments industry finally lost faith in Germany's victory." He argued that during the last phases of the war, large numbers of designers unemployed by the destruction of aviation factories tried to escape last-minute conscription by inventing valuable projects.[172] While capturing the mood of 1945, this argument does not explain why so many projects appeared in 1944, while aircraft factories still functioned, or why many of these projects were fully supported or even initiated by the RLM. Furthermore, the *Entwicklungs-Hauptkommission* (Main Development Committee) — which included top representatives from the aviation industry and from the RLM — discussed in detail many of these projects. Committee members included prominent designers, executives and officials like Kurt Tank of Focke-Wulf, Walter Blume of Arado, Heinrich Hertel of Junkers, Roluf Lucht and Siegfried Knemeyer of the RLM. They served high enough in the hierarchy to feel safe from being drafted for front-line service. In March 1945 they still met regularly to discuss future projects, like a new jet-propelled all-weather and night fighter.[173] This business-as-normal attitude of these top-ranking experts makes the question of proliferarating new projects even more interesting. Experienced designers and top executives like Kurt Tank knew very well that it took 3.5 to 4 years to develop a conventional piston engine aircraft and make it ready for operational use.[174] So why were they still at work on such projects so late in the war? Why did the RLM, the Jägerstab and the Rüstungsstab fail to impose stricter control on research and development activities at times of growing rationalization?

The amount of investment in research and development in the last two years of the war is particularly striking. Between 1943 and early 1945 the budget allocated to these activities increased from around 340 million RM to around 500 million RM. The number of personnel allocated to research and development also increased significantly at the same time: from around 7,000 in 1943 to around 8,000 in early 1945.[175] Large percentages of these people worked in fairly large firm-based research and development teams. Next to the aircraft production boom in the 1930s there was a marked increase of patents submitted by aviation firms. The firms invested much of their increased profit in development work. Therefore, research and development became an integral part of the expansion of the aviation industry and this trend continued during the war.[176] Messerschmitt's research and development department, which moved in June 1943 from the firm's main offices at the Augsburg factory to a new facility at Oberammergau in the Bavarian Alps, numbered some 1,400 workers — around 600 of them involved directly in research and development. British intelligence officers who inspected the facility immediately after the war noted the strong team spirit prevailing among members of the staff there.[177] This "dream team" was one of the strongest firm-based research and development teams, and it was responsible for numerous wartime developments — including some late-war advanced "fantasy" projects. Kurt Tank, who functioned in parallel as Focke-Wulf's general director, chief designer and senior test pilot, led another strong research and development team based at Bad Eilsen. The mere existence of such large teams dedicated to research and development contributed to the proliferation of advanced development programs. The firms continued investing in these teams because their work sometimes brought large contracts. Henschel's *Abteilung F*, for example, had since the late 1930s been developing different guided weapons using rocket engines. It was relatively low-risk development work compared to the development of manned rocket aircraft, and by summer 1941 the result of this work — the Hs 293 radio-guided missile — was ready for production. By the end of 1943 Henschel delivered 24,720 missiles of this type, worth almost 200 million RM. This was the largest single contract in Henschel's history and it helped secure the firm's financial status after years of insecurity.[178] This and other teams continued their work almost undisturbed up to the last days of World War II, and therefore kept streaming proposals and blueprints.

Other factors contributing to this influx of projects were also deeply rooted in the structure and institutional culture of the German aviation industry, and were not related only to the late-war period. The story of some projects connected to the Ar 234 jet bomber exemplifies these factors quite well. Arado engineers and aerodynamicists Rüdiger Kosin and Walther Lehmann developed in 1942 a new type of crescent-shaped swept-back wing for high-speed flight, but were forced to put it aside because they were busy designing the Ar 234. After finishing their work on the Ar 234 they suggested a heavier four-engine version of this aircraft. This design soon received the internal project number E395, and on 16 January 1944 the RLM gave the project a go-ahead under the impressive priority level *Nationale Aufgabe* (National Task). The aircraft was viewed as a potential replacement for the He 177 as the Luftwaffe's future heavy bomber. Some RLM officials even proposed to terminate the He 177 production immediately in order to free production capacity for the E395. The new aircraft was supposed to fly for the first time at the end of 1944. It should be remembered that by January 1944 only 5 early configuration prototypes of the Ar 234 existed and that the aircraft and its engines were at least 6 months away from series production. Following "Big Week" and Arado's assignment to increased conventional fighter production, the E395 project was abandoned. However, as Kosin later recalled, "One day in May 1944 Siegfried Günther (Heinkel's chief designer) came to my office in Landshut, under orders from the RLM, to obtain the complete preliminary design drawings for the airplane. Heinkel was to design and build it under the designation He 343. We were not angry when this happened, but as far as I know, the project was scraped eventually because its engines never became ready for installation."[179]

Colonel Siegfried Knemeyer, the influential head of the Aircraft Development Department in the Technical Office, in fact asked Heinkel and Junkers in January 1944 to develop a four-engine jet bomber in accord with Hitler's order to concentrate all efforts on jet planes. The project was conceived as a "crash" program, to be completed by early 1945 through a tight interindustrial cooperation. The RLM authorized twenty prototypes and development planes right from the start.[180] In order to cut development time, Heinkel decided to base its design on two existing projects: its own P 1068 jet bomber concept study and the proven Ar 234 design.[181] The easiest way to do it was to take over Arado's E395 project. Günter's visit to Arado took place not in May but on 24 January, after Knemeyer decided to go ahead with Heinkel's design and ordered Günter to personally collect the required material from

Jet bomber. Junker's futuristic Ju 287 won the RLM's heavy jet bomber tender in 1944. This is the first crude prototype/technology demonstrator, created by combining components from different existing aircraft (courtesy U.S. National Archives and Records Administration).

Arado.[182] Further data for the project were obtained from the DFS research establishment, which carried out numerous wind tunnel tests on P 1068 models with different configurations.[183] Heinkel started to construct two prototypes of the new bomber — now designated He 343 — in early summer 1944 and ambitiously planned to produce 200 planes by July 1945, as if the less revolutionary He 177 affair never happened.[184] The construction of the prototypes was underway when the Rüstungsstab decided on 22 November to stop the work and concentrate all the efforts on the Junkers'-built rival, the Ju 287. Heinkel stored and later scrapped all the parts and tooling manufactured for the first two prototypes.[185] The reasons for the termination of the project were probably the desire to free Heinkel's capacity for the new He 162 fighter, and because the RLM and then the Rüstungsstab favored the more complicated Junkers design as the Luftwaffe's future four-engine jet bomber.

Before the He 343 was terminated Kosin and Lehmann took out of the drawer the blueprints of their crescent wing and proposed a version of the aircraft using this wing. Since this was a long-term proposal, it was decided to try the wings first on one of the Ar 234 prototypes. It was never finished, and British troops who took over Arado's facility in Dedelsdorf at the end of the war destroyed it.[186]

These Arado/Heinkel jet bomber projects were among the more rational "fantasy" enterprises undertaken by the Germans, while many others were completely far-fetched considering technical capabilities or production capacity available at that time. However, they clearly demonstrate that many such "fantasy" projects were part of a well-established evolutionary process integral to so many German technological research and development projects. Furthermore, many of the most fantastic schemes involving jet and rocket propulsion were long-term in nature. For instance, most of the further developments projected for the Me 163 rocket fighter were prepared and submitted before the end of 1942 — when the basic design was still in its early fligh-testing phase and long before the desperate days of 1944–1945.[187] It should be remembered, of course, that the Me 163 prototype first flew in 1941 and was based on research that started before the outbreak of World War II.

It is hard to make comparisons between German and Allied research and development processes, but it appears that the Germans tended to invest too much effort in completely new projects at times when streamlining and type reduction was the call of the day. Much effort and many resources were wasted industry-wide on such projects: designers worked on the E395/He 343 in two firms, prototypes and mockups were manufactured, wind tunnel tests were carried out, etc. Furthermore, while this project was well under way, the RLM changed its mind and chose the Ju 287 instead of Heinkel's design.[188] To make matters even more intriguing, while the He 343 was an aircraft of a quite simple design, the Ju 287 was a much more complicated design, using revolutionary forward-swept wings. Therefore, it was a risky choice of a very advanced and untried aviation technology.

Finally, it is clear that even in 1944 and under new management, control of research and development was still loose. Similarly to the E395/He 343 project, other expensive projects were fully or partially initiated by the RLM and received its backing and financing. While not responsible anymore for aircraft production since March 1944, it was still responsible for research and development, and therefore could work with the industry behind the back of the Jägerstab/Rüstungsstab. While these new staffs tried to streamline production, the RLM initiated new projects and new versions of existing types, which were mostly useless and wasteful. Thus, for example, towards the end of 1944 Göring ordered the conversion of the surviving He 177 bombers into "Mistel" air-to-ground missiles. Consequently TLR directed Heinkel to manufacture two prototypes of the weapon. Soon it became clear that it was a completely unrealistic contract because at that time Heinkel worked on the He 162 light fighter project and lacked the required resources to design this complicated conversion.[189]

In several other cases, aircraft firms and designers submitted "fantasy" aircraft proposals in response to competitions for a contract issued by the RLM. This was certainly the case with the single jet engine day fighter competition initiated in late 1944, which resulted in at least 8 separate design submissions. No winner of this contract was ever announced.[190]

The unrealistic attitude prevailing at the Reich's leadership was crucial in pushing forward some development programs. Hitler was supposedly aware of the time it took to develop modern aircraft, especially those using advanced technologies. His orders, however, reflected to a large extent the unrealistic approach prevailing in the aviation industry and in the RLM. In some cases, like with the Me 262, he was severely misled by the experts. In other cases he independently developed his own expectations. In June 1944, for example, he expressed the hope that the development of the Do 335 heavy fighter would go ahead of schedule, because otherwise it would enter series production too late to make any difference.[191] The Do 335 was another advanced design, which took long time to develop and therefore never entered series production. When the Rüstungsstab decided to stop all work on this aircraft in March 1945 and to strike it out of the production plan, Hitler objected. On 22 March 1945 he ordered Saur to continue the trials program of the aircraft and approved the construction of a prototype of a twin-fuselage long-range version — another far-fetched concept.[192]

Even the cold-headed technocrats of the late-war leadership were sometimes extremely unrealistic in their projections and planning. As Junkers received the production contract for the aforementioned Ju 287 in March 1945, it was expected to deliver the first series aircraft by October 1945 and to achieve a monthly output of 25 aircraft by December 1945.[193] Considering the complexity of the Ju 287 and Germany's situation at that time, the production plans for this aircraft were purely imaginary.

Perhaps the most important impulse for the large assortment of advanced project proposals was the availability of an advanced aeronautical research infrastructure in Nazi Germany. Aeronautical research was well established in Germany before the Nazis came to power, but from 1933 German aviation research expanded beyond recognition. It also went through a process termed by historian Helmut Trischler "from aviation research into grand research" ("von Luftfahrtforschung zur Grossforschung"). The last term refers to research projects that demanded huge resources allocation on the national level and the coordination of different activities in many research and industrial centers.[194] The expansion of research facilities in this framework was extensive. During World War II German scientists had 63 wind tunnel complexes at their disposal: 14 of them belonged to universities, 39 to special aeronautical research centers and 10 were owned by aviation firms. An additional 15 wind tunnels were available to the Germans in the occupied countries. Nineteen of the total of 78 wind tunnels available to the Germans were useful for the research of high-speed flight regimes. In contrast, in the USA there were only 3 high-speed tunnels, and none of these was as powerful as the most powerful German tunnels. Consequently, German wind tunnels — including the advanced ones — were more accessible to firms and designers than the American tunnels were.[195]

Messerschmitt rigged this Me 323 transporter at the Obertraubling airfield to drop the fuselage of the fourth Me 262 in order to test its aerodynamic features (courtesy U.S. National Archives and Records Administration).

Large numbers of these tunnels were operated by research organizations dedicated to aeronautical research and development. Messerschmitt's research and development team at Oberammergau enjoyed not only the availability of the firm's own modern research facilities, but also the facilities of the *Luftfahrtforschungsanstalt München* (LFM — Aviation Research Institution Munich) that was established early in the war at Ottobrunn by the state in order to provide research services to aviation firms based in southern Germany.[196] The LFM facilities were designed right from the start to test aviation technologies associated with jet propulsion, and they took advantage of the availability of powerful hydroelectrical stations in nearby Tirol in order to power their multiple modern wind tunnels.[197]

One of the biggest and most influential research and development organizations was the *Deutsche Forschungsanstalt für Segelflug* (DFS — the German Research Institute for Gliding). Professor W.J. Georgii from the Darmstadt Technical University founded this important aeronautical research organization in 1924. As its name implied, it dealt initially only with unpowered flight, but in the Third Reich it expanded immensely and dealt with almost every aspect of aeronautical research. It was largely financed by the RLM and by the aviation industry, which used its services and implemented many of its developments.[198] Some research organizations outside the aeronautical research establishment also contributed significantly to aeronautical research. One of the most important among them was the *Hamburgische Schiffbau Versuchsanstalt* (the Hamburg Experimental Shipbuilding Institution) in Hamburg.

This maritime research and development center operated several model test basins and other facilities, which were also very useful for aerodynamics research.[199]

The DFS, LFM and other organizations dedicated to pure research enabled scientists and designers to systematically test theoretic aerodynamic and propulsion concepts before implementing them on full-scale test aircraft. This testing process was especially critical for the development of high-speed flying machines, which required a huge amount of data collection and data processing. It was impossible to develop such advanced concepts with the old-fashioned "trial and error" process that characterized the earliest ages of aviation. The advantage offered by these research complexes to advanced aerodynamic research is clearly visible in the case of swept-back wing research, done mainly at the *Aerodynamische Versuchanstalt* (AVA — Aerodynamics Research Institution) in Göttingen.[200]

Some firms operated their own highly developed research facilities. Junkers, for example, operated in Dessau a research complex of two large high-speed wind tunnels that had been completed in 1943. This complex contributed significantly to the research and development activities of this firm in the late years of the war.[201] Among other projects, it enabled Junkers to develop the swept-forward wing design for the aforementioned Ju 287 and to successfully fly it on a full-scale prototype before the end of the war.[202] Kosin and Lehmann's sophisticated crescent-shaped wing also benefited from the advanced research facilities available to them within Arado.

Even some secondary aircraft producers operated wind tunnel and research complexes. The Blohm & Voss firm operated in its development facility in Finkenwerder (not far from the current huge Airbus factory near Hamburg) a small wind tunnel and a small smoke tunnel in addition to other research equipment.[203] These in-house research facilities enabled Blohm & Voss to develop some advanced technology products, like the BV 155 high-altitude fighter and guided bombs.

The Luftwaffe and the RLM also operated several test centers. Although structured differently than the pure research and development centers, the test centers contributed immensely to this activity. The best-known and the most important test centers were Rechlin, where new aircraft types were tested, and Peenemünde-West — a large development and test complex for missiles, rocket planes and guided weapons. Next to them there were smaller test centers dedicated to the testing of aerial weapons, seaplanes and instruments.[204]

In 1943, in line with recent rationalization drives, the RLM made an effort to improve control of the maze of research and development activities. A committee called *Forschungsführung* (Research Leadership) was established and placed directly under the Supreme Commander of the Luftwaffe, namely Göring. Its four members were leading scientists, and one of them, the prominent Prof. Ludwig Prandtl, functioned as a chairman.[205] The committee was put in charge of planning and coordinating all aeronautical research activities. Through its work, coordination improved, but although the *Forschungsführung* determined policy and assigned projects, the research organizations under its control preserved a large degree of independence and were able to initiate or continue their own projects. Furthermore, the *Forschungsführung* also promoted multiplication of projects by initiating competition between the organizations under its control.[206]

More chaos was caused in the field of research and development by faulty decision-making in the higher echelons. An example for this and other problems described here is the story of the Horten flying wings. The two Horten brothers were a garage startup team operating outside the main aviation establishment. Their gliders and proof-of-concept aircraft were based on flying swept-back wing designs, similar to a concept developed at the same time in the USA by legendary designer Jack Northrop. Göring regarded the Horten brothers as genius independent designers and decided to support their work. In a meeting on 28 September 1943 at the RLM he expressed his full support of their flying wing development and ordered the RLM to fully support their work.[207]

The Hortens succeeded in making good progress with their rudimentary flying-wing concept thanks to access they were granted to an advanced supersonic wind tunnel at the AVA in Göttingen in early 1944. This tunnel was one of only four supersonic wind tunnels available in Germany[208] and it helped them to apply their design concept, which was restricted until then to low-speed flight, to a full-scale jet fighter. They offered Göring this design and he decided to allow them to work on it. He based his decision to support this far-fetched jet fighter project on a loose appraisal of the Horten brothers' design capabilities and on his own dislike of firm-based developers at this stage of the war. There are indications that some of these established developers fully realized the impracticability of a flying-wing fighter.[209] Nevertheless, with Göring's backing these outsiders were authorized to produce several prototypes of their advanced fighter. At least two prototypes made largely of wood were completed before the end of the war. Series production of this "wooden wonder" by the Gothaer Waggonfabrik firm was planned even before the first prototype took to the air, but never started.[210]

Flying wing. The Horten brother's Ho IXV2 jet fighter prototype before its first flight in Oranienburg on 2 February 1945. Although it was a design far ahead of its time, Göring authorized its serial production by the Gotha firm even before the first flight (courtesy National Air and Space Museum, Smithsonian Institution, SI 77-14586).

The Horten flying-wing jet fighter is another example of a project made possible by advanced research facilities and a gradual evolution process. It is also another "fantasy" project, which had its roots many years before the desperate days of 1944–1945,[211] and that was pursued by higher authority at a stage when resources were scarce and the number of types was constantly reduced. Furthermore, it represented again the incredible optimism prevailing at the German aviation establishment regarding the quick ripening of sophisticated technical developments and their implementation in operational military aircraft.[212]

One last explanation for the proliferation of advanced projects in 1944–1945 may be found in the structural changes that took place in the management of aviation production after "Big Week." As the RLM was gradually pushed out from control of aviation production, one of the only practical functions left under its authority was research and development. After the establishment of the Rüstungsstab this function was transferred to the new TLR division, placed within the Luftwaffe's High Command (OKL).[213] Since both the RLM and the OKL had nothing else to do in the field of aircraft production, they promoted new studies and projects. Although sidelined in many respects, Göring still had a say when it came to research and development and procurement of completely new weapons. This shift of power explains why many of those projects described above received their initial approval from the RLM and the TLR, and not from the Jägerstab or from the Rüstungsstab. Speer tried to solve this structural problem by establishing with Göring's approval the Main Development Committee on 15 September 1944. Its chairman was Roluf Lucht, chief engineer of the Luftwaffe. The deputy chairman was Karl Frydag, who doubled as chairman of the Airframe Main Committee. The new committee was expected to "centralize and direct the development of aircraft in the spirit of the Führer's order from 19 June 1944 regarding the concentration of armaments and war production. It is also being tasked to quickly complete tasks related to it."[214] The committee sought in the next months to centralize control of the development of new types, but research activities stayed largely beyond its domain.

The streamlining of the German aircraft industry began in mid-1943 by concentrating most of the production capacity on the production of a few important types. Under a new leadership this trend continued in 1944 by terminating most types that were previously produced in small numbers. In this regard, therefore, the Jägerstab simply continued what had started under Milch. The merciless termination of types in 1944 was similar to what Great Britain had done back in 1940. At the height of the crisis, after the fall of France, British production was concentrated only on the so-called "5 types": Hurricane, Spitfire, Blenheim, Wellington and Whitley.[215] This and other measures described earlier gave the British aviation industry great momentum. As a result, even though in 1939 Germany possessed a larger industrial base than Great Britain's, by the end of 1943 Great Britain and its Commonwealth produced a total of 102,018 aircraft, while Germany produced only 70,534.[216] These figures point to the huge — and somewhat little-known — achievement of Great Britain and its war industry. They also point to the grave failure of German aviation production until 1944.

The rationalization of 1944 succeeded in cutting the number of types, but it failed to introduce into service some badly needed next-generation aircraft, such as the Ta 154. It also failed in reducing subtypes proliferation. Although certain designers, like Prof. Hertel of Junkers, agreed to "freeze" variants development,[217] most of the aviation firms continued to develop and produce new variants. Besides failing to reduce the number of types until a late stage, the Germans also failed to eliminate costly research and development projects that obviously were not going to mature in the near future. In this regard, by 1944 the Germans should have learned the lesson from their bitter experience with the development of jet-propelled aircraft, and from several conventional development programs. Furthermore, It seems that German developers and industrialists were preoccupied foremost with problems of research and development rather than with the daily run of production lines. U.S. intelligence officers remarked strikingly after the war: "The great reliance on the development of new weapon types and on research was evident in all plants. This research was concentrated to a great degree on new designs and new weapons. There was little to indicate similar effort in studying the problems of production, with the possible exception of machine tool design."[218] The result of non-standardization and preoccupation with research and development therefore caused much wastage right at a time when the rationalization drive finally reached its peak efficiency.

Dispersal

One of the most important changes in the wartime structure of the German industry was the dispersal and relocation of key factories in order to reduce the effect of bombing raids on the production of certain items. The largest relocation and dispersal took place in 1944 and hastily put thousands of men and machines on the move. The aviation industry was the most massively dispersed industry due to its importance and due to the fact that it was heavily targeted by the Allies since mid–1943. Its dispersal not only affected the way aircraft were produced, it also changed social structures and daily life in the factories and around them. It turned a centralized industry based on well-established factories located near large population centers into something completely different.

It is interesting to note that the first general dispersal of an aviation industry has taken place in Great Britain in 1940 under similar circumstances. After the Luftwaffe attacked several aviation factories during the early stages of the Battle of Britain, almost all firms were ordered to disperse their production in order to minimize further damage. British dispersal was largely based on the prewar "Shadow Factories" scheme, through which non-aviation firms, mainly car and electric equipment manufacturers, undertook to run aviation production facilities financed by the government. These factories were to manufacture designs supplied by established airframe and engine producers. The facilities were constructed during the last two years of peace for immediate operation on the outbreak of hostilities.[219] Therefore, in 1940 these "shadow" factories were activated and formed the nucleus of massive expansion and dispersal of the British aviation industry. One of the biggest dispersals was of the two main Supermarine factories at Castle Bromwich and in Southampton, which produced the important Spitfire fighter. After the Germans heavily bombed them, these factories were dispersed to more than 50 different locations. All the new sites were located within a 50-miles radius of the original factories and a fleet of trucks was employed to transport raw materials, subassemblies and components from one place to the other. The Rolls-Royce aero-engine firm also dispersed its 3 main factories to 72 smaller factories and workshops around Derby, Crewe and Glasgow. It also subcontracted 45 percent of its production. Other firms followed suit in similar patterns, but since the Germans failed to press home their attacks on the British aircraft industry after the initial waves of attack, the dispersal scheme was largely stopped in 1941.[220]

Early RAF bombing of Germany was largely inaccurate and failed to pose a serious threat to the aviation industry. The Focke-Wulf firm, and especially its general director Kurt Tank, nonetheless viewed minor British night raids in the area of Bremen in 1940–1941 as a bad omen and obtained permission from the RLM to disperse some of the firm's production capacity eastwards to four smaller locations in Silesia, East Prussia and the eastern Wartheland region. Most of the new facilities were established in former textile factories. Tank claimed that he was aware of the range of both American and British bombers and purposely chose locations beyond the range of such bombers operating from Great Britain. Spare machine tools stored in the main Bremen complex were used in the new factories, so output of the main factory was not affected at all.[221] The firm's large design and administrative staff with around 2,000 workers was also moved in late 1941 to the small resort town of Bad-Eilsen, practically overwhelming its quiet community of around 900 people. The new offices were established in two big hotels and the workers were lodged in different houses and villas in the vicinity.[222] Although Tank allegedly recommended the RLM to commence general dispersal of the aviation industry, nothing was done and other firms took no similar measures.

Following the beginning of heavier bombing raids in mid–1942 and the arrival of the USAAF, the Planning Office of the RLM prepared a general dispersal plan for the entire aviation industry. In autumn 1942 it ordered aviation firms to submit detailed preliminary dispersal plans to their respective Special Committees for the so-called *einzigartige Fertigungen* (unique productions — main production programs).[223] These dispersals were supposed to take place in case the main factories were damaged by air attacks. This plan foresaw the future, because up to this point aircraft factories had suffered damage only as a side effect of RAF carpet-bombing of cities.

The main mode of dispersal foreseen in this plan was an internal dispersal. It sketched the move of production machines into unused facilities owned by the firms. In many cases such facilities existed within the perimeter of main factories but were not used, mainly because of the inability of most factories to fully exploit their production capacity. In other cases damaged factories were supposed to relocate their capacity to other firms producing the same type of aircraft and possessing idle facilities. AGO's FW 190 production center at Oschersleben, for example, was supposed to move to unused facilities in Focke-Wulf's Bremen main factory in case of bomb damage. This type of relocation enabled the firm whose factory was damaged to resume production after a relatively short time, and at the same time it did not affect production of the host factory. Complete relocations requiring massive use of transportation were supposed to take place, according to the plan, only in case a factory was hit and 100 percent damaged.[224] The defensive posture of this scheme, however, caused a principled rejection by higher authorities and its costs deterred the industrialists. Nevertheless, in April 1942 the RLM ordered the aviation industry to disperse at least its material stores.[225]

It was not until mid–1943 that accurate USAAF daylight bombing began to pose a direct threat to the aviation industry. Single firms started to relocate, on their own initiative, some of their factories to the East. These relocations were done independently, but with the RLM's approval and financial support. The Vereinigte Deutsche Metallwerke (VDM), the biggest manufacturer of propellers, was one of those firms that moved eastwards. Motivated by the destruction of its Hamburg factory in the heavy and particularly deadly raids on the city in August 1943, the firm dispersed to 3 new factories in Silesia.[226]

The Wiener Neustadter Flugzeugwerke (WNF) of Austria started to plan the dispersal of its main factory after it was heavily attacked by the U.S. Fifteenth Air Force on 1 October 1943. WNF officials held extended discussions with the Wehrmacht's armament officials in the Vienna Military District and made tentative arrangements for dispersal. The main obstacle was the inadequacy of most of the alternative locations. Some of them were factories engaged in different non-aviation production programs, and some were unsuitable for aircraft production. Therefore it was impossible to perform the planned dispersal within a reasonable time frame, and the scheme dragged on well into early 1944.[227]

Messerschmitt, which bore the brunt of the air raids on aviation factories in 1943, reacted more quickly. In the aftermath of the August 1943 attack on its Regensburg factory, the firm relocated two fighter assembly lines to its Werk II at the Obertraubling airfield. By the end of 1943 the workforce at this location comprised 11,463 workers.[228] It became such a well-established factory that during "Big Week" it was bombed twice.

In late 1943 the RLM became especially concerned about the safety of its new technology production centers. The Junkers aero-engine division (Jumo), which tooled up for the production of the Jumo 004 jet engine, was ordered to move its production line to the army barracks at Zittau, where series production of the engine was supposed to start in January 1944. A second production line was then opened in a former chocolate factory near Magdeburg. While visiting Junkers headquarters on 4 November 1943, Göring told Jumo's chief Walter Cambeis to establish the firm's next new factory underground, preferably in an unused railway tunnel or a cave in order to save construction time.[229] Production of the Ar 234 jet bomber was also supposed to move to a new factory at Alt Lönnewitz.[230] It is clear that at that time both Göring and Milch generally considered dispersal of several important production programs. During the above-mentioned visit to Junkers they even discussed relocating some of the day-fighter production to factories in North Italy.[231]

Therefore, at the beginning of 1944 a dispersal of at least portion of the aviation industry was underway. Göring's order to move the most important production programs to underground factories issued on 26 January 1944 represented a major strategic shift. It was the first step in implementing Hitler's vision of moving the entire war industry to bombproof facilities. The underground relocation was planned as a long-term set-piece operation directed by the RLM. Göring's order, however, included at its end the following sentence: "Interference of current running relocations and decentralization should not be affected by the relocation to bombproof spaces. On the contrary, it should be done in a way that this aspect of the air-raid protection measures would also be executed to the outmost."[232] It is therefore obvious that Göring and Milch initiated a general dispersal of the aviation industry before "Big Week." Furthermore, they intended to take only the most important production programs underground, while the rest were expected to continue in dispersed locations.

It was, however, "Big Week" and its shock that really set things in motion and forced the Germans to increase the pace of the dispersal. The Germans were well aware of the priority the Allies allocated to destroying their aviation industry and feared continued waves of attacks on this sector. The dispersal plan therefore became one of the Jägerstab's top priorities.[233]

Beside dispersal other means were used in order to restore production in the available factories and to make them less vulnerable to bombing. The Jägerstab quickly organized special repair teams, which went to the damaged factories and salvaged whatever they could from the ruins. The massive repair was made possible by putting the bulk of the Reich's construction capacity at the disposal of the aviation industry.[234] The least damaged structures were hastily repaired and new or salvaged production machines were brought in. In some ruined structures tarpaulins were hung above the machinery so there was no urgent need to fix the roof.

Repair of the damaged factories was made easier by several protective methods implemented after the attacks of 1943 in order to make production less vulnerable. Some factories made preparations for "internal" dispersal, similarly to what the RLM suggested in 1942. In these factories spare machine tools were kept on standby in a different place and were brought in after the factory was bombed. This method was used, for example, at BMW's Allach factory and proved to be effective. The Americans found out after the war that "the plant was never idle for more than 24 hours because of the availability of 'stand-by' machines."[235] In many factories, blast walls were constructed around vital machinery in order to protect it. This method was used to protect machine tools, generators, climate control devices, conveyor belts and parking spots for finished aircraft. According to Junkers officials these protection measures were extremely successful.[236] Saur estimated that these walls reduced the loss of production machines in bombing raids from 30–35 percent to just 5 percent.[237] Even if a plant was bombed, these walls usually protected the machinery and enabled the Germans to salvage it and use it again.[238] Generally the unused capacity of the aviation industry up to this point was characterized by the availability of idle machine tools stored in different locations. These were mostly produced before the war or during its early period, but were stored because the industry never geared up to its full capacity. Now the availability of these spare production machines proved invaluable to the rapid reconstruction of damaged factories and to the dispersal scheme.[239]

The repair and salvage activities were only short-term solutions. General dispersal was viewed as the main long-term solution and underground dispersal formed the ultimate solution. The principle behind the dispersal scheme was simple: the establishment of multiple sources for each part, component and even complete aircraft in smaller locations unknown to the Allies. The dispersal scheme was also an opportunity to implement lessons learned from previous air attacks and to make the new factories less vulnerable to air attacks. It was done, for example, by not placing factories inside cities or even close to them, by putting them away from prominent landmarks, and by keeping their buildings widely apart. Careful camouflage was implemented during the construction phase in order not to attract the attention of aerial reconnaissance. Firms were ordered to make use of all sorts of camouflage methods to hide completed facilities. Since roads inside the factories were thought to be especially conspicuous from the air, their camouflage was considered as highly important, but it was also suggested to avoid camouflaging smaller existing buildings, because it was feared that the use of camouflage would indicate to the enemy that they are now being used for important purposes. Through less use of inflammable materials it was also hoped to make new aviation factories less vulnerable to fires.[240]

Such a comprehensive relocation posed a major security headache to the Germans, and several measures were taken in order to keep the operation as secret as possible. The main fear was espionage — especially by foreigners working in the factories and in possession of secret radios.[241] No mention of firm names was allowed on road signs pointing to the new plants, and workers were warned not to discuss the location and the purpose of their factory.

The Armaments Ministry also issued a special order that strictly prohibited wearing uniforms in dispersed factories in order to not compromise the fact that the plants were involved in military production. Only Luftwaffe and SS personnel involved in so-called *Sonderaktionen* (Special Activities) were allowed to wear uniforms on such locations. Uniformed Jägerstab members were ordered to wear civilian clothes while visiting or inspecting new factories.[242]

Saur and his staff thus started the gigantic task that eventually dispersed 27 main aircraft factories into 729 smaller plants. The aero-engine industry also dispersed in the process from its existing 51 main and secondary plants to 249 new locations. The dispersal was to take place between April and August 1944. The firms were instructed to look for alternative locations and to arrange the dispersal of their plants in accordance with the RLM's pre–"Big Week" plan.[243] The main challenge facing firm mangers was how to complete the dispersal with a minimum negative impact on output.

The dispersal was therefore divided into several overlapping phases, which were supposed to secure the continuation of production with a minimum loss of output.

Phase 1 — Representatives of the firms went out and looked for complexes suitable for the construction of production lines. Since the fighter production program was allocated the highest national priority, the firms were allowed to acquire and even confiscate any structure deemed suitable. With such authorization the aviation industry practically "invaded" the consumer goods industry and converted many of its factories into aviation factories. This overtaking was an indirect way to reduce less important sectors of the German industry. A wide array of facilities was confiscated for the relocation scheme: from glass and porcelain factories to paint and furniture factories.[244] In May 1944 BMW even moved some of its oil pump manufacturing to a cookie and chocolate factory in Landshut.[245] In several cases a competition for the best available locations had developed as firms tried to push their way ahead of other firms and disperse as fast as possible.[246] Sometimes completely new complexes were constructed for the dispersed factories. Four of the 21 dispersed facilities of the Gothaer Waggonfabrik firm around Gotha were newly constructed.[247]

Phase 2 — An advance party was sent to the new location and began preparing it, while production was still running in the original location.

Phase 3 — A production line was brought to the new location. In some cases the existing production line, with all the production machines, conveyor belts, jigs, etc., was dismantled and loaded on trucks or trains. In other cases new tooling was delivered from the machine tools manufacturer or from the machinery depots to the new location in order to create a completely new production line. In these cases the new location formed a branch of the original factory.[248] The efficiency of the transportation system and its density, especially of the railway network, was a central factor in the accomplishment of a quick and successful relocation of most factories.[249]

Phase 4 — Upon arrival at the new location the production line was assembled or reassembled and made ready for operation. Workers were then brought to the new place and started production.

The dispersal of the aviation industry was an expensive undertaking. It was estimated that it cost the Junkers Aircraft Company 15,000,000 RM and Junkers Motors 35,000,000 RM to relocate and disperse their production facilities. These estimates exclude the costs of moving some production facilities to underground locations. The costs were paid or reimbursed by the RLM.[250] Furthermore, the dispersal decreased the efficiency of manpower use by increasing the number of workers engaged in tasks other than production. Relocation demanded massive administration, paperwork, organization, survey, preparation, and the move itself. All these tasks had to be performed by the firm's own workers in addition to or instead of other duties. Junkers, for example, was able to keep its normal output almost throughout the dispersal process, but was forced to increase its workforce by 25–30 percent in order to accomplish regular production and dispersal at the same time.[251] Even after the completion of the initial moves, firms spent quite a lot of man-hours on dispersal-related tasks. In November 1944 employees of the Fieseler firm worked a total of 2,100,749 hours; of these, 20,810 hours were spent on clearing and repairing air-raid damage and 138,097 hours were spent on dispersal-related tasks.[252] Dr. Karl Frydag, head of the Airframes Main Committee, estimated that the number of workers engaged indirectly in production rose from 40 percent to 50 percent following the dispersal and relocation scheme.[253] Then came an extra important cost: even with meticulous planning it was difficult to accomplish such massive relocation plan without causing a major disruption to production. The phased process did not completely prevent disruption of production, and in many cases long disruptions occurred. Long disruptions were especially common when bombed factories were dispersed.

It should be remembered that workers were also relocated in most cases along the production lines. Relocating the workforce posed another organizational challenge to all involved parties. Firms like Junkers found it necessary to construct new houses or barracks for the relocated workforce and to initiate new transportation schemes in order to bring the workers to the factories, which in some cases were located far from the normal transportation network. The move could be further complicated by the difficult transportation situation. Local transportation was often disturbed by air raids and became clogged and overburdened.[254]

The Germans sought to reduce these problems at least to some extent by relocating the plants in a regional manner and by clustering new factories around the original factories. Messerschmitt's and BMW's factories were relocated mainly to new locations in Bavaria, where their main factories were originally located. Most of Junkers' new factories were also clustered around the original factory complex between the cities of Magdeburg and Leipzig. Arado's main dispersal area was around Berlin and in the Brandenburg region, where its main prewar factories were located.[255] Gothaer Waggonfabrik's furthest dispersal was 48 km away from the original factory, while most of the other locations were only 14–23 km away. All these dispersed factories were located next to rail lines that formed part of the Gotha-Mülhausen regional railway network. Reliance on local railways made transportation between the dispersals and the main factory independent of marshalling yards and main rail lines, therefore making it less vulnerable to wartime disruptions.[256] This system worked well even away from the traditional centers of the aviation industry. The Junkers aero-engine factory in Strasbourg was dispersed to 5 smaller locations in textile mills and other premises following a destructive bombing of the main factory in May 1944. Some 60 percent of the machine tools were dug out from the ruins, repaired and shipped to the new factories. All the new dispersed factories were within 48 km of Strasbourg and were connected to the excellent local rail network and waterways of this region.[257]

These patterns of dispersal spared much travel for both the equipment and the personnel. They also kept the new factories close to their parent factories, on which they were still dependent for raw material supply, administration and workforce allocation. Being

Production of Me 109 fuselages inside a wooden hut at the Gauting forest factory. After the dispersal of 1944, completed components were usually delivered painted and marked to the final assembly plants. Note the cramped conditions (courtesy U.S. National Archives and Records Administration).

close to the parent factory was practical also because many dispersed factories supplied sub-components and parts for final assembly at their parent location. But there was a flip side to this advantage.

The dispersal complicated the workflow of aircraft production. The production of the Me 109 fighter was practically dispersed all over Germany and some places in the East. This widespread production network looked like a logistical and organizational nightmare. However, regional clustering was again an efficient way to alleviate some of the problem associated with such widespread dispersal. The Erla Maschinenwerk firm, located originally in Leipzig, was one of those firms licensed to produce the Me 109. By the end of 1944 some 20 dispersed factories of this firm were involved in the production of parts, components and final assembly of this aircraft (sub-contractors not included!). Most of them were located within a 145-km radius of Leipzig and Chemnitz. The production process was composed of 4 main stages, which included basic subassembly production, component construction, component assembly and final assembly. Almost all these stages were performed at dispersed locations. The last stage — final assembly — was carried out in only 3 plants, which were completely dependent on the production of other factories in the chain. It was a very delicate network and a glitch in one primary factory, or in the transportation system between the factories, could have caused a delay in the final assembly at all three locations.[258]

Nevertheless, it seems that dispersal saved the Me 109 production. At the end of 1943, 66.7 percent of the factory floor space available in 1942 for the production of this important type was destroyed in air raids. However, in the first quarter of 1944 the amount of production space increased by 77.4 percent through substitute space and relocation, thus not only replacing all the destroyed space but also increasing the amount of available space by 10.7 percent. This trend continued also during the next two quarters.[259] The Austrian WNF firm, another licensed producer of the Me 109, dispersed in late spring 1944 all its production to 24 smaller production centers mainly in Czechoslovakia. It also made long-term plans to relocate most of its production to a large underground factory in Hungary. This factory was supposed to produce 500 aircraft per month, some of them for the Hungarian air force. In May 1944 the Jägerstab pinned high hopes on the ability of this factory to secure fighter production during the second half of 1944.[260] Hitler personally approved the plan for this 200,000–300,000-square-meter factory on 30 April 1944, because it involved international cooperation.[261] Although this factory was never completed, the other dispersed factories enabled WNF to continue high-rate production to the last months of World War II.[262] Even though bombing raids on the Me 109 production centers continued after "Big Week," they had little effect on its production rate. By July 1944 a large portion of the Me 109 production capacity was dispersed, so when the Regensburg factory and its Obertraubling branch were bombed on 21 July 1944 it did not affect the production of this type at all, because Regensburg's lost production capacity was simply taken over immediately by its dispersed plants.[263]

Although the dispersal scheme was pushed forward most energetically, there were some dissenting opinions considering the logic of the move. By June 1944 the relocation plan was largely completed, but some people in the RLM wondered about the usefulness of this scheme, especially because of the disruption it caused to production. On 30 June 1944 the Technical Office submitted a memo in which it was estimated that reconcentrating production in a few large factories would be the best way to get higher output figures. It acknowledged the fact that concentration would also risk the production of certain types due to bombings, but pointed out that if the most important aim was output increase, concentration would result in a monthly increase of 900 aircraft.[264] Speer also argued after the war that he generally opposed dispersal, unless absolutely necessary, on the same grounds. He preferred to repair damaged factories and increase their survivability.[265]

Considering these objections, it can be said that under the circumstances of early 1944 dispersal was a reasonable solution, even if it badly decentralized production. As we saw, regional dispersal sought to minimize the problems associated with such decentralization.

The aforementioned RLM report implied, however, that the main risk to dispersed production was its dependence on a reliable transportation network. Even with regional clustering, only efficient and reliable transportation could guarantee the flow of materials, parts and components between factories. Even minor producers like Fieseler, whose Kassel complex was dispersed to 15 places around the city, required an average monthly transportation capacity of 800 train carloads as well as many truckloads.[266] This dependency on a functioning transportation system soon turned out to be a major Achilles' heel in the infrastructure of the German aviation industry.

A dominant and very successful form of dispersal was the *Waldwerke* (forest factories) scheme. The idea behind these factories was to create completely new and well-camouflaged factories in woods and forests. Since most of the facilities of the forest factories were placed

A 24 July 1944 American aerial photograph of the Leipheim airfield with its adjacent Messerschmitt factory. The airfield and the factory are heavily damaged. Allied photograph interpreters missed the camouflaged forest factory seen on the top, right below the highway (courtesy U.S. National Archives and Records Administration).

in simple wooden barracks and huts, they were easy and cheap to construct. There are indications that the move to forest factories was based on local initiatives of specific firms, particularly of Messerschmitt. The firm started contemplating moving some of its facilities to forests in the vicinity of its main factories during the "Big Week" attacks. On the morning of 25 February — the last day of "Big Week" — the director of the Regensburg plant, Roluf

Lucht, and his chief of administration, Dr. Wedemeyer, flew in a Fieseler Storch light plane to look for suitable places for forest factories.

According to Wedemeyer, at that time Lucht also initiated the creation of special trains, equipped to move personnel and equipment to the location of the new factories.[267] Messerschmitt constructed its first forest factory in the framework of the general dispersal scheme. It was an Me 109 factory constructed in a forest near Gauting, southwest of Munich. It was a typical example of forest factories constructed elsewhere. Its construction was completed within two months. It started producing wing sets and fuselages for Me 109 fighters in June 1944. The construction started with concrete bases used to support the conveyor belts and the jig fixtures. All the structures of this factory were prefabricated wooden huts. Their size was kept to the absolute minimum dictated by the size of components produced in each of them. Living barracks for the workers were also provided on the grounds of this factory and the canteen was placed right next to the main assembly buildings. Interestingly, this factory reflected the prewar "workers-friendly" aircraft factory in having a vegetable garden next to its kitchen, providing it with a daily supply of fresh field crops.[268] Slave workers allocated to this factory also lived in a nearby labor camp. Finished aircraft with their wings separated, were towed by trucks at night to the Obertraubling airfield and were assembled and flown out from there. From early 1945 this factory also produced parts for the Me 262.[269]

Messerschmitt chose to disperse most of Me 262 production to forest factories after only 15 prototypes and pre-production Me 262s were manufactured in its main conventional Augsburg and Regensburg factories.[270] The Me 262 production lines of the bombed Regensburg and Augsburg factories were dispersed following "Big Week" to 5 forest factories in Bavaria.[271] The decision to produce Me 262 in forest factories resulted from the positive experience with the construction of the Gauting factory and from delays in the construction of underground and bunker factories — especially the "Weingut II" factory near Kaufering. One of the new five factories was a complex of two factories constructed near Horgau and was another typical factory of this type. Its construction had started in mid–September 1944 and it was completed by 1 March 1945. Frequent transportation breakdowns caused by Allied bombings delayed its construction by approximately 3.5 months.[272] Jigs and conveyor belts were brought in on 15 January 1945 and production began immediately after the factory was completed. A team of 80 workers — only 10 of them skilled — performed the entire construction work.[273] The plant comprised 21 prefabricated wooden structures with a total floor space of approximately 4,700 square meters. One of them was a long and narrow structure, in which the main production line was placed. Narrow roads hidden among the trees interconnected all the structures.[274] Around 845 workers lived on location in 7 wooden barracks, thus solving the housing and transportation problem. A canteen and a recreation hall were conveniently constructed right next to the main production buildings. All the structures were painted green and camouflage netting covered the complex. Just like the Gauting factory, a labor camp was constructed nearby. Luftwaffe soldiers unfit for front service watched the inmates of this camp.[275]

Production was carried out in two separate locations within the complex. One section produced wing, tail and nose assemblies, which were then transported to the second section for final assembly. The factory was situated in a forest some 35 km west of Augsburg and right next to the Augsburg-Ulm highway. This location was extremely practical, because

3. Reorganization of Aircraft Production

Unpainted Me 262s roll out of a makeshift final assembly hall in one of the forest factories. The production hall is basically a long tent constructed over a wooden frame. Note the heavy camouflage netting over the exit area (courtesy National Air and Space Museum, Smithsonian Institution, SI 79-10097).

test and delivery flights could be performed from a modified stretch of the nearby highway.[276] Dispersed factories increasingly used highway strips for these purposes. This practice became crucial as the transportation situation in Germany deteriorated and normal airfields were frequently attacked by the Allies towards the end of World War II. The smaller Me 262 forest factory near the Schwäbisch Hall airfield (code named "Miessgeldingen") was even simpler and consisted of 10 large tents reinforced by wooden structures and several simple wooden barracks. The nearby airfield served as a factory airfield for test and delivery flights, but it also attracted the attention of Allied air power.[277]

The construction of this fairly simple factory suffered a severe delay due to the refusal of OT to provide wood for the project when it started in late June 1944. Like in other forest factory construction projects, Messerschmitt hired a construction firm to carry out the work. It failed, though, to fillout the required paperwork in order to obtain the wood and as a result the factory was completed only in early 1945.[278]

As the Americans argued after the war, forest factories were the most efficient means of dispersal because none of them were ever found by Allied intelligence before their capture. Although aerial photo interpreters located some of the adjacent highway strips, and Allied fighters frequently attacked the completed aircraft parked along them, the factories themselves remained hidden.[279] Even factories located next to large exposed airfields and other prominent facilities were not discovered and consequently were never attacked.[280] Among them was the "Stauffen" Me 262 final assembly factory located approximately 5 km from the Obertraubling airfield. Although this plant included three relatively large assembly halls and a widespread network of roads, and although it was located not far from an important airfield, it was never discovered. "Stauffen" therefore produced 335 aircraft between October 1944 and April 1945 with no disruption.[281]

Even though the German leadership was well aware of the advantages offered by the forest factories,[282] it was fixated on a third form of dispersal, which formed the ultimate solution to war production under strategic bombing. Even Hitler, who acknowledged the cost-effective nature of the forest factories in terms of construction materials and construction time, viewed them solely as a provisional solution to the problem of protecting important war production from air attacks.[283]

Moving Underground

While relocating the aviation industry to numerous new locations, the Germans also started constructing better-protected factories. There were two types of such factories: underground factories and bunker factories, some of which were constructed partially underground. The main difference between normal dispersal and dispersal to underground and bunker factories should be clear. Normal dispersal schemes sought to relocate existing plants to new places, where basic production halls and other facilities already existed or could be easily constructed within a short time. As was already mentioned, firms were empowered to confiscate every facility they considered suitable for their relocation plan. In contrast, relocation to underground facilities or to other kinds of protected plant usually meant a large construction project, either in order to create the required facility or to adapt it for its new role. The amount of construction involved in this type of relocation was a crucial element. It demanded a large workforce for the excavation and construction. It also required massive cement castings and other heavy-duty construction. These characteristics made these construction sites extremely suitable for slave workers under SS custody. Furthermore, it meant that production in this type of plant could not commence a short time after the relocation; it usually took several months before these factories were ready. Some of them were never finished before war's end.

Immediately after the war Willy Messerschmitt boasted to his American interrogators that he sought to place his factory underground as early as 1935, and that these initiatives received no support.[284] This claim is probably an exaggeration, like other exaggerations made by Messerschmitt in his interviews, but by the late 1930s the fear of strategic air attacks brought the first notions of placing key facilities underground. After the war some BMW executives told the Americans that in 1937, when the firm planned its new aero-engine plants, BMW's engineering group pointed out to the RLM the dangers of bombing raids and suggested placing the factories entirely underground in converted salt mines or alternately in a forest near Munich. These factories were intended to serve mainly as "shadow" factories, to be used only in case of war in order to expand production. Eventually it was decided to construct the new factory in the Allach Forest, not far from the firm's main factory in Munich.[285] When the Germans conquered France in 1940 they were surprised to find an underground aircraft factory at Creil near Paris, where light bombers of the firm Lioré et Olivier were produced.[286] It is possible that this discovery, as well as a survey of another plant of engine manufacturer Gnôme et Rhône near Paris, influenced BMW's general director Franz Josef Popp. In November 1940 he submitted to the RLM a memo in which he discussed the advantages of windowless hardened factories compared to underground factories.[287] The RLM took no action after receiving Popp's memo.

The Germans first gained experience with giant fortified structures while constructing concrete submarine pens along the French Atlantic coast in 1941–1942. These bunkers originally had a reinforced concrete roof of between 3.5 and 7.5 meters thick, but later in the war additional layers of concrete were added to many of them.[288] Factories were not fortified at this stage, but curiously the first documented move of an aircraft factory to an underground facility occurred on the French Atlantic coast. The managers of the SNCASO firm in Bordeaux, being an important subcontractor of Focke-Wulf at that time, feared Allied bombings already in 1942 and sought to move some parts of their main factory to a chalk caves system in a hillside near St. Astier. It seems that this move was also viewed as an opportunity to ease congestion in the main factory and to expand its capacity. Excavations and further tunneling in the caves began in late 1942 and the floor of the first cave was ready by July 1943. The initiative was fortunate, because the USAAF heavily bombed the Bordeaux factory on 17 May 1943. Only one cave with a floor area of 147,828 square meters was used. In order to improve working conditions for the workers, a canteen was also constructed in the cave and offered the workers low-priced hot meals. Focke-Wulf planned an eventual monthly output of 400 rear fuselage and tail units for the FW 190 fighter in this factory. Delays caused by sabotage and by the uncooperative attitude of the firm's French management, among others causes, severely hampered production in the St. Astier factory. Nevertheless, even as late as July 1944 — after D-Day — Focke-Wulf and the Jägerstab sought to enlarge the cave, bring in more machine tools, and start a two-shift daily schedule in order to make full use of the capacity of this factory. Achieving this goal proved to be very difficult due to French partisans operating in the area and the generally difficult supply and transportation situation.[289] By the time of the liberation, the St. Astier factory produced only 20 tail units. After the war it was found out that the factory was getting enough fresh air and that a constant temperature of 12° Celsius was maintained inside the caves without artificial assistance. Furthermore, "it was stated that the chalk dust had no adverse effect on the moving parts of machine tools."[290]

The Germans started contemplating constructing bombproof factories following the first wave of American daylight raids on aviation-related factories in early 1943. On 11 April, Hitler demanded during a meeting with Speer and other armaments officials to prepare the relocation of the production of some key items, like crankshafts and bevel wheels, to fortified factories.[291] Following this conference the RLM assigned a civil engineer named Bilfinger to head a special staff and recommend the best way to protect the existing factories. Bilfinger's *Sonderstab Höhlen-Bau* (Special Staff Caves-Construction) was composed of engineers, geologists and construction experts. Its initial recommendation was to encase important factories in concrete bunkers. This scheme was totally unrealistic due to the amount of work and resources required in order to complete it, and it was not pursued further.

It is assumed that the first concrete initiative to move vital factories to a protected underground facility came after the RAF and the USAAF bombed several factories associated with the V-2/A-4 missile production in June and August 1943. The production of this missile was supposed to be carried out in 3 main centers. The first was the highly secret

Peenemünde complex, where the A-4 and the Fi 103 (V-1) were developed and tested, and where the prototype series of the A-4 was already produced. The second factory was the Henschel-owned Rax Works in Wiener-Neustadt, Austria. It was an old locomotive factory, now converted into a missile factory. The third center was the old Zeppelin Luftschiffbau factory in Friedrichshafen, which once constructed airships and now produced radar antennas and other light metal products. The RAF attacked the Zeppelin factory on the night of 21–22 June 1943. The factory was targeted because the British received intelligence indicating that it produced radar equipment. The USAAF attacked the Rax factory on 13 August 1943 because it was thought to be a Me 109 factory. These attacks unintentionally hit at the heart of the V-2 production even before it started. The third strike, however, was fully intentional and was based on precise information gathered by British intelligence. On the night of 17–18 August 1943, the RAF heavily attacked the V-2 development center at Peenemünde. The attack missed the most important facilities, but it shocked the German establishment, which was sure until then that the base was safe.[292] On the day after the attack Speer suggested to Hitler to move the entire ballistic missile project underground. Hitler agreed and ordered the allocation of the required manpower for the construction work from Himmler's concentration camps reservoir.[293] Soon afterwards Special Committee A-4 chose an underground fuel storage facility in the Harz Mountains, near Nordhausen and Niedersachswerfen as the site of the main A-4 production plant. Construction of the missile plant began under the direction of Dr. Hans Kammler, an engineer and a rising star in the SS hierarchy. The SS cooperated closely in the construction of the factory with the Wirtschaftliche Forschungsgesellschaft (WIFO), which was originally responsible for the site.[294] This underground plant was largely completed by the end of 1943 and became infamous as the Mittelbau-Dora, or the Mittelwerk complex.

The Mittelbau-Dora project set the stage for later similar projects. The SS became deeply involved in it right from the beginning mainly because it was the only agency able to provide the required manpower for its construction. The entire project relied on SS-supplied manpower, and all this manpower came from the concentration camps reservoir. Himmler initially based his growing involvement in the V-2 production on the premises that the SS would be in a better position to protect the V-2 facilities if it became more involved in the program. In this regard slave labor seemed to offer an excellent solution to several security problems.[295] The relocation of the V-2 program to a single massive underground facility was the first of its kind in Germany during World War II. It was done after other measures to avoid bombings — like dispersal and camouflage — seemed to fail. It was a total solution, driven mainly by the shock caused by the bombing of Peenemünde and other production centers within a short time period, but it was a sign for the type of reaction expected in similar cases of Allied action.

In autumn 1943 Hitler and Speer also discussed the move of certain other factories to bombproof underground locations. The aviation industry figured prominently in the following discussions. On 2 October 1943 Xaver Dorsch, the head of *Organisation Todt*, offered to build for Göring a giant underground factory, capable of producing 500 fighters or 200 bombers monthly as an alternative to the "Ultra" above-ground factory. The output of this factory was obviously far from "Ultra's" projected "thousand bombers" goal, but it was better protected. Dorsch was consulted earlier about the construction of "Ultra" and saw here an opportunity to broaden the responsibilities of his organization.[296] Furthermore, due to earlier allocations of responsibilities, the OT was allowed to operate only in the occupied countries and not on the Reich's territory. Construction projects inside the Reich were managed by the Construction Office of the Armaments Ministry or by the military. Construction of bombproof aircraft factories therefore offered Dorsch the opportunity to operate his organization inside the Reich.[297] Several days after Dorsch's offer, Göring, who became especially concerned for the safety of aircraft production following recent damage caused by daylight bombing, wrote to Speer and pointed out the urgent need to protect his factories. He suggested allocating the highest priority to the aero-engine industry and to factories producing other bottleneck items. These factories were supposed to relocate to existing underground facilities, like tunnels, mines, etc., and if these were not available, Göring asked to encase the crucial factories in concrete.[298] Preparations for the move of some of the most important aero-engine production began soon afterwards, and Bilfinger's staff, now bearing the title Cave Commissioner (*Höhlenoberkommissar*), was ordered to prepare preliminary surveys of possible underground locations.[299]

At this point BMW and Junkers stood at the focal point of the scheme. According to Junkers executives interrogated by USSBS teams after the war, the firm drew plans to move engine production underground already in early 1943.[300] However, during a visit to Junkers in Dessau at the beginning of November 1943 Göring expressed again his intention to move the most important aviation factories underground. First to go were several aero-engine production lines, including Junkers' own modern engine production, as well as BMW's Allach piston engine production. Then the next step, according to Göring, was to move underground the most important airframe factories.[301] At that time, however, BMW moved ahead with its own plans. In October 1943 Wilhelm Werner, chairman of the Aero Engine Main Committee, recommended the relocation of the entire Allach production to bombproof underground facilities. The RLM offered BMW several caves in West Germany and France. Since relocation to these caves meant a huge investment and posed a range of organizational challenges, BMW and the RLM decided eventually to construct a large fortified production hall at Allach instead. Initially it was decided to construct a long arch-form bunker just south of the Allach complex. Later Speer decided to construct a dome-shaped bunker that was considered more complicated to construct. The Construction Office of the Armaments Ministry reversed the decision in mid–February 1944 and ordered the construction of a simpler cube-like bunker. Following "Big Week" it was decided to construct a smaller two-storied bunker of 32,000 square meters floor space. After more deliberations between OT, BMW, the Jägerstab and the Armaments Ministry it was decided not construct the bunker, but to cover one of the existing production halls with a 60-cm layer of cement. This project was also never completed.[302] Instead, BMW was incorporated in several underground factory projects that started to form in spring 1944.

At the end of 1943 Himmler ordered underground spaces to be dug in some former SS quarries in order to use them for different war production purposes. This order set in motion several tunneling projects next to different concentration camps.[303] Some of these projects soon came up as potential locations for some of Göring's production programs.

In early December, Himmler suggested to Milch to produce the V-1 in some free underground space in the Mittelwerk complex. Nothing came out of this idea in that stage because Volkswagen, the prime contractor of the V-1 program, started working its own plan to move some of the missile production to an underground facility at Tiercelet.[304] The free space at Mittelwerk came up at approximately the same time in another context. In the search for a suitable place for the Junkers plants, it was found out that the Mittelwerk factory complex could house more production lines. It was therefore decided to relocate some of Junkers' aero-engine plants involved in the production of the Jumo 004 jet engine and of the Jumo 213 piston engine to a new tunnel complex code named "Anhydrit." WIFO submitted concrete plans for the new facility in January 1944, and in early February the RLM and the Armaments Ministry ordered Kammler to carry out the project. The SS allocated Kammler 10,000 concentration camp inmates for this project.[305] At the beginning of February, Junkers also proposed to broaden the scope of its underground relocation and include in the plan airframe production facilities in an unused potash mine near Neusollstedt and in the Heimkehle cavern near Rottleberode.[306]

In the meantime Göring and the RLM decided to begin the relocation of the aviation industry to underground facilities. The decision was a culmination of a ripening process, which started after the deadly attacks of 1943, and following the preliminary studies done by the Cave Commissioner staff. The formal order was issued on 26 January 1944 and was signed by Milch: "The Reichsmarschall issued a basic order regarding the immediate relocation of important Luftwaffe production programs to bombproof spaces."[307] This general order was followed up on 1 February 1944 by a detailed order printed on a unified template sent to most of the important aviation firms. In the blank places each firm was allocated its underground location and the type of aircraft to be produced in them. Focke-Wulf, for example, was ordered to relocate parts of its FW 190 and Ta 154 production to the Prinz Adelbert mine (40,000 square meters) and to a potash mine in Meimershausen.[308]

A special RLM staff called *GL/A Sonderstab H* (*Höhlenbau*—Cave Construction), headed by *Regierungsoberbaurat* Treiber (an engineer), was appointed to direct the operation. Each firm was ordered to submit initial plans for the relocation to its allocated locations by 15 February 1944. Members of the Cave Commissioner staff helped the firms to prepare their plans. Dr. Solle, a professional geologist from the RLM, conducted on 14 February a survey of caves along the Nekar Valley and suggested the relocation of some Daimler-Benz press workshops into them. The firm integrated his report in its own plans.[309] Other firms followed the same procedure, but it seems that this was not done under an atmosphere of extreme urgency. A week before "Big Week" Göring wrote Himmler and asked his help in the relocation of the aviation industry to subterranean locations. Himmler's positive answer came almost one month later, as the relocation was already in full swing after "Big Week": "The move of manufacturing plants of the aviation industry to subterranean locations requires further employment of about 100,000 inmates. Plans for their employment on the basis of your letter of 14 February 1944 are already underway."[310]

This correspondence proves that Göring clearly foresaw the need to move his key factories underground before the decisive effect of "Big Week," and that he was looking for unorthodox ways to rush this relocation.

Following "Big Week" and the establishment of the *Jägerstab*, Göring's visions regarding the move underground received an enormous boost. On 5 March 1944, Hitler declared that the underground relocation should be viewed not as a stopgap solution, but as a long-term total solution, with the final aim of moving the entire German war industry underground.[311] There were some deliberations regarding what scheme of underground relocation should get precedence. Relocation into existing tunnels and caves could be accomplished within shorter times and with much less expenditure. The main disadvantage of these sites was the restricted size of their tunnels and entrances. Completely new bunkers could be constructed according to specific needs and therefore better suit the requirements of specific production lines. Hitler left the decision to Speer, who favored bunkers, probably because of the architectural challenges involved in their construction.[312] After further discussions, which continued until late April, it was decided to construct 6 bunker factories for several highest priority productions and to relocate all the rest to already-existing underground spaces.

The underground relocation required massive and comprehensive administration. Immediately after the establishment of the Jägerstab, each firm allocated a special executive to survey underground facilities in its region in order to determine which of them was suitable for the construction of underground factory.[313] The *Rüstungskontor*, a state-owned firm which financed different armament projects (Mittelwerk was one of its subsidiaries), administered the purchase of the properties located by the firms. Next it handed over the property to the firm, which carried out the required conversion work. The owners were offered alternatively to keep ownership of the property by financing its adaptation into an armaments plant. In this case the finished facility was rented to the firm.[314] The move underground was administered mainly by the concerned firms and financed by the Armaments Ministry and by the RLM. Kammler and his staff generally supervised the biggest underground relocations. Those comprised 20 projects that were divided into two categories. Category A projects included relocation to existing underground facilities, offering a total floor space of 240,000 square meters. Category B projects were more ambitious new tunnel complexes, with a total floor space of 500,000 square meters.[315] They were largely modeled on the Mittelwerk project. Each category included 10 projects. All Category A facilities were allocated to firms dealing in one way or another with aircraft production, but one of the Category B factories (B6 at Leitmeritz) was intended for tank motor production. Later two of the projects (B1 and B6) were terminated or allocated to other organizations, but in May three were added (B11–13 — all expansions of Mittelwerk) and others were enlarged, therefore bringing the total floor space of the Category B factories to almost 1 million square meters.[316]

Most of the daily work on the simpler relocation sites was done on lower levels using normal business procedures. Aviation firms usually contracted mining firms, who made initial geological surveys of the sites and then carried out the required construction under the overall supervision of OT. In case of a move into an existing mine the Jägerstab ordered each firm to appoint a representative who worked with the operating mining firm and made the necessary coordination and arrangements.[317] When significant work was required the involved organizations contacted the SS and contracted it to provide the workforce.

Some firms like Daimler-Benz formed direct contacts with the SS at early stages and asked it to support the relocation. Most of these contacts ended eventually on the desk of Kammler and his Special Staff. In the case of the relocation of Daimler-Benz's Genshagen factory, however, the SS played a central role in every aspect of the operation — from the allocation of slave workers, to planning and conducting the required construction work.[318]

These sorts of local arrangements proved apparently unsatisfactory, and in October 1944 the Armaments Ministry founded a special subsidiary firm to administer and finance the entire underground relocation. This firm was called Industriekontor GmbH, and among its functions was the acceptance of the underground floor space requirement from the ministry, buying the land, drawing up contracts between construction firms and OT or the SS, financing the construction, and renting the spaces once constructed to tenant firms.

OT or the SS conducted technical control and planning of the projects while the construction firms managed the work on-site.[319] Similarily to the above-ground dispersal, the Armaments Ministry issued special regulations considering the need to conceal the underground factories and their construction sites from the Allies. Generally, OT and its local project directorates were responsible for concealment. The entrances to the underground facilities were heavily camouflaged and the structures used by the constructors were placed in an inconspicuous way around the place. Aerial photos of the construction sites were taken regularly in order to ascertain the quality of their camouflage.[320] Information regarding the true nature of the construction site was withheld even from those directly involved in the work. In many cases OT did not inform the firms hired to conduct the project about the intended use of the factory they constructed. Plans were simply handed over to them and they were asked to accomplish the construction according to them.[321]

There were, however, some firms that expressed reservations regarding the underground relocation scheme and its rationale. Some of them flatly refused to follow their relocation orders due to the expenses and due to misgivings regarding the functionality of underground factories. Henschel, for example, was able to refuse relocation to a potash mine near Stassfurt and get away with it — probably because other firms agreed to move in. Resistance to the underground relocation was so widespread that it was suggested to punish acts of resistance with concentration camp internment or with hanging. One BMW executive was sent to a concentration camp at least temporarily in May 1944 because he delayed the underground relocation of his factory for a couple of weeks.[322] In general, however, most firms were more or less cooperative and the relocation went ahead with full energy.

Junkers and the Mittelwerk complex continued to figure prominently in subsequent projects. Four days after the establishment of the Jägerstab, Hitler demanded to check if the underground spaces of Mittelwerk could also accommodate production lines dedicated to fighter production.[323] In mid-March Milch and Saur considered not only moving the production of the Jumo 004 to the Mittelwerk complex, but also the production of the Me 262 airframe. At that time the Jägerstab considered, next to the general dispersal scheme, some big bunkered or underground factories, in which complete production of individual aircraft would take place. Besides the jet fighter, it was also planned to concentrate the production of the Ta 152 piston-engine fighter, which should have replaced all other single-engine fighters, in a single underground plant.[324] The final decision to move the Jumo 004 production to Mittelwerk was made, however, only at the beginning of April 1944 in the framework of the general relocation plan. Hitler approved this decision, which was viewed as a solution to the jet engine production bottleneck, thought to be the main obstacle to the Me 262 program.[325] The Jumo production line was constructed eventually in a separate part of the Mittelwerk tunnel system, called "Nordwerk" and operated by the Nordwerke firm — a subsidiary of Junkers. It consisted of 20 underground galleries with 44,000 square meters of floor space. Some of the Jumo 213 production facilities of the Köthen and Magdeburg engine plants were moved to Nordwerk in early May 1944. Production started slowly because the move was not completed and because frequent use of explosives in the expansion work of the Mittelwerk complex constantly disrupted work in Nordwerk.[326] All the production machinery of the Jumo 004 engine with around 4,500 workers was moved from the Magdeburg plant into Nordwerk in July/August 1944.[327] This move, combined with the fact that the Jumo 004 engine was still under development, caused a delay of at least 3 months in the production of these engines in the underground factory.[328] However, when it finally started, series production at Nordwerk appears to have been quite effective, with a monthly output of approximately 500 Jumo 213 and 550 Jumo 004 engines. In March–April 1945 it was planed to start production of the second standard jet engine — the BMW 003 — in the original Mittelwerk factory, but the construction of this production line was never completed before the end of the war.

The Mittelbau-Dora complex was supposed to expand even further and to encompass at least five high-tech weaponry underground factories and a couple of underground oil refineries, including the 123,000 square meters "Anhydrit" complex — also known as Construction Project B3, dedicated to the production of the Jumo 004 engine.[329] Tunneling for this facility began only in mid–April 1944 in the side of the Woffleben Hill — only 2 kilometers north of Mittelbau-Dora and northeast of the original plant. Yet, there is no evidence that Junkers ever used this complex. By mid–September only 15 percent of the facility was completed, while the construction of an additional extension of the project has not even started.[330] It appears that the project was largely stopped in October 1944.[331] In February 1945 a missile development section of the Henschel firm was moved from the firm's development center in Berlin-Schönefeld to the uncompleted tunnels in Woffleben. There is evidence that development work of air-to-air, surface-to-air and antitank missiles was carried out there, as well as the pre-production series of the Hs 117 surface-to-air missile. None of the missiles developed there ever came even close to entering operational service.[332]

Although the Mittelwerk complex was the biggest underground factory project, most facilities were moved to much smaller existing tunnels, caves, quarries, storage spaces, mines, church cellars and even beer storage cellars — as happened in Munich.[333] These subterranean spaces were located by different military and civilian survey teams according to different criteria: size of the space, proximity to supply sources, availability of power sources, temperature, dampness and the amount of work required in order to convert them into a functioning factory. Usually as soon as a location was found, the Jägerstab decided which factory would move into it, but in some cases the relocation was decided upon at the local level and was just approved by the Jägerstab. Expansion and conversion of existing underground spaces were carried out quite fast. Up to the end of September 1944 most of these projects — including Kammler's Category A facilities — were largely completed and started producing.[334]

Among the key programs that were relocated in this framework was the Me 262 production. After Messerschmitt's main production center at Regensburg suffered crippling air raid damage in 1943, the RLM contemplated moving the Me 262 production to underground facilities. The first schemes for this move were apparently conceived already after the August 1943 attacks. By the end of October 1943 the Cave Commissioner staff submitted plans to construct production lines in at least 3 road and rail tunnels: in Eschenlohe, in Obergreinau and in the Engelberg highway twin tunnel near the town of Leonberg (it was originally intended for ball-bearing production). By using existing tunnels it was hoped to cut short the time needed for underground relocation, and the Eschenlohe rail tunnel was supposed to be readied in 6 weeks.[335] It took, however, several months to complete the conversion of the tunnels. In the aftermath of "Big Week" the Leonberg factory (code named "Leo"), which made the best progress, was still not ready.[336] Nevertheless, on 25 March 1944 the Jägerstab hastily ordered the movement of the necessary machine tools into its tunnels.

The factory was constructed by German workers as well as by a large contingent of slave workers. Inmates were first brought to a camp constructed next to the tunnels' entrances on 10 April 1944. Most of those early inmates, as well as later transports, arrived from different Dachau subcamps. At least 3,329 inmates were allocated to this factory at one time or another.[337] Eventually only Me 262 wings were produced at the 14,750-square-meter Leonberg tunnel factory — mostly by concentration camp inmates. These wing sets were delivered by train for final assembly in a forest factory near the Schwäbisch Hall airfield and to other final assembly factories.[338]

Other aircraft manufacturers, even those not engaged in the highest-priority projects, also moved to different underground facilities. Blohm & Voss moved some workshops of its main factory complex in Hamburg to the cellars of the local Elbschloss Brewery in Altona.[339] Junkers produced some components in a salt mine it shared with BMW at Stassfurt, and landing gear for Ju 88 and Ju 188 aircraft in the Heimkehle gypsum cave near Nordhausen.[340] Henschel produced subassemblies for Ju 88 nightfighters in an unfinished subway tunnel in the outskirts of Berlin.[341] Even the rather minor Me 110 heavy-fighter manufacturer Gotha moved one of its machine shops into a railroad tunnel near Friedrichroda.[342]

The production of the V-2's cheaper rival, the V-1, was also moved to several underground locations. In autumn 1944, after the USAAF attacked Volkswagen's Fallersleben factory four times, the Armaments Ministry decided to move large parts of its V-1 production capacity to the expending Mittelbau-Dora complex. A production line was installed in the main tunnel complex, where V-2 missiles and Jumo engines were already produced. It was estimated that until the end of the war 10,600 V-1 missiles were produced in Mittelbau-Dora — mostly by slave workers.[343]

As demonstrated by the case of Junkers, the aero-engine industry was also included in the underground relocation scheme. Earlier fears of devastating bottlenecks in engine supply in case the Allies concentrated their attack on the aero-engine industry was the main reason for this emphasis.[344] Daimler-Benz moved its main aero-engine plant at Genshagen, south of Berlin, into a gypsum mine near Heidelberg. The plant, code named "Goldfisch," manufactured parts and components for the DB605 piston engine.[345] BMW moved one of its main factories to a railroad tunnel near Markirch in Alsace and some of its Allach parts manufacture to the cellars of the famous Franziskaner (code named appropriately "Franz II") and Hofbräu breweries in Munich.[346] Most important, however, was the massive dispersal of the BMW 003 jet engine production line to salt mines in Heiligenrode, Abterode, Ploemintz and Stassfurt (codenamed "Ludwig II").[347]

In some instances several firms shared the same underground location, therefore cutting costs. BMW shared the huge Franziskaner Brewery cellar complex in Munich with a Dornier workshop, which manufactured some parts and piping for its Do 335 fast bomber, and with a branch of the Süd Deutsche Bremse Werk, which manufactured fuel transfer pumps for aero-engines. A total of 886 laborers worked in this underground complex, which was well lit and well ventilated.[348] BMW, ATG and Junkers shared an underground factory constructed in two salt mines in central Germany, which were initially intended for ATG.[349] Although underground factories were intended mostly for fighter production or fighter-related production, in late May plans were made to move the production of the He 177 bomber and of the new Ju 287 jet bomber to bombproof facilities. The concentrated bombing of the Oranienburg and Brandenburg He 177 factories on 18 April 1944 was the main

Incomplete V-1 fuselages found by U.S. troops in the Mittelbau-Dora underground factory in 1945. This photograph illustrates the cramped nature of an underground factory — and this was one of the largest of its kind (courtesy Yad Vashem Archives).

incentive for this plan. These raids practically shut down the aircraft's already struggling production. Milch estimated that the He 177 production would require at least 120,000 square meters of underground floor space.[350] Producing such a large aircraft underground posed completely new technical and organizational challenges to the Germans. The termination of the He 177 soon afterwards and the delays with the development of the Ju 287 doomed this ambitious plan.

Some of the relocation projects, particularly those of Category B, required massive tunneling in soft rock. Conventional mining techniques were normally used in these projects. Various hand-held jackhammers and augur drills driven by compressed air or electricity were used to make holes at the rock face for blasting charges. After these were detonated the loosened rocks were removed and loaded on light railway cars, which took them out of the tunnel. The normal practice was to dig a pilot tunnel big enough to accommodate a light railway for the removal of rocks and then to enlarge it. Up to four pilot tunnels were excavated in parallel during the excavation of bigger tunnels. Tunneling machines were sometimes used, but only in limited numbers and in specific places. Soil and rock debris from the tunnels were usually extracted by light railway cars or by mechanized scoops. In several cases, spoil spread in the countryside, and a rail line entering a hillside formed an excellent indication of an underground factory construction site for Allied aerial photo interpreters.

An American intelligence report summarized the issue of underground construction:

> More technical ingenuity was expended on expediting the removal of excavated rock from the tunnel face, than on the actual boring of the tunnel. The application of these principles reflected the general position of Germany in 1944: an abundance of slave labor, a large supply of mining tools, a shortage of modern tunneling machinery (although a rotary tunneling machine and a small compressed air driven power shovel had been developed during the war and were encountered in underground factories), and finally a desperate need for speed which led to the abandonment of planned attempt at concealing or camouflaging the new projects.[351]

Messerschmitt's "Seelachs" plant at Kematen was a "spontaneous" project requiring massive tunneling. Located in a picturesque valley in Tirol, it was originally conventional factory producing parts for different aircraft. Between January and July 1944 a series of tunnels were dug into a nearby mountainside to serve as an air-raid shelter. In mid–1944 it was decided locally to enlarge these tunnels and to turn them into an underground factory. Work began in August 1944 with 200 workers. Two months later the number of workers was doubled and the first tunnel was completed on 8 December 1944.[352] The following day the welding department and the control office of the original factory were moved in. Tunnel 2 was finished in February 1945 and Tunnel 3 was ready on 15 March 1945. Different parts for the Me 262 were produced in these tunnels. Because the original factory was so close

Dornier's cutting edge. The U.S. Army found this Do 335 heavy fighter at the Lechfeld airfield in April 1945. This particular aircraft was assigned to *Erprobungskommando 335*, which was established in order to perform operational tests of the aircraft (courtesy U.S. National Archives and Records Administration).

to the underground facility, here the move underground caused almost no disruption of production. Five more tunnels were not completed before the end of the war, largely because of missing air-conditioning equipment.[353]

Another advanced project relocated largely to underground facilities was the Do 335 fighter-bomber. It was a high-speed fighter-bomber powered by two piston engines in an unconventional tandem arrangement. The Do 335 was another of those development programs which took a long time to mature. Although it first flew in October 1943 and was allocated high priority in April 1944, second only to the jets,[354] only a few examples were produced before the end of the war and none entered operational service. At least two underground factories were constructed for the planned series production of this aircraft. One, in Überlingen, was dug into a sandstone hill, but was never completed. The other was a well-concealed factory constructed in the cellars of a brewery in a wooded hill near Ravensburg. The American technical intelligence team that visited this site in early June 1945 was impressed by the location and by its machinery. Although this factory was completed, only a few aircraft were completed there and none was ever delivered.[355]

The underground relocation project progressed quite well, despite its complexity. By the end of May 1944 there were already 200,000 square meters of underground production space ready for use by the aviation industry. Between June and August this figure increased fivefold. By the end of 1944 some 2.5 million square meters of underground production space were available for aviation production purposes.[356]

However, moving underground proved to be anything but a simple operation. In autumn 1943, as Göring first discussed plans to move factories underground, Milch objected to the scheme and expressed his concerns regarding working conditions in mines and tunnels. Göring dismissed these concerns and saw no real problem in working underground besides "some little cold air."[357] In reality, cold air was not the only problem involved in underground relocation. Although Category A and other similar relocation projects were completed relatively fast, their facilities were particularly troublesome. Mines, for instance, were limited by the size and capacity of their elevators. The small size of many elevators and their slowness seriously delayed traffic in and out of the factory, and limited the size of machine tools that could be brought in. These problems were especially apparent in the move of BMW's factories into mines in Stassfurt and Springen.[358] Another underground factory which suffered from this kind of a bottleneck was the dispersed AGO factory constructed in a potash mine near Oschersleben. This factory was served by a single vertical shaft typical to mines. This single shaft limited the amount of floor place that could be used, and because of the limited capacity of the elevator it took around 3 hours to change the 12-hour shift of around 2,000 workers. It was reported that in the similar underground factory at Bartensleben, where Askania produced different instruments, it took 4 hours to change the shift.[359] The elevators and their prominent towers were also vulnerable to air attacks and their destruction could paralyze the factory underneath. In at least one case plans were made to encompass the elevator tower of a mine in a bombproof bunker. The scarcity of material and labor made it impossible.[360]

Another difficulty, which affected almost every underground factory, was ventilation and supply of fresh air. Adequate ventilation was essential both during construction and even more after the factory started producing.[361] This problem had to be solved during the construction phase. In some instances the construction firms were aided by mining experts in order to solve the problem of ventilation.

The high level of dampness in many underground facilities was troublesome both for the workers and for the equipment. Frequent flooding occurred in many places and had to be dealt with by installing pumps and drainage systems. Some of the converted mines were especially problematic because of the additional effect of the minerals. In July 1944, as the Soviet army advanced into Poland, the Germans tried to relocate the Mielec Heinkel factory to a salt mine in the vicinity of Wieliczka, west of Cracow. Workers and inmates were brought there from the original factory and for at least two weeks they moved production machines and equipment into the mine. It was soon discovered, however, that the salty and humid air inside the tunnels rapidly corroded the equipment and the Germans were forced to abandon the plan.[362] Another move into a salt mine — this time of a Büssing aero-engine plant, producing DB605 engines for the Me 109 fighter — also had to be canceled after salty humidity ruined equipment brought in.[363]

Another cause of problems was fine dust in some mines. Some 80 percent of the flight instruments produced by the Askania factory in the Bartensleben salt mine were rejected because of damage caused by dust to their delicate machinery during manufacture. This Askania factory suffered generally from the poor conditions, and its output was only 20–30 percent of that anticipated.[364]

Many underground factories were constructed in relatively soft or fragile ground, which sometimes caused dangerous collapses during their construction and even after they were finished. A substantial part of the roof of the Heimkehle cave came down shortly after the end of the war and demolished a workshop.[365]

Finally there was the effect of limited space on efficiency. American engineers and technicians surveying several underground factories after the war noted that in many of them space limitations imposed by the shape and the size of the tunnel dictated inefficient factory layouts. The resulting confused workflow, combined with the other difficulties, resulted in increased production costs.[366]

Daimler-Benz's "Goldfisch" factory, constructed in a former gypsum mine near Heidelberg, demonstrates not only the plethora of problems affecting underground factories, but also their dangerous nature. In early August 1944, while the plant was still not completed, several ceilings collapsed and damaged machinery. This accident was only a prelude, and soon came a series of fatal collapses. In September 1944 it was reported in a Rüstungsstab meeting that 195 people were killed within 4 weeks in "Goldfisch" in different accidents — mostly caused by collapses due to drying up of the earth and rock layer in the tunnels. These accidents happened at the worst time: just as the plant was about to start producing engines. These cave-ins forced the management of Daimler-Benz to suspend the work and bring in mining experts as consultants.

Further problems in "Goldfisch" were caused by the high level of humidity in the tunnels. Besides being unhealthy for the workers — as SS hygienic research had determined — it also caused corrosion of machinery and equipment. Installing special warming boilers, which warmed the air sucked in by the ventilation system, solved the problem. This installation, however, dried the air going into the tunnels and consequently caused more collapses.[367] The factory management was forced to find a compromise, which it was never able to accomplish satisfactorily. As American surveyors remarked after the war, many facilities in "Goldfisch" were either primitive or were never completed. Its ventilation system was inadequate and it suffered frequent power failures due to the fact that it lacked its own power supply. There was only one entrance to the factory, which was very narrow, and it could be reached only by a primitive road. These problems caused unbearable working conditions, mainly due to high humidity over long periods. The continued dampness in the tunnels caused high rates of sickness among the workers and damaged the machinery. The Americans determined that "under such conditions it was almost impossible to carry on precision work on aircraft engine parts."[368] The factory manager also admitted in a contemporary report that, during the winter, extreme cold inside the work halls made precision work—so important in engine manufacture—almost impossible.[369] Even in some better-equipped underground factories, like Kematen, workers complained of headaches during their first 7 to 10 days underground. It was believed that this illness was caused by higher barometric pressure underground.[370]

The troubles with the Category A facilities and with other smaller relocation sites disappointed the industry and the Armaments Ministry. From autumn 1944 they started to shift their attention and hopes to the larger and better-constructed Category B projects.[371] The biggest problem with these projects was the long time it took to complete them. Indeed, most of them were never completed before the war's end.

The underground relocation scheme was highly ambitious. However, the most ambitious protected-factories schemes involved the production of the Me 262 in several completely new underground or bunkered facilities. The concept behind these factories was different from all other forms of dispersal. The initiative to move the entire production of the Me 262 underground came on the day its main dispersed final assembly plant at Leipheim was heavily damaged by an American bombing on 25 April 1944. Nine almost completed aircraft were also destroyed in this raid. The attack was viewed as proof that Allied intelligence located the dispersed above-ground factories of the Me 262 program due to security breaches. It was therefore thought to be a prelude for more attacks. Later in the same day Milch pointed out during a Jägerstab meeting that large parts of the Me 262 production facilities were already bombed and that the dispersal of its production became crucial.[372] Major Dr. Krome, the Jägerstab member responsible for jet aircraft, urged the movement of the Me 262 underground at all costs and even at the expense of the A-4 program.[373] Such relocation was also viewed as a way to recentralize the production of this important aircraft at a time when transportation difficulties began to plague dispersed production programs.[374]

The idea to centralize certain production programs in bombproof bunker factories had already been discussed earlier by the Jägerstab. Saur remarked in a meeting on 6 March 1944 that while it was intended to produce a maximum of only 100 fighters per month in the above-ground factories, it was planned to reach output of up to 1,000 in big bunker or underground factories. Factories with floor space of 500,000–600,000 square meters were mentioned, because Hitler demanded them in earlier meetings with Speer and Milch.[375] These fantastic plans obviously had their origins in "Ultra"—the "thousand bombers" factory concept of 1942–43—and in turn in what the Germans learned earlier about wartime aircraft production in the USA. The amount of work and resources involved in the bunker factories initially forced the Jägerstab to restrict them to only two projects. According to the original plan they were located only 60km apart, one near Kaufering and one near Ottmershausen.[376] Originally they were supposed to offer only 60,000 square meters of production space, but the head of the Construction Office in the Armaments Ministry, Carl Stobbe-Dethleffsen, argued that it would be possible to expand them to a total of 600,000 square meters in accordance to Hitler's wish. One of these giant factories was intended for the Me 262 production, while the second was to be given to Erla, which became the main contractor of the new Ta 152 fighter. These factories were so big that they were intended to produce not only airframes, but also their engines.[377] In any case, the Me 262 enjoyed the highest priority and it was decided to dedicate the biggest tunneling and bunker construction projects to its production.

During the next couple of weeks the German leadership decided to expand the bombproof factories scheme and to add four more projects to the original two. Hitler and Göring showed great interest in the bombproof factories. On 14 April they discussed with Dorsch their construction and the schedules for their completion. On the following day Dorsch presented draft plans of the factories to Göring and in the evening of the same day he was summoned to present them to Hitler. Hitler, who was always interested in architecture and construction projects, made several suggestions for changes. Dorsch also used the opportunity to privately discuss with Hitler the legal problem of OT operations inside the Reich. On the following day Dorsch presented to Hitler the new drafts. Dorsch discussed again with Hitler the operation of OT inside the Reich's borders. The redrafted plans impressed Hitler and he said that only OT could execute this construction project, therefore fulfilling Dorsch's wish.[378] This decision was formally announced on 21 April 1944, when Hitler appointed Dorsch and OT to carry out the construction of the bunker factories.[379] On the following May Day, Hitler appointed Dorsch as the head of the Construction Office in the Armaments Ministry, therefore broadening his powers to other construction projects.

From now on OT dealt not only with construction of aircraft factories, but also with other construction projects within the Reich. This appointment practically ended a long and bitter rivalry between OT and the Armaments Ministry. In his new position Dorsch was also able to intervene in some construction projects of the SS, especially in matters related to contracting and raw materials allocations.[380] At that time none of the bunker factory construction had started, although detailed survey work was already done. In the following weeks the location of the new factories was often discussed in the Jägerstab meetings. There is some confusion considering the location of the 6 fortified factories in the existing documentation and literature. The main problem is that many sources, including contemporary, tend to mix up underground factories constructed in existing tunnels or in expanded tunnel systems with the bunker factories. The situation became even more confused because each agency allocated underground and bunker factories its own code name. The Armaments Ministry used to allocate each project a code name, while the SS allocated the same project its own code. The tenant firm sometimes also referred to the same factory with yet another name. As a result, even an insider like Saur got it wrong when asked after the war to name the locations of the bunker factories. He named the Leonberg tunnel and a Hungarian underground factory as part of this scheme.[381]

The first bunker factory was in fact a massive tunnel complex. The 50,000-square-meter "Bergkristal-Esche II" plant (also known as Project B8 by the SS) was constructed in a massive tunnel system near St. Georgen in Austria. It was a cooperative enterprise of the SS — represented by its construction enterprise Deutsche Erde- und Steinwerke (DESt), the Luftwaffe and Messerschmitt. Slave workers provided by the SS from the nearby Mauthausen-Gusen concentration camp complex started working on the site in March 1944. The SS also constructed a large concentration camp — Gusen II — specifically for this project.

It was opened on 9 March 1944 and at the end of 1944 its population amounted to 12,000 inmates.[382] Although at the beginning it was not intended for Me 262 production, it was intended subsequently to perform there Me 262 final assembly. The factory was also supposed to produce fuselages and the entire slats supply for the Me 262 program.[383] Production at "Bergkristal-Esche II" started in early 1945, although the installation was never fully completed. The entire tunnel complex was completely reliant on rail communication. Raw materials were brought in on trains from all over the Reich, and completed fuselages were transported to at least 8 other final assembly plants.[384] Some 1,000 fuselages were produced at the "Bergkristal-Esche II" factory before the end of World War II, as well as a large number of other parts — largely by slave workers.[385]

Samuel G. Wilson was a USAAF technical officer who visited the plant little more than a month after it was captured by U.S. troops. He provided a detailed description of this plant:

> Entrance was through any one of five portals. Tunnels had been driven horizontally in a grid with the same size drifts (at right angles to the tunnel) on about 250 feet (76 meter) centers. Tunnel and drifts were about 20 feet (6 meters) high and 20 feet wide. Wall, floor and ceiling were of concrete. The place was lit by Mercury vapor lamps, walls were all painted white. We saw 30 completed Me 262 fuselages on the production lines. Assembly began from the rear of the tunnels and progressed out horizontally to level outside ground. At the farthest point in each tunnel were key sub-assemblies that were added to the assembly line as it progressed towards the tunnel portals. There was an underground 480KV diesel generator for standby, normal electrical power being provided from the regional grid. There were also complete testing laboratories. Huge bins were filled with [everything] from rivets, bolts and nuts to instruments ... all to make up complete fuselages as the end product. The units holding the frames for the fuselages were portable and in some cases made of wood. Railroad cars were apparently used for final assembly of the fuselage.... I noticed many of the big machine tools were manufactured in Cincinnati, Detroit, Chicago, etc.... Big ventilation ducts were made by putting wood sheet ceilings in, they were suspended from the arched ceilings of the tunnels. The intervening space was large enough to get fresh air to all parts of the factory. There were elaborate wash rooms, all large. Trains could back into the main tunnel for the purpose of shipping completed fuselages or delivering heavy equipment and materials. Most of the ventilation was natural.... The plant was never bombed but it would have been a waste of bombs. As I recall the tunnels were 50 to 100 feet below ground and had heavy iron doors for blast protection at the portals.... Production at the end of the war was ten fuselages per day and there were plans to increase rate to 50 fuselages per day.

Beside the detailed description of the plant, Wilson also mentioned the workforce used in it:

> Labor was of many nationalities. Some of it is said to be slave labor from the nearby concentration camp at Gusen. People when worn out by work and poor food were shipped to Mauthausen camp (also nearby) and replaced by new intake from Gusen. There were signs in the plant in German, French, Czech and Italian. I saw no sign in English or Russian.... Much of the labor was skilled and most of all plentiful and at nearly zero cost to the manufacturer, Messerschmitt.[386]

The aforementioned Kematen factory also became part of the recentralization scheme. It was planned to move a full Me 262 assembly line into its tunnels in May 1945, but this never happened.[387]

"Bergkristal" was a normal underground factory and not a bunker factory, but it formed one of the most important efforts to secure the Me 262 production from Allied bombing

The "Weingut II" bunker factory. Its half-buried long round form is visible here (courtesy U.S. National Archives and Records Administration).

raids. Of even greater magnitude was project "Ringeltaube," in which 3 separate large bunkers of identical design were planned in the area of Landsberg, Bavaria under the code names "Weingut II," "Walnuss II" and "Diana II." In many respects these three factories formed the cornerstone of the bunker factory scheme. They were supposed to produce different components for the Me 262. The Jägerstab established a special firm, Weingut-Betriebsgesellschaft, to manage the operation of the "Ringeltaube" factories. The firm planned to operate 3 shifts of 10,000 workers in each of the three factories. By employing a total of 90,000 people (30,000 of them were supposed to be concentration camp inmates provided by the SS) in the area of Landsberg, the new firm was about to turn this region into one the biggest centers of the German aviation industry. The three factories were expected to turn out 900 Me 262 fighters each month. In order to accommodate the inmates used in this project, the SS constructed eleven labor camps in the general Kaufering region. Around 30,000 inmates, mostly Jewish, lived in these camps throughout their existence.

Due to shortage of cement and steel only "Weingut II" was ever constructed.[388] This bunker was constructed in the Ingling Forest near Kaufering and was supposed to replace the capacity of Messerschmitt's battered Augsburg factory. Its construction was in line with a general design used in most bunker factories: a long and narrow arc-shaped concrete structure.

It was at least five stories high (28.4 meters), 400 meters long, its floor space was 95,000 square meters and its roof was 3 meters thick. A 5-meters-thick roof was planned but its thickness was reduced due to raw materials shortage. Nevertheless extra protection was provided by burying 40 percent of the bunker in the ground. Just for comparison, the 3 main dispersal locations of the Augsburg plant had a combined floor space of only 12,700 square meters.[389] "Weingut II's" semi-buried structure was covered with a layer of earth, in which trees were planted in order to improve its camouflage.

The factory was constructed under the management of the Leonhard Moll construction firm from Munich, beginning in late May 1944. The constructors used a unique construction method. First, the huge carapace concrete roof was poured on molds shaped from giant piles of gravel. After it was completed and the cement had dried, the molds and an additional layer of earth were dug from beneath the concrete arches. Precast concrete beams were then used for the interior floor construction. This construction method was especially efficient in making the construction site less vulnerable to air attacks. Normal construction of large concrete structures was dependent on large wooden casting forms and their associated scaffolding. These were very conspicuous and could easily be damaged by air attacks.[390]

A former construction superintendent told the Americans immediately after the war that the workforce employed in the construction of "Weingut II" consisted of 600–800 Germans, 300 Russians and 1,500 concentration camp inmates.[391] However, it is now assumed that around 10,000 Hungarian Jews worked at one time or another in the construction site. At the time the U.S. Army captured the site, the excavation beneath the concrete roof was far from being completed, but some jigs and production machines were already in place.[392]

A fourth similar bunker, code named "Weingut I," was constructed in a similar way to "Weingut II" near Mühldorf am Inn, east of Munich, by the Polensky & Zöllner engineering firm. Construction here started in June 1944 and most of the workforce for this project came from the nearby Dachau concentration camp. The airfield in Mettenheim, located right next to the bunker, was supposed to serve as a fly-out field, sparing the need to transport the completed aircraft on trains. It was planned to use this bunker especially for pressed parts manufacture. Production was supposed to start on 15 June 1945.[393] Although initially the construction of "Weingut I" progressed according to the plan, harsh conditions during the winter slowed down the work significantly. Then a general shortage of materials and workforce disrupted work on the construction site. In January 1945 only 57 percent of the project was completed and by March only 69 percent was completed. As a result production facilities were never brought in before the U.S. Army overran the site on 2 May 1945.[394]

The last of the six bombproof factories was the REIMAHG factory in Kahla. On 8 March 1944, Fritz Sauckel, the Reich Plenipotentiary for Labor Mobilization and *Gauleiter* of Thuringia, suggested to Göring the construction of a massive complex of bombproof factories in the Kahla-Pössneck area, south of Jena. Armaments giant Wilhelm Gustloff Werke teamed with OT and contracted various mining firms in order to construct these factories. Gustloff purchased aircraft producer AGO, whose Oschersleben factory was largely destroyed by American bombing. Then it established a subsidiary firm named REIMAHG (after Reichsmarshall *H*ermann *G*öring) to take over production of FW 190 and Ta 152 fighters formerly produced in Oschersleben.[395] Construction began in a former china clay mine dug in a hillside on 11 April 1944. The mine needed relatively little adaptation, including strengthening of weak points, flooring with cement, spraying with concrete against dust, and finally whitewashing. The Jägerstab planned to commence production in August 1944, but in October 1944 the facility was still not ready. The delay has been attributed to the failure to construct a branch railway to the site at an early stage and to Sauckel's dilettante interventions in the planning and construction process. These allegedly included the removal of the ducted air conditioning system installed by the contractors.[396] By that time Allied intelligence had already identified the factory and determined that it was supposed to produce jets.[397]

Hitler was briefed about the progress of the construction on 21–23 September, and he suggested that a BMW engine production line also be moved to the tunnels [398] Göring and Saur visited the site on 10 October 1944 and two days later Hitler ordered production switched to Me 262. Messerschmitt hurriedly moved production facilities to the factory and began producing there some subcomponents. Since the tunnels were still unfinished this initial production was done in makeshift bunkers constructed on the southern side of the complex. Final assembly began only after some of the tunnels were finished in January 1945. The factory comprised 75 tunnels totaling 32 km in length, offering approximately 30,000 square meters of floor space, and four bombproof buildings with 2-meter-thick walls. Final assembly was supposed to be carried out in one of the bunkers. A special elevator was supposed to deliver completed aircraft to a 1,250-m-long runway on top of the complex, used for test flying and for delivery flights. During early 1945 around 15,000 workers — two-thirds of them slave and foreign workers — worked in the Kahla REIMAHG complex. The slave workers were provided and guarded by an SS watch. At least 3 labor camps were constructed for them around the site.[399] Production, however, never really picked up the pace at REIMAHG and it is estimated that REIMAHG completed only between 15 and 26 aircraft by the time the U.S. Army captured the place on 12 April 1945.[400]

Thus, despite all the monetary investment (estimated at around 30,000,000 RM) and resources invested in this ambitious project, practically nothing came of it. In addition, its unique construction made it very conspicuous to aerial reconnaissance and Allied intelligence easily found it.[401] As the first USSBS team to visit the place summed up, this factory "must have been one of Germany's less efficient industrial enterprises."[402] However, REIMAHG definitely represented prevailing late-war thinking regarding armaments production in Nazi Germany. It was also another example of multiple partners' cooperation in the field of aircraft production. Fritz Sauckel played here a central role in his double function as a local *Gauleiter* and as the Reich Plenipotentiary for Labor Mobilization, therefore controlling both the territory on which the complex was constructed and the workforce required for its construction.

The underground relocation of large parts of the aviation industry was one of the biggest armament projects the Germans undertook in 1944. It was a project of national magnitude, which involved hundreds of firms and numerous state and military organizations. The only comparable underground dispersal program was the relocation of the oil-industry. This project was also triggered mainly by a general bombing offensive on this industry in April–May 1944. Initial Allied bombings on oil refineries proved to be exceptionally devastating. In contrast to the aviation factories, repairing the damaged oil-producing facilities was a long and difficult process. Once repaired these factories were easily damaged again by

Saukel's project. An aerial photograph of the REIMAHG bombproof factory near Kahla, 26 December 1944. The photograph was included in the target sheet file of this factory. Allied interpreters marked the workshops and tunnel entrance area. Visible at the top is the landing strip constructed on top of the complex (courtesy U.S. National Archives and Records Administration).

bombings. At the end of May 1944, Speer appointed Edmund Geilenberg, a munitions production expert, as a *Generalkommissar für die Sofortmassnahmen* (General Commissioner for the Immediate Measures) responsible for the underground dispersal of the oil industry. Just like Saur three months earlier, Geilenberg was also provided with all the authority and resources he needed. In fact, some resources previously allocated to the Jägerstab were now redirected to Geilenberg. Several conflicts soon appeared between Kammler's staff and Geileinberg's regarding the allocation of specific underground spaces, previously promised to the aviation industry. Two of the initial five underground refineries were supposed to be constructed in tunnels allocated to Junkers within the Mittelwerk complex and the A5 project. Kammler was able to prevent Geilenberg's takeover of these facilities, but the conflict delayed their completion.[403]

Geilenberg's task was truly monumental and the prospects for its completion were slim. The construction of a normal oil-producing plant was not an easy task, but doing it underground and under a worsening war situation proved to be extremely difficult. As a result very little was accomplished by the end of the war. The fate of the underground aviation fuel plants scheme demonstrates well this lack of progress. Eight underground aviation fuel factories were supposed to produce 130,000 tons of aviation fuel each month — an output that formed 82 percent of the total aviation fuel output in January 1944. Regular production was expected to start in these factories in the summer of 1945. The hydrogenation plants were large and therefore extensive excavation was required in order to construct them underground. The scarcity of workforce and materials, the loss of construction sites to the advancing Allies, and changing conditions compelled the Germans to change their plans repeatedly. As a result only two of the planned aviation fuel plants were under construction by the end of the war. None of them was even close to completion.[404] Just like the dispersed aviation industry, the oil industry was also heavily dependent on efficient transportation and simply fell apart once it was heavily disrupted.

The move of many factories to underground facilities was an extreme measure taken by the Germans under extreme circumstances. The main motivating force behind it was the need to protect important production from strategic bombing. Some firms supported the Reich's leadership's decision to move their facilities to underground facilities because they hoped to preserve this way their valuable production tooling for the postwar era.[405] Another reason was the willingness to re-centralize at least some of the production in single well-defended underground factories. Their construction involved, however, huge investments of money, materials, time and workforce. As we saw, in many underground factories working conditions were poor; in some of them they were unbearable. The bunker factories represented in this context an even more extreme solution, by trying to create artificial underground-like facilities. These projects demanded an even larger investment and most of them were not completed before the war's end. The forlorn remains of some of these concrete monsters can still be found in several places in Germany.

It is puzzling, therefore, that so much energy and resources were invested in these projects while at the same time a much better solution was apparently available in the form of the forest factories. The German leadership viewed these makeshift factories only as an interim solution,[406] but they offered several important advantages. First of all, they were much cheaper to construct and they were completed within a short time. It took, for instance, only two months and 10 work hours per square meter to complete the forest factory in Gauting. Additionally, it cost only 700,000 RM to construct this completely new factory. In contrast, it took 10 months, 100 work hours per square meter and 4,076,000 RM to construct Messerschmitt's underground factory in Kematen, which was never completed.[407]

Perhaps the best feature of the forest factories was that they were well hidden. The Allies quickly noticed the early dispersal drive, and the Germans were surprised to learn how much Allied intelligence knew about their dispersal after they recovered some documents from a shot-down aircraft. Hitler was sure that this knowledge could only be explained by some high-ranking treason and ordered an immediate investigation of the matter.[408] Apparently most of the information Allied intelligence gathered came from aerial photo interpretation, and was cross-referenced in at least some cases with other sources.

Allied intelligence was also well aware of the existence of underground aircraft factories and listed them regularly in intelligence summaries. On 24 October 1944 the British air intelligence listed four existing underground aircraft factories and noticed that they "are being specially examined for targeting and suitable method of attack, but in view of their importance it was considered advisable to give this preliminary notice of their existence."[409] Listed were the factories at Leonberg, Kahla, Eschenlohe (converted rail tunnel), and the FW 190 factory at Langenstein. Further information was received as German underground relocation progressed, enabling the Allies to get a quite accurate picture of these activities and about the main factories. The relocation of the aero-engine industry was also closely followed by different western Allied intelligence agencies, and close to the end of the war the German Economic Department of the British Foreign Office was able to submit a detailed report about this industry, in which both its underground and above-ground dispersal were detailed. The report even mentioned the dispersal locations of some smaller subcontractors of the aero-engine industry.[410]

Immediately after the war, special teams were sent to locate the underground factories and to confirm their existence.[411] One team visited 12 underground factories in central Germany and noted that 10 of them had been located and identified during the war by Allied aerial photo interpreters. Air intelligence was quick to publish updated manuals for aerial photo interpreters in order to help them recognize underground factories, and this reference material eased their work.[412] In most cases the interpreters were unable to precisely determine what was produced in these factories, but they were able to put them on the map and in some cases even to relate them to aviation production. In any case, the Allies were well aware that important production was done in underground facilities and therefore kept a watchful eye on every newly discovered underground activity.[413]

Even several aboveground factories relocated to industrial complexes never associated before with the aviation industry had been correctly identified by Allied intelligence. Examples of such successful discoveries were Messerschmitt's factory in Telfs near Innsbruck (in a former textile factory) and Focke-Wulf's factory in Rathsdamnitz (in a former paper factory).[414] Interestingly, in several cases former workers of the aviation industry, who were drafted and then captured, disclosed initial information about dispersal factories during their interrogation. A former Junkers employee worked for the firm for eleven years until he was drafted in September 1944. He was captured, and during his interrogation by British intelligence officers, he disclosed almost the complete dispersal layout of the Junkers Dessau complex.[415] As we have already seen, the forest factories were not compromised, although the Allies received hints regarding their existence through POW interrogations.[416] For instance, a former Messerschmitt employee provided his interrogators the exact location of the forest factory located near the Leipheim factory, but it seems that the Allies never realized the importance of this forest factory for the Me 262 production.[417]

The effectiveness of the forest factories is also proved by their productivity. As mentioned before, besides 15 prototypes and preproduction Me 262, most of the 1,433 aircraft of this type to be finished were assembled in dispersed forest factories.[418] Series production in underground factories also suffered qualitatively. The mix of a largely slave workforce with poor working conditions underground badly affected not only quantity, but also quality. This was the case in Junkers' Gross Schierstadt underground factory, located in a potassium mine, where aft fuselages for bombers were produced. Mostly slave labor — including some 800 British POWs — was employed here. Former executives estimated that approximately 40 percent of its output was of unacceptable quality.[419]

The only advantage of the underground and bunker factories was their invulnerability to normal bombing, as the Allies acknowledged.[420] In this regard it is interesting to note that an underground factory could become a death trap if its entrance was hit. The Germans were extremely concerned about possible precision strikes against shafts or entries. Gabel, the Jägerstab representative for underground relocation, declared in late May 1944 that securing the entrances was the most important task he was dealing with, because failure to do that could turn the underground factories into death traps.[421] There is no indication that the Allies ever tried such a strike against underground aircraft factories, but they successfully struck other hardened targets. From mid–1944 the RAF attacked several massive bombproof or underground targets in France using massive 5.5-ton "Tallboy" earth penetrating bombs. These were heavy bombs with thick casing, which exploded only after penetrating a thick layer of earth or cement. They destroyed their targets mostly by causing strong underground concussion waves. The Germans noticed this development, which had already brought their V-3 bunker project to an abrupt end. Consequently at least one German observer suggested at the end of October 1944 the abandonment of the aboveground bunker scheme because the thickness of the roofs always lagged behind the latest developments in bunker-busting bombs. He claimed that the only protection is to place the factories at least 50 meters underground.[422]

During 1944 Allied intelligence collected detailed information about Mittelwerk from different sources. It also prepared a large-scale model of the factory and its surroundings. In December 1944 all the material was added to a target folder with the obvious intention to attack the plant.[423] In February 1945 the RAF made initial plans to attack Mittelwerk with special bombs filled with inflammable fluid. This plan was put aside after it was pointed out that it would not destroy the machinery and would only kill the slave workers working in the plant.[424] However, at the end of the war the development of earth-penetrating bombs reached its culmination with the British 10-ton "Grand Slam" bomb, which entered operational service with the RAF in mid–March 1945. With it the Allies possessed a potentially formidable anti-underground facilities weapon. Even earlier it was found out that conventional bombing was also effective against hardened targets, especially during their construction phase. The campaign conducted by Allied air power against V-weapons bunkers in 1944 resulted in the Germans' abandoning them or never completing them. The main effect of these attacks was not damage to the structures themselves, but by peripheral damage, which prevented their completion and rendered them unusable.

It seems that one of the main explanations for the bunker factory projects had to do with the general atmosphere of extreme solutions prevailing in the German leadership at that time. As was the case with the V-weapons, the jets and the new submarines, the huge concrete factories represented a "wonder weapon," which was supposed to help Germany win the war despite the heavy odds. As Albert Speer has put it in one of his postwar books, it was the prevailing mentality of *Höhlenphantasie* (cave fantasy) that kept the Reich's leadership fixated on the underground relocation master plan.[425] Rational decision-making processes and strategic planning should have stopped their construction at least towards the end of 1944, after the forest factories proved their effectiveness. This lack of realism — even among the construction professionals — is particularly evident in the planned expansion of the Mittelwerk complex. In January 1945 the SS submitted a plan for the construction of the B3, B11 and B12 projects with a total floor space of 440,000 square meters. It seems that the planners totally ignored the fact that it took WIFO seven years to dig the original 110,000-square-meter tunnel complex of Mittelwerk.[426]

This sort of unrealistic planning formed a central feature of the entire reorganization and extreme measures of 1944. At the time the reshuffle drive started in the follow-up of "Big Week," Germany's situation was bad, but it was on the verge of a great production boost. The Germans themselves failed initially to fully appreciate the limits of Allied air power and its effects. Most of the measures they initiated proved to be effective and fulfilled the production goals set by the Reich's leadership. The main problems that doomed German air power were in two areas. First was the failure of the Luftwaffe to efficiently use the equipment delivered by the industry. The second problem was the earlier failures to streamline production and production programs. The Germans started World War II with good aircraft but totally failed in the timely development of the next generation of military aircraft. Although the RLM fully appreciated the value of jet technology and ordered its high priority development from early on, it failed to realize the time it will take it to mature. Consequently it failed to supervise the development of stopgap aircraft. Instead it ventured into highly complicated upgrade scheme of existing aircraft, and as a result the Germans invested too much time and resources in producing a maze of variants and subvariants. The existence of many short- and long-term development programs and futuristic research and development projects aggravated this wastage. Although many of these programs were far-fetched "paper projects," others were real and turned into serious projects. They could have been very reasonable projects, but not in 1943–44, when streamlining and rationalization were the call of the day. This lack of realistic planning and management ironically happened exactly when the management of the aviation industry was put in the hands of the most efficient body to ever control it. On the one hand it can be said that years of neglect and mismanagement created a sort of managerial culture that was hard to change under the new leadership. On the other hand, as we saw, even under Saur and the SS, an unrealistic mentality prevailed. The conclusion is therefore that not only did the RLM and the aviation firms suffer from a poor managerial culture, but also the supposedly rational and cold-headed "non-flyers" that stood at the helm from March 1944.

4

From Technological Expertise to Slave Labor

Technological, industrial and organizational aspects of the German aviation industry has stood so far at the focal point of this research, but now it is time to turn to some social issues. One of the most striking features of late-war German aviation industry was a sharp shift from the industry's traditional manpower. Large proportions of highly proficient German personnel has been replaced by late 1944 with foreign manpower composed of, among others, large numbers of Jewish housewives and teenage schoolgirls picked up in places like Auschwitz. This change formed a major break from the national-community-building aspirations so widespread in the aviation industry before the war. It was also a major break from the guild-like character of aircraft manufacture. The massive use of slave labor became within months a central characteristic of this branch. Historians, like Constanze Werner, noted: "Therefore, forced labor and concentration camp inmates were employed so massively, so early and so unscrupulously in the aviation industry like in no other industral branch."[1]

Even before modern historians started pondering over this peculiar phenomenon, it bewildered American technicians and engineers surveying the German aviation industry after the war. One of the intelligence teams described it in a technical report dealing with production techniques of the German aviation industry:

> An outstanding feature of German production methods was the extensive use of slave labor. Among these people were relatively few possessing skills. The "Auslander" (*sic*), as they were called, were handled in various ways, ranging from extreme cruelty and neglect to attempts to secure cooperation and good work by more favorable treatment. In the larger government-controlled plants, such as the Mitteldeutsch (*sic*) Motoren Werke and the Nordhausen V-plants, the slave laborers were badly mistreated; in the small, privately-controlled plants their lot was comparatively good and efforts were made to properly feed and house them. In smaller plants, some elementary training was provided, particularly in the instrument manufacturing works.[2]

A maze of general structural factors and firm specific interests contributed to the deeply rooted practice of slave labor in Germany's most technologically advanced industry. In order to understand the background of this change and the way it enabled the Germans to increase their production rates in 1944, we need to look first at the way modernity affected industrial production.

Basically, industrial modernization, as applied by German industrialists, could appear in two forms, which were different types of production lines:

1. Automating the production process by replacing humans with production machines and automated production lines. That method usually reduced manpower, but still required skilled workers to operate the complicated machinery in a proper way. This was basically the system used in the modern car industry and implemented partially unsuccessfully in aircraft production — mainly in the USA.

2. Employing output-multiplying machines or processes, which still needed the human touch. Here workforce was not necessarily reduced; the output of each worker was simply multiplied.[3]

The main difference between these methods, as applied to aircraft production, can be exemplified by the type of riveting machines used in them. In the first method, automatic and complicated riveting machines were used. In the second method, improved hand-held riveting machines enabled workers to complete more riveting than with older machines or by using manual riveting.

Both methods relied basically on the mechanization of the production process, but the first method was difficult to implement in the production of complicated machines like aircraft and their engines. Therefore, the second method was more suitable for a rapidly expanding industry, which employed a large number of employees with little or no training. This method also perfectly represented production-line-based industries, such as started to appear in Germany from 1942. Ford used this strategy in his "Willow Run" factory to produce bombers. Ford's engineers used machinery and fixtures that had accuracy built into them. The workers simply needed to load the machines and quickly attach the parts.[4] Automatic production machines replaced in several cases complete production processes by combining in their mechanism several different functions of the manufacturing processes. Such complicated machines were, however, rare even in the American aviation industry, and most available machines were just designed to improve the performance at certain stations along the conveyor belt. Hand-held riveting machines had not changed much, but became easier to operate and enabled faster riveting. Their operation also required less training than was required for the operation of complicated automatic machines. These types of output-multiplying machines and the production method associated with them perfectly suited the type of workers that started to appear on German production lines in 1942.

These changes in the production process enabled the Germans to seek some extraordinary solutions to a growing manpower shortage. Initially, German industrialists viewed the conveyor belt system merely as an interim solution to the demand for an ever-increasing output. Few viewed it as a way to solve the manpower shortage affecting all of German industry, and particularly the aviation industry.[5] This shortage represented one of the biggest structural problems of the German aviation industry. Solving this shortage represented one of the only ways to boost aviation production, particularly after the outbreak of World War II. National-Socialist ideology, which defined certain human groups as inferior, provided the ideological motivation for a radical solution of this industrial problem.

Germany's Manpower Crisis

Although in 1939 the German aviation industry possessed large unused production capacity in terms of production machinery and floor space, it suffered from an acute manpower shortage. By definition, the highly modern aviation industry relied on quite narrow and professional manpower reservoirs, which were already stretched to the limits by the rapid prewar expansion. Skilled technicians were available in limited numbers, mainly because the German apprentice system required four years of professional training, supplemented afterwards by frequent qualifying tests. With the outbreak of the war the training period was cut to 3 years, but it was still relatively long and therefore failed to solve the shortage of skilled workers.[6] Basically, the RLM planned an immediate expansion of the aviation industry in case of war on the erroneous premise that an increased single shift would suffice to provide the required output, and that in an emergency it would be no problem to add a second shift, using partly trained workers from other, less important branches of industry on modern production lines.[7] However, even before the outbreak of World War II it became clear that it would be difficult to recruit this extra manpower in case of a war.

Contingency plans for war foresaw a significant increase in the workforce of each factory and firm. The Focke-Wulf firm reported in April 1938 that in case of war it would need to increase its workforce from 7,547 men and women to a total of 11,436.[8] The comprehensive prewar expansion of the military-industrial complex, as well as the massive reduction of unemployment in Germany, starkly reduced the available manpower required for the implementation of such ambitious mobilization plans. It was difficult enough to find skilled workers for the new prewar factories. The difficulties were particularly acute in areas where several aviation factories were clustered, like greater Berlin. According to Heinkel's mobilization plans, 10,500 workers were supposed to work in Heinkel's Oranienburg factory in case of war. In April 1940—more then seven months after the outbreak of World War II—only 7,585 men and women worked in this important factory.[9] At that time the aviation industry was already in possession of potential overcapacity, created mainly through massive prewar expansion of fixed assets and by capital increase in most firms. This increase could not be matched by a similar rate of manpower increase, and as a result the extra capacity stayed largely unused. This discrepancy resulted in empty structures within the compound of various factories and in idle production machines, put mostly in long-term storage. Although immediately after the outbreak of the war extra manpower was recruited, increased military conscription soon reduced the available manpower.[10]

One of the results of this manpower shortage was the industry's inability to run more than one daily shift or to increase it meaningfully. In order to somewhat compensate for the lack of manpower and yet increase production, most firms increased the workers' weekly working hours from the prewar standard of 53 hours to 56 hours in 1939–40, and then to 58 hours in 1940–41.[11] This increase brought a limited improvement in productivity, but conscription continued to dry up the manpower reservoir. Generally, at that time the Reich's economy started to gradually convert parts of the consumer goods industry into armaments production. By late 1940 most of the consumer branches already devoted 40–50 percent of their output to the military. In the following years the consumer goods industry was further reduced, freeing more capacity and manpower to the war industry.[12]

Yet, aviation manufacturers continued to suffer from acute manpower shortage. In June 1942 Focke-Wulf's general director Kurt Tank complained bitterly in a letter to the Industrial Committee for Production of Luftwaffe Equipment (*Industrierat des Reichsmarschalls für die Fertigung* Luftwaffe*ngerät*) about the manpower crisis created by increased call-up of mostly skilled workers. While his firm demanded 4,250 additional workers for its expanded production, only 320 new workers were allocated, and at the same time the firm lost 1,434 men to the Wehrmacht.[13] Plainly said, firms were unable to increase their manpower in order to fulfill their expanded production plans. When in 1942 the aviation industry finally geared up for total war production, the lack of essential manpower was the main obstacle to the industry-wide introduction of a badly needed second daily shift. Especially missing were foremen and engineers to supervise a multi-shift work schedule.[14]

Taking this limitation into account, it is still amazing that in June 1942, when the Reich's leadership realized Germany's worsening strategic situation, most factories still operated on a one-shift basis. Within Focke-Wulf, for instance, only the main Bremen factory operated an emergency second shift, which was manned by only 70 workers (7,447 workers manned the first shift).[15] Heinkel also initiated a firm-wide second full shift only in summer 1942, and only because of the need to complete urgent tasks associated with the He 219 night-fighter project.[16]

The general manpower shortage became more and more desperate in 1942, mainly due to the enormous losses on the Eastern Front. The need to replace these losses led to an increased call-up of older age groups and of men previously exempted from the draft. The aviation industry gradually lost an increasing number of workers to the Wehrmacht and no German replacements were found.[17] Engine producer MIMO, for example, up to the end of 1941 lost only 285 employees to the Wehrmacht. By the end of 1942 this number has increased to 1,182 — 1,024 of whom were skilled workers. During 1942 some 73 of MIMO's workers were lost forever because they were killed while serving in the Wehrmacht.[18] As Milch pointed out in a meeting of the Armaments Ministry's Central Planning Board at the end of October 1942, lack of manpower was the main obstacle the industry faced in fulfilling its production plans.[19]

Earlier in the war Great Britain had faced a similar problem. The British aviation industry expanded significantly immediately before the outbreak of the war and afterwards. The small peacetime British Army also suffered heavy losses in the early campaigns and had to replace them while increasing its size almost ten times. Just like in Germany, these developments badly affected manpower availability to the aviation industry. The British solution was straightforward: cadres from the main factories were sent to new and/or dispersed factories. They formed a professional nucleus for additional workforce, which was mostly recruited locally by closing or trimming down non-essential businesses and domestic services. Most significant, however, was the relatively short training required for most production tasks on Britain's modern production lines. Most of the new workers could be trained to perform some of the simple tasks in 3–4 weeks, and only a few technical skills required more than six months' training. American officials stationed in London in 1944 reported about the revolution in British aviation production: "Within a year many of the small shops were operating with a new labor force except for supervisors and two or three veteran assistants from the home works."[20]

The British relied on massive mobilization of women in order to replace drafted men. In contrast to a widespread belief, women had worked in the German armaments industry, and specifically in the aviation industry, since before the outbreak of the war. Hiring more women was in fact one of the solutions sought by the industry in order to solve its chronic manpower shortage before the war.[21] Furthermore, a massive recruitment of women was planned in the framework of prewar contingency plans for wartime expansion. Focke-Wulf, for example, planned to increase the number of its female employees from 378 (5 percent of its total workforce) to 4,276 (27 percent if its increased total workforce).[22] This plan also foresaw a major change in the pattern of female employment within the firm. While before the war, women performed mostly administrative and logistic work, and only a few were directly involved in production, Focke-Wulf planned to change that pattern. In April 1938 only 3.5 percent of its technical workforce were women, but the firm intended to increase their share to 39.5 percent in case of war. These plans clearly show that women were viewed at least by the aviation industry as an important potential workforce if Germany were to be involved again in a large-scale conflict.

In practice, these plans remained largely unfulfilled. Firstly, recruitment of women never really picked up pace. Secondly, during the war, or at least initially, only a relatively small number of women became directly involved in production. Most of them fall into the unflattering internal category of *Unproduktiv*. In August 1943 the Focke-Achgelis firm, one of the first helicopter producers in the world, listed 672 employees on its payroll, 109 of them were women; 87 of them were part of the design or administrative staff, and the rest were categorized as unskilled workers, which suggests that only a few of them, if any, were engaged in production.[23] This is a micro example of the situation prevailing within this industrial sector. The total number of women employed by the aviation industry nevertheless increased in the immediate prewar period and in the early years of World War II, as presented in the following statistics of the Mansfeld firm, which produced different subcomponents for Heinkel and other aircraft producers.

	21/12/38	21/7/39	25/11/40	20/9/41
Total no. of workers	356	489	1015	1093
Women	38 (12%)	73 (17%)	273 (30%)	278 (28.8%)

Number of women employed by the Mansfeld firm, 1938–1941.[24]

This trend was not a general one, and in accordance with the general trend prevailing in the German industry between 1939 and 1941,[25] in some firms the number of women actually dropped. The most extreme example is probably Arado, which lost in 1940 74 percent of the women working in its Brandenburg-Neuendorf plant. The share of women within Arado's total workforce dropped from 19.9 percent in 1939 to 15.6 percent in 1940 and to 15.1 percent in 1941. The drop here and in other places was probably caused by poor working conditions in the wartime industry and by the fact that married women received up to 85 percent of the former wages of their drafted husbands, thus providing them a respectable income without the need to work.[26] In other cases, although the number of women increased, their share of the entire workforce decreased. In 1939 some 19 percent of the workers of the Henschel aero-engine company were women. Their total number increased during the war, but their share dropped gradually to only 11 percent in 1944.[27] The number of BMW's female employees more than doubled from 2,186 in 1939 to 5,354 in 1944, but their total share dropped from 20.6 percent to 17.6 percent.[28]

Earlier, Messerschmitt had become the most distinguished firm that tried to compensate for the loss of German men by increasing recruitment of women. By early 1942, 24.6 percent of the firm's workforce was composed of women. It is clear that by relying on increased numbers of women, Messerschmitt required only a small number of foreigners. While at that time 11.3 percent of the workers of the aviation industry were foreigners, only 5 percent of Messerschmitt's workforce was composed of foreigners.[29] The share of women continued to increase in the following year. Milch reported in late October 1943 that 25 percent of the 1,852,000 workers of the aviation industry were German women and 7 percent foreign women.[30]

It seems, therefore, that at least in some places true efforts were made to bring more women into the aviation industry. The employment office in Munich regularly allocated women to the BMW factories in and around the city. During April 1943 a total of 100 German women were allocated to BMW; some of them were trained immediately to perform technical work.[31] Focke-Wulf employed in June 1943 a total of 1,949 women — 115 of foreign origin — in its Bremen main factory. This number formed 27.9 percent of its total workforce of 6,991.[32] The number of women employed at Messerschmitt's Kematen factory in Tirol also increased in mid–1943, and then again at the end of that year. In any case, the number of women employed in this factory towards the end of the war was almost five times higher than at the beginning of 1941.[33] In the Siebel light-aircraft plant in Halle, 60 percent of the 9,000 workers were women, but this seems to be an extraordinary case and it is not clear how many of them were foreigners or slave workers.[34] On 19 March 1945 the Junkers firm, still one of the largest aviation concerns in Germany, listed a total of 9,646 German women on its payroll. The number of German male employees at that time was 20,899; therefore almost one third of the German workforce was composed of women.[35]

The manpower shortage became especially critical after the Battle of Stalingrad and the total-war emergency measures declared in its aftermath. The Wehrmacht urgently needed an influx of manpower in order to replace its huge losses,[36] while the armaments industry needed extra workers in order to increase its output. Some high-priority programs, like the He 177, were allocated special status, which exempted men involved in them from the draft. In March 1943 the firms involved in the production of this aircraft received a special status that protected their workers from conscription.[37] Such cases were rare, however, and during 1943 most firms lost large numbers of German and foreign workers. MIMO, an important producer of aero-engines, lost 1,585 of a total of 8,033 workers (19.7 percent) during 1943, mainly due to conscription of Germans and of foreigners choosing not to renew their contracts. This loss happened after MIMO's manpower statistics peaked at the beginning of 1943 and in spite of the massive anti–Bolshevik propaganda drive used to recruit foreigners from Western Europe after the Stalingrad catastrophe.[38] BMW also lost a large number of workers as a result of the Stalingrad defeat. In early 1943 a total of 6,189 workers — 11.5 percent of BMW's entire workforce — were called up by the Wehrmacht. Most of these workers were both skilled and experienced, and their loss was immediately felt.[39]

The problem of foreign workers choosing not to renew their contracts became a major problem in summer 1942. The annual contracts of numerous foreigners, who had volunteered for work in Germany in the wake of the previous year's propaganda campaign, started to run out, and many chose not to renew them. At a meeting in the RLM on 26 August 1942, Karl Frydag reported that the Junkers aero-engine division lost 38 percent of its foreigners, BMW 24 percent, Arado 18 percent, Daimler-Benz 26 percent and Heinkel 10 percent. Two solutions were suggested: Göring suggested forcing the foreigners to renew their contracts by law. Hitler, on the other hand, wanted to let them go and replace them with Soviet POWs and other foreign civilians.[40]

Since late 1942 all the workforce requirements for war-related production had been formulated by the Armaments Ministry and submitted to the office of the Reich Plenipotentiary for Labor Mobilization (GBA), headed by Fritz Sauckel. This arrangement made the RLM and its industry dependent more than ever on other organizations in matters of workforce allocation. The situation was further complicated by the emergence of two central competing factions within the armaments industry and its management: Speer and the army's Armaments Office on one side, and Milch and the Luftwaffe on the other side. Speer and Milch already cooperated closely in the reorganization of the aviation industry, but Speer represented to a large extent the army and its interests. The navy was also a competing agency, but on a smaller scale, and in any case it decided to hand over the management of naval production to Speer in July 1943. Milch became desperate to get the workers he needed. As Saur told his interrogators after the war, at least on one occasion he yelled at him violently during a telephone conversation and accused him of taking away all the manpower required for the aviation industry and allocating it to the army's production programs.[41]

In mid-1943 the Planning Office of the RLM warned Milch that the development and production of aerial weapons would suffer heavily as result of the continued drain on the industry's manpower through military call-up.[42] Göring and Milch tried to avert this trend in a cooperative manner but largely failed. The problem became simple: the aviation firms lacked the required workforce and therefore were unable to fulfill the output figures of the RLM's production programs. Mismanagement on behalf of the RLM made the situation even worse, because in various cases it failed to appropriately prioritize production programs. In December 1943 Arado estimated that it needed 9,180 employees to produce the new jet-propelled Ar 234, but it could spare only 750. At the same time, 12,293 Arado workers were involved in the production of the troubled and long-overdue He 177 bomber.[43]

This state of affairs underscores two inherent problems of German aircraft production: lack of manpower and confused prioritization. Arado would have been able to produce the Ar 234 with the available manpower only if the already outdated and troublesome He 177 was stricken from the production programs. The RLM was required to make the choice between allocating more manpower and eliminating aircraft types from production. Since it failed to a large extent to eliminate troublesome and unnecessary projects, it was obliged to find the extra manpower required to replace German workers and to support the production of high-priority types.

Another pressing matter, which exemplifies well the RLM's dilemma, was allocation of workers to two of the Luftwaffe's prime high-tech projects: the Me 262 jet fighter and the V-1 cruise missile. The preparation for the Me 262 series production was a difficult task that required a large number of skilled workers for manufacturing production tooling. Speer later estimated that 6,000 skilled workers were employed in preparing the production facilities and tooling for this aircraft.[44] These extra workers were required particularly in order to man Messerschmitt's new factory at Kottern that was established specifically in order to produce for the firm the tooling it needed. As it turned out, finding a workforce for this factory became a major bottleneck of the Me 262 production in the next months.[45]

After winning a high-priority contract from the RLM to produce this aircraft, Willy Messerschmitt employed tactics bordering on blackmail in order to secure the workforce his firm required. He did not hesitate to use his high status and his political influence in order to get what he wanted. In 1943, immediately after the deadly attacks on Hamburg, he went to Hitler and demanded either massive expansion of the V-weapons program or the allocation of all available manpower to the buildup of the Luftwaffe; in other words, to

The meddler. Willy Messerschmitt congratulates company test pilot Fritz Wendel upon completion of a record-breaking flight in the Me 209V1 prototype on 26 April 1939 (U.S. Air Force via National Air and Space Museum, USAF-A19341AC).

the production of his jet fighter. Hitler concurred and ordered increased aircraft production, but at approximately the same time he also authorized high-rate production of the A-4 missile.[46] Armed with this Führer decision he kept demanding from the RLM more draftsmen and jig-makers for the Me 262's production. By October 1943 the RLM was winding up other companies in order to free workforce for the Me 262 production. Messerschmitt subsequently demanded 4,000 additional workers directly from Göring, warning that the Me 262 would be up to six months late if he did not get them.⁴⁷ Messerschmitt now became almost intolerable because of his nagging demands for more workers and for his lobbying for the Me 209 conventional fighter. In November 1943 Messerschmitt demanded 1,000 additional workers for the Me 209 program, which was canceled earlier by Milch and was revived after Messerschmitt's intervention with Hitler. This conventional fighter was supposed to enter operational service only at the end of 1945 or in early 1946 — much later than the more advanced Me 262. It is a clear example of the way Messerschmitt's meddling behind the scenes caused the waste of time and resources on a completely unnecessary project.⁴⁸ This kind of behavior caused Göring and Milch to prefer working with Focke-Wulf's general director Kurt Tank, whom they also considered a better designer and a fighter aircraft expert, largely because he personally tested his aircraft.⁴⁹ Göring mistrusted Messerschmitt so much at this stage that he refused to speak with him without a stenographer present to record exactly what was said in the discussion.⁵⁰

As mentioned earlier, at that time the RLM became heavily dependent on the Armaments Ministry for manpower allocations, and Göring and Milch could do little to fulfill Messerschmitt's demands without severely curtailing other RLM programs. As relations with the Armaments Ministry grew colder, Speer's staff increasingly refused to allocate resources to the RLM, including manpower. Speer was convinced that aircraft production was crucial because regaining control of the air over the Reich was the only way to prevent his war industry from being bombed, but he had other obligations.⁵¹ Therefore, he tried to find a compromise solution, as evident from his contemporary personal correspondence with Milch. In November 1943 he offered to initiate the transfer of 3,000 Italian POWs from agricultural work to the Me 262 program in order to fulfill at least partially the demand for 6,000 workers. However, he was unable to provide the required skilled workers and asked Milch to comb the Luftwaffe's own technical services and production centers in order to obtain them.⁵² Although Milch doubted the productivity of Italian foreign workers and military internees, and had recently ordered to beat them if they would not work satisfactorily,⁵³ he had no choice but to accept the offer. Speer set the wheels in motion in order to facilitate the allocation of Italian POWs and of some craftsmen from north Italian aviation factories.⁵⁴

In mid–December a Führer conference discussed the question of the missing 3,000 skilled workers for the machine tooling and jig-making of the new aircraft's production program. Hitler ordered for them to be available for four months from the Wehrmacht's manpower reservoir. This measure was acknowledged as a temporary solution, and the transcript states that after 4 months the status of these workers would be reconsidered. They were then either to stay in the aviation industry or be redrafted.⁵⁵ This kind of arrangement was quite common in wartime Germany, but was usually discussed at lower levels. Soldiers or officers possessing special civilian professions urgently required in the home front were temporarily released from the service upon specific request. They worked in the firm or organization requesting them and usually their special status, known as *UK-Stelle* (UK— *Unabkömmlich*— indispensable position), was reconsidered after a predetermined period. If it was found out that they were not needed anymore they were redrafted. In some cases soldiers were released from service under these conditions if they went to work in an industry related to their branch of service — like Luftwaffe soldiers being released in order to fill key positions in the aviation industry. By the end of October 1943, some 435,000 of the 817,000 German men working in the aviation industry were *UK-Stelle* personnel.[56] Theoretically they could have been redrafted at any moment, as frequently happened.

Besides its temporary nature, this arrangement was also risky, because letting such workers go back to the front posed a security risk. Due to their professional status they possessed much knowledge and if captured, they could provide their captors valuable information. Such was the case of a man released from military service on the Eastern front in March 1944 and sent to work as a technical drawings librarian at Messerschmitt's Leipheim factory. He was redrafted to the army in October 1944 and was soon captured on the Western front. Because of his central position he knew a lot about the Leipheim factory and generally about the production program of the Me 262. His interrogators thus obtained from him minute details about the aircraft and its production program, including the exact layout of the Leipheim plant, technical description of parts and components, further development of the Me 262, a list of subcontractors, etc.[57]

Allocation of workforce also became a burning issue of the V-1 program. In December 1942 it became clear that the developer of the weapon, the Fieseler firm of Kassel, lacked enough capacity to maintain series production of the missile on its own. Additional capacity for this unique project could not be found within the aviation industry. Part of the problem was that a Fieseler factory that was supposed to manufacture 1,000 V-1s per month was reallocated to fighter production. Some Volkswagen executives heard about this problem and offered to produce the flying bomb for the RLM. At that time Volkswagen possessed extra free capacity, which the firm was eager to use in order to expand its aviation production. The RLM was also eager to commence series production as soon as possible, especially in light of the looming rivalry with the army's A-4 missile. It quickly decided to contract Volkswagen to produce the V-1 in its giant Fallersleben factory complex. This contract turned Volkswagen overnight from a minor aviation contractor to an important member of the aviation industry.

Although commonly associated with car production, in 1940 Volkswagen established an aviation division and became an overhaul contractor of the Ju 88 bomber. It also produced wing sets and control surfaces for this aircraft, and later for the more advanced Ju 188 and Ju 388. Volkswagen's *Sonderfertigungen* (Special Productions) Division, which also produced bombs, external drop tanks, fuses and torpedo components, was tasked to produce the V-1.[58] The firm suffered from manpower shortage and relied heavily on foreigners. The GBA was able to fulfill a demand for 2,511 foreigners for June 1943 — when the firm was supposed to start preparing for series production. However, none of the 480 skilled workers demanded at the same time had arrived. Shortage of German workforce was especially critical for this program. Because of its highly classified status only Germans were allowed to work on the V-1 final assembly lines and on production lines of highly classified components, like the Argus engine and the Askania guidance system.[59] By employing foreigners in other, less classified production tasks, the involved firms were able to free at least some of the required German workforce. It was only barely enough and with projected output being constantly increased, further manpower shortages undermined Volkswagen's plan to reach a monthly output of 100 missiles in August 1943.[60] The firm demanded subsequently some 2,990 workers in order to sustain the projected monthly output of 2,000 missiles by December 1943, and 5,000 missiles by June 1944. These output figures eventually proved to be completely unattainable. They form one of the best examples of over-optimistic planning by the RLM for a revolutionary weapon system that was still in the development phase.[61] Anyway, it was difficult to obtain the required workers and from the 2,180 workers that were formally authorized for the project by the end of July 1943, only 1,427 had arrived.[62]

Workforce shortage continued to plague the V-1 production after the August 1943 bombing crisis and consecutive decisions regarding the priority level allocated to the V-1. On 8 September, Volkswagen urgently requested 122 German workers in order to fulfill its September output quota. From the 300 German workers that were eventually authorized, only one arrived, and he turned out to be a professional tailor. By the end of September the firm had received a total of 100 German workers.[63] The Fallersleben factory failed to make the designated production pace, and in October 1943 it was still short 250 skilled workers, although by that time 394 foreigners and slave workers had been brought in. This failure formed a serious check to the series production of the V-1, because on 22 October 1943 Fieseler's works in Kassel were heavily damaged during a British nocturnal carpet-bombing of the city. It was an accidental lucky strike, since the RAF targeted the entire city and not specifically Fieseler's factory. As a result both the factory and the homes of its workers were hard hit. Around 50 percent of the factory was destroyed in the raid, including Fieseler's main design office.[64] The damage was so severe that it was decided not to reinstitute production in Kassel. This decision practically left Volkswagen as the only V-1 producer until Fieseler completed the move of its V-1 production to a new branch factory in Cham, Bavaria. This move was prolonged by Hitler's order not to employ foreigners in the V-1 production. He issued this order because it was thought that foreign workers had informed the British about the Kassel factory and that the bombers had attacked Kassel specifically in order to destroy it. As a result of Hitler's order, Fieseler was forced to leave behind 45 percent of the workforce intended for Cham. This setback caused in turn a delay in opening the Cham factory.[65]

Shortage of skilled workers continued to form the biggest manpower problem. On top of its highly classified status, the V-1 was a unique weapon system that required a relatively high number of skilled workers during its evolving production process. In November 1943 some 8,417 workers were involved in V-1 production, but the involved firms were still around 2,000 short of the number demanded from the GBA in October 1943. Since most of these vacant 2,000 jobs were technical and directly related to production, it was particularly difficult to man them.[66] Despite the efforts to locate workers at the beginning of 1944, Volkswagen was still short 428 Germans and 677 skilled foreigners. This shortage and the imminent production of the final series version of the flying bomb led to intensive correspondence and negotiations between the RLM and the Armaments Ministry.[67] Milch's situation became extremely difficult. On the one hand he was competing for skilled manpower with the army, and especially with its A-4/V-2 program. On the other hand, he desperately sought ways to divide his meager total allocation between different air armament production programs — foremost among them the V-1, conventional fighters and jet aircraft.

Under such pressing circumstances, in early January 1944 the German leadership finally realized that the Armaments Ministry, which now enjoyed unprecedented authority, must set clear priorities. On 17 January 1944 Speer issued a directive in which the Me 262 and Ar 234 were listed as the highest-priority armament programs in terms of manpower allocation.[68] Two days later, under the impact of a recent wave of Allied bombings, Milch decided to grant fighter production, including Me 262 production, the highest priority within the aviation industry.[69] Nevertheless, these clear-cut decisions failed to solve the manpower shortage plaguing the Me 262 and the Ar 234 programs at this stage. Real efforts were made on the regional level and by the relevant Special Committees to solve at least specific cases of manpower shortage. Just as Great Britain had done in 1940, in early 1944 Special Committee F2 — responsible for Messerschmitt — and *Rüstungskommando Augsburg* (the regional armaments command of the Wehrmacht in Augsburg) sought to pull out manpower from lower-priority economic sectors in the Augsburg region and redirect it to Messerschmitt and to other firms associated with the Me 262 program. By 20 January 1944, *Rüstungskommando Augsburg* located and ordered the transfer of 550 civilian workers from different firms. Some of these workers came from other aviation firms like Gotha, Weserflug and Henschel, which were involved in less-important production programs.[70] In other cases, soldiers — mostly from the Luftwaffe — based in different rear-area bases and depots were loaned to Messerschmitt in order to provide it at least with a short-term solution. Milch sent his personal thanks to *Rüstungskommando Augsburg* at the end of January 1944 for assigning 100 idle soldiers from a local *Genesenden Kompanie* (convalesce company) to work for Messerschmitt.[71] Unfortunately for the Germans, these efforts to scratch out every bit of available manpower were largely useless, because they located mostly German manpower, which now become scarcer than ever in all industrial sectors. Furthermore, fewer and fewer soldiers sat idle in Germany waiting to be spared for industrial work.

Sauckel also failed to deliver the foreign workers he was supposed to deliver. In the beginning of February 1944 Speer complained to Sauckel that his request from December 1943 to transfer Italian workers from agriculture to advanced aircraft programs was only partially fulfilled. He emphasized again the importance of the Me 262 and Ar 234 programs in order to influence Sauckel to act.[72] At that time the production plan of the Me 262's Jumo 004 engine was also lagging behind, partially due to the lack of skilled construction workers for Junkers' new engine factory at Zittau.[73]

The manpower shortage affecting the Me 262 program was not solved even after the Jägerstab took over. Friedrich Wilhelm Seiler, Messerschmitt's chairman, later estimated that at that time only 30 percent of the necessary manpower was supplied, but this figure looks far too low.[74] At the beginning of March 1944 it was reported in a meeting in Lechfeld that the Me 262 program still required an extra 500 German workers. Since it looked like it would be impossible to obtain them from other branches of industry, it was suggested again to use a large number of Italian POWs in German captivity.[75] However, as Fritz Schmelter, the Jägerstab member responsible for labor assignment, testified in Milch's postwar trial, Sauckel's ability to provide workers, especially skilled ones, declined considerably in 1944. He also testified that even after the creation of the Jägerstab, Sauckel's support of the aviation industry continued to be halfhearted, and that in many cases he tended to disapprove requests for manpower. As a result, most of the efforts to receive additional manpower concentrated on transferring skilled workers from other industries under the jurisdiction of the Armaments Ministry.[76] In desperation Saur even suggested in April 1944 making Sauckel a Jägerstab member, thereby making him committed to the cause of aviation production. Milch declined and offered instead to use his high military rank to influence Sauckel.[77]

By mid–March 1944 the situation of manpower allocation to the Me 262 production generally improved, but the Germans were in the position of someone trying to cover himself with a short blanket on a cold winter night. By the allocation of extra personnel to this program, other programs suffered. The Luftwaffe's overhaul facilities, for example, complained that the Me 262 was draining them of skilled manpower. As a result, Nobel — the Jägerstab member responsible for overhaul and repair — asked on 17 March 1944 for clearance for foreigners to work in these facilities. Milch approved the request but remarked sarcastically, "Nicht als Flugzeugführer!" ("Not as pilots!"). Frydag added that foreigners employed as ground personnel should not be allowed to enter aircraft without a German supervisor.[78] These worries were not baseless, because in mid February 1944 two Soviet POWs tried to escape from the Regensburg factory airfield in a stolen Me 109 fighter.

Conflicting needs and conscription continued to disrupt aviation production, and particularly the Me 262 production, until the end of the war. When at last it was about to commence in early summer 1944, engine production required extra workers. In May 1,500 workers were demanded for the initial series production, mainly at Junkers; in June 1,500; more and then an extra 500. It was hoped again to cover these urgent needs with Italian manpower.[79]

Surprisingly, even with its ultra-high status, the Me 262 program could not escape emergency call-up drives initiated in late 1944 in order to fill the ever dwindling ranks of the Wehrmacht. Conscription was not the only reason for the loss of workers. By May 1944 Messerschmitt declared the loss of around 500 workers since 25 February 1944 due to *Feindwirkung* (enemy action), health problems and social needs. Only a limited number of replacements arrived to replace them. Especially female workers for administrative tasks were in short supply.[80] The Kottern factory and smaller tooling manufacturers continued to experience severe manpower shortage, even though Kottern received an emergency allocation of Luftwaffe soldiers detached on a temporarily basis.[81] The crisis at the tooling factories and elsewhere was only aggravated in the following months by emergency drafting of workers from all sectors of the industry. However, in late September Saur made it clear that only tooling and machine tool specialists were perfectly safe from the draft.[82] During the same month Messerschmitt lost several hundred of its workers to the Wehrmacht and was left short of the 2,700 workers it required, especially for Me 262 production. This requirement was only partially covered by the delivery of 1,400 concentration camp inmates. In any case, Messerschmitt was left with a very thin layer of German technical and supervisory staff and had to operate its production lines with ever-growing numbers of foreigners and concentration camp inmates.[83]

Different attempts were made in the last months of 1944 to increase somehow the number of German workers by applying for special work leaves for specific drafted workers (*UK-Stellungen*). At this stage such cases were rare and never even came close to covering the need. Firms tried subsequently to get hold of any skilled German at hand. Messerschmitt even tried to transfer a Luftwaffe soldier employed as a guard in a labor camp near its Augsburg plant, after it was found out that he was a trained technician.[84] Towards the end, a large proportion of the German workforce was unskilled too, composed of elderly men unfit for military service, unskilled women of all ages, and *Hitler-Jugend* boys, as one slave worker observed at Junkers' jet-engine factory in Zittau.[85]

The struggle for manpower allocations shows that even intervention by higher authority could not solve the escalating manpower crisis. Milch, Sauckel, Speer, Saur and all the other bosses of Germany's war economy fought a losing battle trying to secure workforce for the high-priority programs. Power struggles and conflicts between different ministries and organizations complicated matters, but the core problem was real and difficult to solve. Industry executives aggravated the difficulties when they regularly demanded more manpower than they really needed, and for unnecessary purposes — as happened when Messerschmitt resurrected his Me 209 and demanded a workforce for its unauthorized production. This tendency originated probably from initial experiences, in which firms demanded realistic numbers of workers but were allocated only a fraction of them. By inflating the numbers they hoped to receive the number of workers they really needed. This practice became so widespread that Milch and the Jägerstab launched frequent investigations of manpower allocation demands. Offenders were prosecuted if the demands were found to be exaggerated. It appears, though, that in most cases special war tribunals acquitted the suspects. By May 1944 Messerschmitt became particularly notorious for its exaggerated manpower demands. During an early Jägerstab visit to Messerschmitt, a firm representative demanded a certain number of extra workers. Schmelter answered that the firm would get the workers only after his officials made sure the demand was a reliable one. If not, he promised to prosecute the representative. Upon hearing the threat, the man reformulated his demand. After hearing Schmelter's report about the incident, Milch remarked acidly that if he really wanted to counter the phenomenon, "You must lock up the entire Messerschmitt firm in a concentration camp."[86]

The after-effects of "Big Week" influenced the manpower situation also through the general dispersal and relocation of factories. The multiplication of factories, as well as the construction of new production spaces — especially underground and bombproof facilities — required a massive influx of additional workers. The demands of massive construction stretched the Reich's human resources far beyond the limits. The chronic manpower shortage was a central structural problem that had plagued the German aviation industry since the beginning of World War II. It originated from its rapid and massive expansion process and was aggravated by faulty human resources mobilization and allocation policies during the war. As war dragged on, this problem became increasingly urgent. This shortage of workforce was the main reason why managers and leaders sought radical solutions outside the traditional workforce reservoirs used until the outbreak of the war.

Foreign Workers and Slave Labor in the Aviation Industry

The most obvious solution to the manpower shortage was to recruit non–German workers. Although Germany's general manpower crisis came to a head after Stalingrad, the aviation industry started looking for alternative workforce long before the 6th Army was lost. The manpower shortage was generally the main reason for the use of foreign and slave labor, but the context was more complex than just this single problem. Business considerations and corporate interests played a role in the decision-making process leading to a widespread introduction of slave labor. Structural developments within the Nazi system were yet another important element contributing to the emergence of this trend.

There are several central questions regarding the use of slave labor in the aviation industry. Why, for instance, did specific firms choose to use slave workers earlier than others, and why did one specific firm, namely Heinkel, lead the way in this regard? Were there any alternatives to the use slave labor or could foreign labor alone have fulfilled the demands? How did this practice spread further within the industry? Finally, how much was initiated by top-level decision-makers and how much was initiated from below by the industry?

After the war, Milch, a key figure in this story, denied any involvement in the allocation of forced labor to the aviation industry. He also argued that this was agreed upon by Hitler and Himmler in 1944 independently of the Jägerstab and that subsequent slave labor allocations were carried out exclusively by the SS.[87] At the time of Milch's trial in Nuremberg, his interrogators knew little about the early use of slave labor and restricted their questioning to the events of 1944. This neglect improved Milch's situation, because he could easily shift the blame to others by arguing that he was sidelined and had nothing to do with workforce allocations. With hindsight, Milch's role was crucial, but he was not alone. In order to understand the heavy late-war reliance on foreign and slave labor, it is necessary to take a closer look at the origins of this phenomenon and at the way it evolved.

Early Enterprises

As the aviation industry began to expand after Hitler came to power, the RLM established in 1934 a special office to organize the training of aviation industry workers. This office, the *Büro für Luftfahrtindustriepersonal* (Office for Aviation Industry Personnel — BfL) was directed by Otto Mooyer and organized the professional training of new workers required by the expanding industry. This "employment office of the aviation industry," as Göring referred to it in 1938, sought upon the outbreak of the war to solve the manpower shortage also by studying new manpower resources. After the outbreak of World War II, Mooyer became well aware of the fact that the traditional manpower reserves available to the aviation industry were largely exhausted. In April 1940, he informed industry chiefs for the first time that massive use of foreigners was going to be the solution to the manpower crisis. At that time the use of foreigners as industrial workers was only a faint idea in the mind of several Nazi leaders. The Germans employed large numbers of Polish civilians and POWs after the conquest of Poland, but only as agricultural workers.[88] Mooyer's note should be therefore regarded as a farsighted notion. General change of policy indeed occurred after the conquest of France and the Low Countries in May-June 1940 as the Germans sought a solution to a general manpower shortage in the industrial sector. After the victory in the 1940 campaign in the West, Mooyer's office listed potential groups of foreign workers and offered them to the industry. In July and August 1940 Heinkel and other firms were offered a choice of foreigners from a catalog-like list: "Scores of diamond cutters from Antwerp, native Flemish men (Belgians); 200 white French POWs" or "280 untrained Danish workers."[89] Other firms also received the same offers and were asked to select the groups they liked.

As a result of this initiative, foreigners — both recently captured POWs and civilians — started to enter the aviation industry in increasing numbers. This move was an exception at a time when most foreign workers and POWs were employed in non-industrial sectors. In December 1940, 54 percent of the foreigners worked in the agriculture, 23.4 percent in construction and 2 percent in mining.[90] Besides Heinkel, Junkers also accepted a large number of foreigners. By September 1940, Junkers employed some 3,100 foreigners, among them 2,100 Dutch and Flemish, 415 Poles, 360 Danes, 75 Frenchmen, 50 Italians and 100 men of other nationalities.[91] One year later the number of foreigners employed by Junkers grew to around 7,400 — 8.8 percent of the firm's entire workforce. In August 1941 alone Junkers received 1,300 foreign workers, including 457 Flemish trainees, 52 Frenchmen, 188 Italians, 454 Hungarians, 33 Croatians and 6 Danes — all registered as skilled workers. Ninety Croatian women also arrived.[92] Most of these workers came voluntarily to Germany and enjoyed different rights. We do not know how Junkers selected its foreign workers, but we know that in summer 1940 Heinkel selected right away the POWs it was offered, because it was simpler to arrange living quarters for them and because they were immediately available.[93]

This sort of practical reasoning influenced the decisions of other firms. The Mansfeld firm, a rather small but expanding aviation systems manufacturer, which produced (among other devices) landing gear for the He 111 and He 177 bombers, also chose POWs. This firm rapidly expanded its business following increasing orders. One way to obtain extra workers was through a training scheme for the unskilled. The other way was through the employment of POWs. By February 1941 around 100 Belgian POWs — some of them listed as skilled workers — were employed in the firm's Prenzlau factory. In a report prepared for the firm's directorate in September 1941 it was noted: "The experience with the prisoners can be described without exception as good. Since the employment office cannot allocate us other workforce anymore and since we are ready to receive more workers after finishing some construction work, we will employ more POWs. We expect to obtain shortly an allocation of around 300 POWs and hope that their employment will prove itself, as happened with those employed by us in the last 6 months."[94]

This early trend set the tone for a common *modus operandi* in the next years: professional qualifications played a minor role in the choice of workforce because firms quickly found out that numerous production tasks could be performed by unskilled workers. The most important criteria in choosing foreign workers were the size of the offered group and the amount of organizational and logistical effort required from the firm in order to arrange their employment. In August 1941, a couple of months after the invasion of the Soviet Union, Göring ordered the allocation of 100,000 French POWs to the aviation industry. This was the biggest single allocation of POWs so far. Soviet POWs were supposed to be allocated to other sectors of the economy, especially agriculture, as a compensation for the lost Frenchmen. The RLM was responsible for the detailed allocation of French POWs to firms and factories, although the Reich's Labor Ministry made the initial general allocation.[95] By the end of 1941 the widespread employment of POWs, especially French, became the norm in the aviation industry and continued until the end of the war.[96]

At the same time the recruitment of civilian foreign workers from the occupied countries was also underway. In 1940–1941 Heinkel employed more and more foreigners in addition to POWs. They came from different nations and most came voluntarily to Germany. In the winter of 1941–42 Heinkel finally filled the wartime quota of 10,500 workers required for the Oranienburg factory, but still required around 2,000 additional workers for its other factories.[97] By that time Heinkel even developed a screening process, which enabled the firm to allocate the foreigners to the task for which they were most adequately suited. Besides considering former training and employment history, the screeners tested the foreigners' general intelligence and learning abilities, orderliness, level of hand-eye combination, technical abilities, and ability to calculate. At the end of 1942 the firm noted that most of its foreigners' past work experience was limited to agriculture.[98]

It is noteworthy that this massive influx of POWs and foreign workers forced the RLM to finance a new wave of housing projects around some of the factories. Some 62 barracks and one storage house were constructed near Henschel's factory in Kassel, costing 1,307,000 RM. The labor camp at Riesenfeld and the old workers' settlement in Harthof near BMW's Munich complex were enlarged at the cost of 884,000 RM. New barracks camps were constructed near BMW's aero-engine factory in Spandau and near Junkers' aero-engine factory in Prague.[99] Studying aerial photos of various factories reveals that by 1944 almost every major aviation factory had a barracks camp for its foreign workers in its vicinity.[100] Since simple barracks were considered as adequate for the new workers, this new construction drive was simpler and cheaper than the one carried out before the war. Furthermore, since wartime construction bore no "national community" building pretensions, there was no need to include luxury facilities and landscaping in the new housing projects.

As the aviation industry lost increasing numbers of workers to the Wehrmacht in the second half of 1941, some firms also remarked a steady loss of contracted foreign workers.[101] In order to deal with this worrying tendency, the Germans launched in Western Europe an advertisement campaign for work in German industry. This campaign used among others the theme of Europe's fight against Bolshevism and Germany's leading role in this regard. This type of recruitment propaganda was partially effective. Between September 1941 and September 1943 Junkers boasted an increase in the number of its Flemish and Dutch workers from around 5,000 to more than 20,000.[102] It was obviously an extreme (and most likely an exaggerated) case, because other firms stayed much below the manpower levels required for their geared-up production. Furthermore, the bottom of the foreign manpower barrel started to become visible. The RLM warned some firms as early as late summer 1941 that the reservoir of foreign workers was not going to grow much in the future and that other workforce sources would be sought.[103] This bleak forecast set the stage for several developments, which brought an even sharper deviation from the traditional workforce employed by the German aviation industry.

A new and promising source of manpower started to appear in winter 1941–1942 in the form of Soviet POWs. Since the invasion of the Soviet Union, the dominant German policy towards Soviet POWs was to let them die from cold and hunger in barren POW camps. This ideologically motivated policy led to the deaths of millions of captured Soviet soldiers and to a huge loss of a potential workforce for Germany's war economy. Change of policy came only gradually and following the realization that the struggle in the East, and therefore the entire war, was going to be a prolonged one.[104] As we saw, the early suggestion to employ Soviet POWs as replacements for agricultural workers transferred to the aviation industry were submitted in August 1941. On 15 October 1941, Hitler ordered the conditional employment of Soviet POWs as laborers in Germany's war economy. At the end of the month he ordered their comprehensive employment, including by the Wehrmacht, in order to free German workers to other tasks.[105] Göring briefed RLM officials about this change of policy on 7 November 1941 and noted that the Soviets should be used only in well-guarded large groups. By that time the German leadership intended to use Soviet POWs foremost in the operational areas in the East and in occupied Poland. There was no plan to use them in the aviation industry, besides possible employment of small numbers in the production of aero-engine parts.[106]

As the German armies suffered their first major defeat in World War II in front of Moscow in the winter of 1941–1942, Hitler issued a new order that reapproved massive employment of Soviet POWs in order to support Germany's war economy.[107] On 18 February

1942 the Armaments and Economy Division of the Wehrmacht issued an order regulating the massive use of Soviet POWs in the armaments industry. Although the Wehrmacht gave highest priority for Soviet POW allocation to factories engaged in oil production and production for the army, it agreed that POWs would also be allocated to factories producing aero-engines, flight instruments and aerial weapons. The scheme was under the authority of the Plenipotentiary of the Four-Year Plan — Göring, in one of his many positions — and the Armaments Ministry.[108]

At the same time discussions were underway considering the use of another human group the Germans viewed as lower forms of life: concentration camp inmates. Next to their ideological and policing function, the concentration camps were an expanding business. Their biggest asset were the people locked in them. In March 1942 some 70,000–80,000 inmates were locked in the main camps of the SS and their subcamps. One year later this figure had increased to 224,000 inmates. By mid–January 1945 the number of inmates under SS custody rose to around 714,000 inmates — around 203,000 of them were women.[109] The sheer number of inmates made them a feasible source of manpower. Furthermore, their cheapness and immediate availability also made them an attractive manpower resource from the businessmen's point of view.

It seems that the first documented employment of slave labor in an aviation-related factory began as a limited local initiative. Austrian armaments manufacturer Steyr-Daimler-Puch AG (SDPAG), which built aero-engines and other products, in early 1941 formed close contacts with the SS in order to find solution to a manpower shortage caused by its rapid expansion. One of the reasons SDPAG turned to the SS in the first place was that its general director Georg Meindl was an old member of the Nazi Party and a member of the SS, who was personally nominated for his job by Göring in 1938. By using his own contacts with Ernst Kaltenbrunner, the Higher SS and Police Leader in the Danube Region, and with local *Gauleiter* Eigruber, Meindl obtained inmates from the Mauthausen concentration camp. Around 300 inmates were transported 30km each day on a train to the city of Steyr, where they were employed in the construction site of a new SDPAG aero-engine factory. At the beginning of 1942 Meindl turned to Kaltenbrunner and asked for several hundred inmates as production workers. In his request Meindl also suggested that the inmates be accommodated in a camp near the new factory in order to spare the long daily trip from Mauthausen and back. The SS responded positively and in March 1942 a labor camp — the first of its kind — was constructed in Steyr-Münichholz specifically for SDPAG's aero-engine plant.[110] SDPAG's arrangement with the SS was based on local initiative and did not indicate a decision from the higher levels. It pointed, however, the way for further initiatives in terms of business models and logistics, which would become widespread in the following years.

An early documented high-level discussion of the possible introduction of slave labor into an aviation factory took place during a visit by Milch and Udet to BMW's new aero-engine factory at Allach on 8 August 1941. Allach was designated as the main production center of the important BMW 801 engine. It was planned to expand the plant constantly in order to keep up with the growing demand for this engine. The plant was designed right from the start for modern conveyor-belt production, using a large number of state-of-the art production machines. Both Milch and Udet expressed during the visit the idea to use concentration camp inmates as replacements for German workers. The idea was discussed again in the RLM the following month. However, at that time the BMW 801 engine still suffered from teething problems and its production in Allach was repeatedly delayed. As a result, the first series BMW 801 engine left the Allach production line only in March 1943. The problems with the engine and organizational difficulties were probably the main reasons why nothing came out of the idea to use inmates at Allach until much later.[111]

It is clear that such a scheme required the cooperation of the SS, and it seems that Milch and Udet were confident that such cooperation is possible. The RLM and the aviation industry were among the first in the armaments industry to form direct contacts with Heinrich Himmler and the SS in early 1942 in order to seek help in solving the manpower shortage troubling this industry. Milch was generally on good personal terms with Himmler. They regularly exchanged birthday greetings and often met for lunch or for afternoon tea.[112] They even used the more familiar "Du" form when talking or writing to each other. Milch once said, "I prefer working with him [Himmler] than with other military authorities."[113] It was therefore probably Milch who first approached the SS and asked for help.

On 24 January 1942 Milch met Himmler in Hitler's headquarters compound and generally discussed with him the employment of concentration camp inmates in the aviation industry. Himmler agreed to cooperate and Milch informed his RLM chiefs three days later that the *Reichsführer SS* announced his willingness to make a large workforce available to the Luftwaffe.[114] Soon afterwards the *Generalluftzeugmeister* staff asked the SS for 11,000 male workers. In March, Oswald Pohl, a former navy paymaster and now the newly appointed head of the new Economics and Administration Main Office (WVHA) of the SS, approved an allocation of only 4,000 men. However, he also announced the planned allocation of 5,000 female workers.[115] In the meanwhile, Milch became eager to obtain more

Cheap labor. Jewish slave workers at a construction site on the premises of BMW's Milbertshofen aero-engine factory, probably in 1942. Most of these Jews were deported from Munich after their houses were confiscated. They were accommodated in a labor camp near the factory, but were deported in summer 1942 (courtesy Yad Vashem Archives).

workers out of this potentially promising workforce reservoir. He asked Himmler to give the aviation industry priority even over the industrial enterprises of the SS itself (mainly quarries and brick-making) in allocations of concentration camp inmates. As he remarked shortly afterwards in front of representatives of Heinkel, in case the industry needed more workers, Himmler would "catch them" ("fängt Himmler noch welchem").[116]

The RLM's initiative to form an industrial partnership with the SS was rather unique. At that time the SS already provided inmates to different firms, but only for employment in construction projects or in workshops located inside concentration camps. Only small numbers of inmates were used for industrial production, and none were used to manufacture sophisticated military hardware. Throughout 1942, firm directorates, as well as some Armaments Ministry officials, showed reluctance to form closer working relations with the SS because they feared it would take over parts of their businesses. Even Saur, at that time the head of the Technical Office in the Armaments Ministry, objected to broader cooperation with the SS on these grounds. The SS, on its behalf, was reluctant to provide inmates to firms operating outside the boundaries of its camp system. It generally stuck to the concept of the concentration camp factories, as suggested later to the RLM.[117]

While those initial discussions were underway, an important reorganization took place within the SS. On 1 February 1942, the WVHA was created, and in April the Inspectorate of the Concentration Camps was incorporated into it.[118] This move created a unified supreme organization within the SS to control its economic and industrial enterprises, as well as their workforce. There is no proof that this reorganization was motivated by the agreement with the RLM and its enormous business potential, but the proximity of the events suggests a possible additional motive. It generally reflected a shift in SS policy towards its camp system. One of the first actions ordered by the WVHA was to step up professional training at the camps. Initially professional training was restricted to stonecutting and masonry, but this order emphasized the need to create and preserve of a reservoir of qualified workforce in the camps.[119] Furthermore, Pohl tended more and more towards partnership with the Armaments Ministry and with the armaments industry, allowing them to use inmates outside the main camps system.

In order to make the introduction of inmates simpler, the RLM and the SS agreed to allocate inmates to factories in the proximity of existing concentration camps. Heinkel's Oranienburg factory, directed by Karl Hayn, became a natural candidate for the scheme because of its proximity to the Sachsenhausen concentration camp. This factory was becoming one of the firm's most important factories because in 1941 it started producing the new He 177 bomber. A new branch factory dedicated to its production was constructed at Germendorf, around 3 km west of the original plant.

As historian Lutz Budrass pointed out, the massive employment of concentration camp slave labor began in the factories of the aviation industry almost six months before Speer reached his first agreement with Pohl regarding the employment of concentration camp inmates in the armaments industry. This arrangement was agreed upon only in September 1942, and at that time the employment of inmates in the aviation industry was already in progress. Unlike previous cases of inmate allocations, this time it was intended to use inmates in the production process of important and complicated military hardware.[120]

It is quite extraordinary that the SS was ready to compromise with the RLM and allowed it to employ inmates outside the concentration camps. This willingness can be explained either through Himmler's enthusiasm towards the prospect of getting a foothold in such a prestigious industry, or through the leverage Milch achieved on Himmler through their close relations.

Consequently, after Oranienburg's director Karl Hayn informed his superiors in February 1942 about difficulties in fulfilling the factory's projected output due to manpower shortage, Milch informed Heinkel's executives about his agreement with Himmler.[121] In March the SS allocated 400 Soviet POWs from Sachsenhausen to the Germendorf factory, located approximately 6 km west of the camp.[122] However, after only one day the factory returned them to the camp because they were too weak for work. Some of them were so weak that they were unable to hold the tools given to them — as Heinkel reported to Milch.[123]

This bad start seems to have discouraged Heinkel and Hayn for a while from further use of concentration camp inmates. Even though Pohl approved the allocation of inmates to the aviation industry, the implementation of the plan dragged on. Probably the main reason for the delay was the failure on the part of the SS to obtain sufficient numbers of Soviet POWs, thereby reducing the size of its manpower reservoir at a critical time. Milch later blamed Heinkel, saying that he missed a great opportunity by being hesitant and that this affair proved again how sluggish large firms can be when it comes to innovation.[124]

There is, however, another explanation for Heinkel's lukewarm interest after March 1942. This explanation is deeply embedded in the troubled history of the He 177 bomber, and it forms a prime example of the way specific aviation projects sometimes encouraged and sometimes delayed the use of slave labor. Heinkel had prepared for the mass production of the He 177 since 1939. This aircraft was his greatest project and with it he hoped to consolidate the status of his firm as the most important German aircraft manufacturer.[125] Manufacture of pre-production He 177A-0 aircraft started in November 1941 at Oranienburg, replacing He 111 and Ju 88 wing set production there, and in Arado's Brandenburg-Neuendorf factory. However, operational trials soon proved that the aircraft was totally unacceptable for frontline service and had to be thoroughly redesigned.[126] The resulting A-1 model was also completely unsatisfactory. Especially troublesome was the aircraft's complicated engine arrangement, which was made of two piston engines coupled together to drive a single large propeller. These engines tended to burst into flames in flight, with catastrophic results. Only 35 He 177A-0 and 130 A-1 were produced, and 8 of each type crashed by the end of February 1943.[127]

Another complete redesign of the aircraft was initiated at around the time those exhausted Soviet POWs were sent to Oranienburg. As a result of the subsequent production halt, the plant stood practically idle during spring 1942. Furthermore, Heinkel was reluctant to use a mostly non–German workforce in the initial production run of a new aircraft that still suffered from many technical problems and demanded constant changes and improvements. In his words, "The series production [of the He 177] becomes prototype production."[128] The Oranienburg complex started to produce the improved A-3 model only in autumn 1942. Even then production progressed at a slow pace. Heinkel still worked out a series of design changes and improvements on the still problematic aircraft.[129] Although monthly output of 70 A-3 aircraft was intended, Oranienburg managed to deliver only a dozen by the end of 1942 — hardly 5 aircraft monthly.[130] Therefore, the continued redesign of this troublesome plane and the low-rate initial production run meant that until autumn 1942 there was not much to do with concentration camp inmates in the Oranienburg plant even if the SS had been able to allocate them.

Heinkel's greatest project. The second pre-series He 177A-02 in flight. This aircraft was used for different trials, and was equipped with, among other innovations, an improved engine cooling system. It helped little, though, because it crashed in May 1942 after both engines caught fire (courtesy National Air and Space Museum, Smithsonian Institution, SI 77-4037).

Developments in the foreign workers "market" also influenced the unfolding events. During spring and summer 1942 Heinkel received additional allocations of Ukrainian, Polish and Russian civilian workers, who were assigned to the Rostock factory, still producing the old He 111 bomber. Most of them possessed useful professions and were a welcome reinforcement, but as Heinkel reported to Milch, their employment arrangements were unsatisfactory. Disagreements regarding their salaries and status eroded their motivation. They were also enraged by the fact that their camp was surrounded by barbed wire. Even more serious was a regulation that dictated the provision of the same food rations to POWs and to civilian workers from the East. Anyhow, Heinkel praised the initial motivation shown by the Eastern workers and viewed them as good replacements. Soon they were also assigned to the Oranienburg plant.[131]

This assignment was important because in early May, as Heinkel finally prepared to run two daily shifts in his factories, Oranienburg was still 2,000 men short of its requirement. The RLM tried to recruit French workers in order to fill the vacancies. The Frenchmen were viewed, however, as a low-grade workforce.[132] Eventually Heinkel received a mixed workforce composed of 400 Russian women, 47 Czechs, 300 Poles, 400 Frenchmen (from unoccupied France) and 400 Dutch skilled workers. The Frenchmen and the Dutch were allocated by the newly appointed Reich Plenipotentiary for Labor Mobilization (GBA).[133]

Hitler nominated Fritz Sauckel, the *Gauleiter* of Thuringia, for this job on 21 March 1942 and entrusted him with supreme control of foreign labor recruitment. This appointment came as a solution to the general mismanagement of workforce mobilization and as part of adjusting Germany's war economy for a long war of attrition following the December 1941 crisis.[134] Since Mooyers' BfL lost most of its functions as a result of Saukel's nomination, the provision of foreign workers to the aviation industry constantly declined, even though by August Sauckel had brought to Germany 1.6 million foreign workers—1.3 million of them from the East.[135] In October 1942 Sauckel also became responsible for the allocation of POWs, thus making the large number of Soviet POWs recently captured during the German summer offensive on the Eastern Front a viable option again. Milch suggested on 30 October 1942, during a Central Planning Office meeting at the Armaments Ministry, making the allocation of POWs more efficient by turning them over to the labor agencies immediately after their capture instead of incarcerating them for long periods in POW camps.[136] By that time, however, the German offensive had stalled, and 3 weeks later the Soviet Army encircled Stalingrad. The Stalingrad defeat virtually ended any prospects of further large intakes of Soviet POWs. Sauckel thus concentrated his main efforts on increased recruitment of foreign civilian workers. This effort continued to bear limited fruits for the aviation industry due to Sauckel's preference for other sectors.

The situation became even more complicated when Sauckel's recruitment drives started to interrupt aviation production carried out in the occupied countries. In September 1942 Sauckel launched a massive recruitment campaign in France, seeking especially skilled industry workers. Around 25,000 French skilled workers — many of them working in the French aviation industry — were recruited and sent to Germany.[137] Their skills and experience were thus largely lost because most of them were not allocated again to work in aviation-related

Eastern workers. Foreign female workers from the Soviet Union manufacturing wooden drop tanks at Volkswagen's Fallersleben factory (courtesy U.S. National Archives and Records Administration).

factories.[138] This uncoordinated activity created an absurd situation, in which foreign contractors producing for the German aviation industry were drained of their skilled workers. The GBA expected the affected French firms to recruit new untrained workers as replacements for the lost workers. Messerschmitt was one of those German firms that tried with little success to protect its French contractors from wild recruitment of skilled workers in early 1943.[139] Focke-Wulf's subcontractors in France also suffered from Sauckel's recruitment drives. In February 1943, for example, 100 workers of the French firm SNCASO in Chatillon were recruited for work in Germany. According to the protest by the RLM's representative in Paris, it was the second time that such a thing had happened and experience from previous recruitment drives proved that these workers were permanently lost for the aviation industry. His intervention failed to prevent their transfer to Germany.[140]

In order to avoid more incidents of this nature the Armaments Ministry initiated in autumn 1943, in cooperation with the French Production Ministry, a general scheme intended to protect vital French factories from wild recruitment. These factories were called *S-Betriebe* (*Speer Betriebe*— Speer factories) and they became safe enclaves of sorts for French workers. Subsequently, several aviation firms, among them those working with Focke-Wulf, were declared *S-Betriebe*. Sauckel objected this scheme and argued that the productivity of French workers in France was lower than their productivity in Germany, where they were better supervised. The argument between the Armaments Ministry and the GBA was never resolved, and although the *S-Betriebe* scheme improved the situation, recruitment drives continued — even in these safe havens.[141]

This pattern repeated itself in June 1944 in Italy, months after the Germans took over the North Italian aviation industry. Even as firms like Focke-Wulf contracted Italian firms to produce parts for them, the GBA relentlessly recruited workers of these firms. These workers, many of them skilled, were taken to Germany and were employed in agriculture. In some cases this type of forced recruitment was done at the factories. This activity, of course, caused further damage, because other workers stopped coming to work, fearing that they would also be sent to Germany.[142] In this way the Germans created new manpower problems while trying to solve the main manpower problem.

Under these circumstances the RLM and SS discussed again in midsummer 1942 the use of concentration camp inmates. The old idea of using inmates in factories near concentration camps was mentioned again. Some officials also started contemplating the establishment of concentration camp factories, dedicated to the employment of inmates. Under the impression of a spiraling manpower crisis, this plan raised high hopes within the aviation industry. Directors at the Oranienburg factory complex, for instance, viewed it as a great experiment that could bring a broader use of inmates into the entire aviation industry.[143] Business considerations again played an important role in this renewed interest. The Oranienburg factory had just started to produce the improved He 177A-3 and Heinkel promised to deliver the first examples in September 1942. In early July, Heinkel and Milch agreed to concentrate the entire production run of the He 177 in Oranienburg and in Arado's factory in Brandenburg-Neuendorf in order to free the Marienehe plant for other tasks. Thereby Oranienburg became Heinkel's main factory.[144] The RLM, and especially Milch, hoped that the use of concentration camp inmates would enable Heinkel to fulfill his promises this time.[145]

Timing was therefore perfect for the upcoming experiment, which started in summer 1942. The Oranienburg-Germendorf factory, producing He 177's wing sets, was reorganized as a concentration camp factory, completely separate from the original plant at Annahof. Production halls 3 (parts manufacture), 4 (wing assembly) and 5 (wings middle section assembly) were surrounded by an electrified barbed-wire perimeter fence and watchtowers equipped with machine guns and searchlights. This section thus became a mini-concentration camp. In December 1942 the production tooling production hall was also included in the perimeter, and the presswork in hall 2 was added in May 1943. The first contingent of prisoners was delivered to the factory in early August 1942, and by the end of September 800 inmates worked in the factory. This early group was made up of different nationalities, but most of the inmates were Russian POWs, Frenchmen, Poles and Germans.[146]

As the inmates started working on the production lines, the pressure on Heinkel grew. During a meeting at the RLM on 4 September 1942, Hayn announced a new plan to produce 106 bombers in Oranienburg and 60 bombers in Brandenburg by the end of 1943. It was much less than what was promised earlier to the RLM. Colonel Wolfgang Vorwald, head of the planning department in the Technical Office, reacted angrily and hinted threateningly at what happened to Willy Messerschmitt after the Me 210 fiasco and at Göring's own dissatisfaction with Heinkel.[147] Furthermore, the protracted development, the long delays and the enormous investment brought Heinkel by the end of 1942 to the brink of a financial collapse.[148] Therefore, by the time slave labor intensified at Oranienburg, Heinkel was under growing pressure to conclude the development work and to start high-pace production of the troubled bomber by any means.

In the meantime the RLM and the SS went forward with the "concentration camp factory" scheme. Milch reported in a meeting on 17 September that the SS agreed to allocate the aviation industry 50,000 additional inmates and that the RLM wished to employ them solely in special "concentration camp factories." Heinkel's Oranienburg factory and BMW's Allach factory near Munich were considered as the first candidates for this scheme because of their respective proximity to the Sachsenhausen and Dachau camps. The "concentration camp factories" were to employ mainly inmates with a small German supervisory cadre. This new type of factory contrasted sharply the prewar ideals associated with the aviation industry and its "new model" factories. The RLM therefore suggested restricting contact between German workers and inmates in these two factories as much as possible.[149]

These discussions correlated with another massive loss of Heinkel employees to the military. In September, Heinkel asked the Employment Office of Rostock for 250 Frenchmen to replace recently drafted workers in the Marienehe factory. Although they were available, the Employment Office refused to allocate them because of a misunderstanding regarding the priority level of the aircraft produced at Marienehe.[150] In Oranienburg, Heinkel reported the loss of 3,286 workers between 1 March and 31 August 1942 — 950 were German males drafted to the Wehrmacht. At the same time the factory received 3,599 new workers — none of them Germans. The net increase of manpower amounted to only 313 workers, and since they were mostly unskilled and undermotivated, productivity dropped.[151] More workers, mostly skilled Germans, were lost during October and their loss affected especially the He 177 program.[152]

No wonder, therefore, that the use of inmates by Heinkel increased steadily in the following months. On 3 November Hayn asked for an immediate allocation of 1,000 additional inmates and announced that this was the only type of workforce he was going to ask for. He also asked for 42 tons of barbed wire in order to facilitate the housing of these inmates, as ordered by the SS.[153] In December 1942 the number of prisoners working in Oranienburg reached 1,900 — 15.2 percent of the workforce. This number grew further in the next months until in June 1944 a total of 6,966 inmates worked in the Oranienburg complex — 48.2 percent of the workforce.[154] In February 1944 the inmates' share in Oranienburg went even higher: around 60 percent (a total of around 6,000 inmates).[155] Most of the inmates in Oranienburg worked next to assembly lines and production machines. Despite earlier hesitation, they were mixed among German workers. It is interesting to note that while no real attempt was made to separate them from German workers, the inmates were separated from other foreign workers — probably in order to avoid information exchange between these two unhappy groups. The inmates lived in barracks within the factory perimeter and mostly next to their allocated work halls. A permanent SS guard detachment from Sachsenhausen watched them both in their barracks and while at work.[156]

Initially the firm tried to give the inmates some training in order to prepare them for their work, but this practice was soon dropped and the inmates were sent to learn their trade through work. Since Heinkel was a pioneer in the use of slave labor in aircraft production, the firm was able to handpick inmates according to their profession, and around 20 percent of them were categorized as technically trained workers. It was naturally easier to integrate these workers into the work schedule.[157]

These rapid developments happened as Heinkel still tried to solve the technical problems with the He 177. Towards the end of 1942 it was clear that the aircraft was still far from perfect. In a meeting on 16 November 1942, Milch criticized the program and the Oranienburg plant in harsh words. Among other matters, he made Heinkel responsible for the 35 lives lost in He 177 accidents so far. He even threatened to bring some of the men responsible for its design in front of a war tribunal and to execute them.[158] Milch hoped to see an improvement with the next version, the He 177A-3, but it also suffered from a series of serious technical problems. By February 1943 the problems were considered to be so severe that production was stopped again and a special high-ranking meeting presided over by Göring took place at the RLM in order to seek a way out of the mess.[159] In March 1943, following the meeting, Göring strongly reprimanded Heinkel.

At the same time Heinkel was also hard-pressed by a manpower shortage in the development department, because the RLM ordered the transfer of 30 designers to Messerschmitt following the cancellation of the firm's He 280 jet fighter project in March 1943. The development division, which already employed a number of foreigners, was thus stretched to the breaking point.[160] During the first half of 1943 Heinkel and the RLM made concentrated efforts to solve the problems of the He 177, including totally redesigning the engine housing. The extra design work caused long breaks in production. The RLM, however, kept drawing optimistic production plans. In February 1943 it still expected to increase the monthly production rate from 40 to 65 aircraft by the end of that year. It was intended to reach a monthly output of 100 by April 1944.[161] This output projection never materialized. German engineers estimated that during a 15-month period in 1943 and 1944 Arado lost 46 percent (181 aircraft) of its He 177 output due to frequent design changes and frequent changes to the production program.[162] Between August and November 1943, only 19 new He 177A-3 bombers were produced in Oranienburg, while at the same time older aircraft were brought back to the factory for a refit.[163]

The first redesigned He 177A-5 version was completed in February 1943. Although some components of the aircraft were still troublesome, it became the standard version, which was produced in Oranienburg from the end of 1943.[164] A total of 565 He 177A-5s were produced in 1944 before its production was terminated in September 1944, bringing the total production of this aircraft to 1,169 or 1,446.[165] The Oranienburg complex was subsequently converted to produce the FW 190 and continued to produce this fighter until the end of the war.[166] There were plans to produce Do 335 fighter-bombers in Oranienburg from January 1945, but they never materialized. Workforce made free by the reduction and termination of the He 177 production was made available to other factories. Around 2,000 inmates were transferred in July 1944 from Oranienburg to different factories involved in the Ju 388 production scheme (ATG, Siebel, and Junkers' own Halberstadt and Aschersleben factories).[167]

As in other places, the employment of concentration camp inmates proved to be successful because it increased the productivity of the Oranienburg plant. The new influx of workers enabled the establishment of 2 or 3 shifts each day and increased the working hours of crucial machinery. In other words, these machines could be manned for longer hours and therefore produce more. The fact that most of the new workers were allocated to the production process, and not to one or another peripheral task, also helped to increase the productivity of this workforce. While the average monthly output of the Germendorf plant in 1941–1942 was 25 wing sets for the Ju 88 light bomber, an output of 50 wing sets for the much larger He 177 was planned for 1944. Before the production of the He 177 was terminated, Oranienburg-Germendorf reached a monthly output of 40 wing sets — a clear increase in productivity.[168] The factory therefore represented the peak of the rationalization process of aircraft production. It solved the manpower shortage problem, it overcame the need to move people around, organize their housing, bring them to work and then back home each day, and it spared the need to pay them and to take care of their welfare. The output increase was a result of intensified work as much as it was the result of better organization of the production process. The Germans thus solved consequently the main bottleneck that prevented them from taking full advantage of the potential production capacity of the aviation industry: the inability to get more productivity from the ever-dwindling traditional workforce and from the foreign workers. Concentration camp inmates were theoretically the ideal workforce: they had no unions and no employees' rights; they were exceptionally cheap; they required only basic housing and minimum food provisions; they could be forced to work for long hours in miserable conditions and under strict discipline. And they could do nothing to avoid it.[169]

During 1942 slave labor was introduced into other Heinkel factories. The firm's business strategy in times of an escalating manpower crisis is crucial to understanding this trend. Towards the end of 1941 Heinkel's management decided to move some production programs to new and safer locations in the East. This move also formed part of the firm's expansion program following the decision to build the He 177 while continuing the production of the He 111. The heavy bombing of Rostock by the RAF in April 1942 underlined the need for these safer factories. One location considered was a vacant facility at the Schwechat airfield near Vienna. Another was a workshop complex in a former PZL aircraft factory in Mielec, in the General Government. The workshops in Mielec were operated since late 1939 by the Vereinigte Ostwerke GmbH, a firm owned by Heinkel since October 1939. In early 1942 Heinkel asked the Mielec management to prepare the place for an expanded operation. In February the management reported an empty former textile factory in nearby Rzeszow that was adaptable for the production of tailplane units for the He 177. It also noted its intention to employ Jews from the ghettos in the area in this factory in order to overcome a sudden shortage of Poles.[170] In April Heinkel sent the director of the Mielec factory to look for other possible locations. Heinkel eventually proposed that the RLM use the Rzeszow factory and a former munitions factory in Budzyn. These 3 locations — called collectively "Block Budzyn" — shared between them the same floor space as Oranienburg (90,000 square meters). By mid-1942 the Mielec factory alone employed around 3,000 workers and produced some parts for the He 177 bomber. Among the workers were 280 Germans and 259 Jews. The rest were Poles.[171] He 111 bombers were also produced and overhauled at Mielec at least until April 1944.[172] In June 1942 Heinkel suggested moving the entire He 111 production to the "Block Budzyn," and in 1943 a production line for the old bomber was established in Mielec.[173]

Heinkel approved the recruitment of Jewish workers from nearby ghettos in February 1942 and the Mielec labor camp was opened on the second week of March.[174] It was the first documented direct employment of Jews in German armaments production. Ernst Heinkel used this initiative in order to explain the rationale behind the expansion to the East in a memo he sent to Milch in June: "It is easier to obtain new workforce in the General Government than anywhere else. Besides Poles, we can obtain above all good workforce from the abundant Jewish population. Our Mielec factory collected very good experience with Jewish workers and intends to meaningfully increase their employment."[175] In August Heinkel commenced production in the "Block Budzyn" factories with around 500 Germans as a supervisory cadre and around 6,000 local workers — 2,100 of whom were Jews from the neighboring ghettos. The German cadre came mainly from Oranienburg-Germendorf, where the conversion into a concentration camp factory freed some German workers for other tasks. The most immediate problem in the new factories was the language barrier. Since most of the local workers spoke no German and most of the Germans spoke no Polish, the Germans were forced to allocate several German-speaking workers, mostly Jews, to work as translators.[176]

The new complex in the General Government was openly viewed as a forced labor operation, and Milch considered sending French workers there as punishment for the low output of the French aviation industry.[177] The new complex suffered some setbacks in the summer of 1942 as thousands of Jews were deported from the General Government to various killing centers. Heinkel's representative in Berlin, Rittmeister von Pfistermeister, reported in September 1942 to Frydag, the chairman of the Airframe Main Committee, about the setbacks caused due to losing part of the workforce in the wake of these deportations.[178] It was another case where an action dictated by Nazi ideology sabotaged German industrial interests.

In parallel to Heinkel, aero-engine producer Daimler-Benz introduced a similar system of slave labor in its factory at Rzeszow (the Germans called it Reichshof), located in the southeastern part of the General Government. The Polish company PZL established this former aero-engine factory in 1937. After the occupation of Poland the plant was taken over by two aero-engine firms: *Flugmotorenwerke Reichshof GmbH* (a subsidiary of Henschel) and Daimler-Benz, who used it as an engine maintenance depot and later for parts manufacture. The plant was used as the main overhaul facility for DB engines in the East and its workload multiplied following the invasion of the Soviet Union. In spring 1942 the firm drew a plan with the *Rüstungsinspektion im Generalgouvernement* (Armaments Inspection in the General Government) to draw Jews from nearby ghettos and to house them in a labor camp next to the factory. In the next couple of months Jews from dissolved ghettos in the region arrived at the factory, and from August they were accommodated in a separate newly established camp called "Lysia Gora." In early summer 1942 several hundred Jews started working at the Rzeszow plant and by the end of the year 500 to 700 Jewish inmates were employed in the place. Just as with Heinkel, the workforce was a mix of Jewish inmates and Poles, supervised by a skeleton German staff. Until September 1943 members of the factory's own *Werkschutz* (factory security) watched the inmates both at work and at their camp. In September the SS took over the camp and the associated guard duties.[179]

Unlike Heinkel's operation in Poland, the Rzeszow factory was a rather minor outpost of the Daimler-Benz empire, and its expansion was driven mainly by the contingencies of the Eastern front. It is interesting to note, though, that Daimler-Benz was closely associated with Heinkel, because it produced the engine of the He 177 — the same engine that caused so much trouble.

Heinkel's expansion in 1942 was motivated by the developed program of another new aircraft and by hopes to produce it in great numbers. In summer 1941 the RLM urgently asked Heinkel and Focke-Wulf to develop a twin-engine night fighter. Heinkel's candidate was already partially developed in Marienehe as a private venture and was rashly submitted. Initially the RLM rejected Heinkel's submission, designated He 219, and preferred the Ta 154, Focke-Wulf's candidate. In contrast, the Luftwaffe liked the He 219 and as a result the RLM formally authorized Heinkel on 29 July 1942 to produce 4 prototypes in its new factory at Schwechat, and then to commence limited series production. In late July 1942 — long before the plane first flew — Heinkel decided to produce the He 219 in Marienehe and Mielec. The RLM opposed the plan. Due to lack of capacity and because of the fear of more attacks on Rostock, it suggested the complete relocation of the He 219 production to Vienna and Mielec.[180] In September, after more RAF attacks, the RLM ordered the termination of most production tasks in Marienehe and to transfer three-quarters of the factory's workforce to the "Block Budzyn" complex. The factory complex in Poland became even more important following Milch's principled decision to use it for the production of the new Ju 188 light bomber, with an expected monthly output of 100 planes.[181]

Soon after the first flight of the He 219 prototype on 6 November 1942 at Marienehe, it became clear that manpower shortage was a major obstacle to its series production. Therefore, the fuselages of the next three prototypes were produced in Mielec. German skilled workers preformed most of the work, as demanded by Heinkel from early on when discussing the manufacture of the prototypes and pre-production airframes. Later, series fuselage production continued at Mielec.[182] Expansion of the "Block Budzyn" complex in preparation for the He 219 and Ju 188 production encountered difficulties due to lack of construction materials and so dragged on into early 1943.[183] In early January 1943 Heinkel submitted a detailed production plan for the He 219. The firm foresaw the employment of 2,706 foreigners and 656 POWs in its production. These groups formed 67.8 percent of the total designated workforce, and only 160 of them required some relevant skills.[184] At the end of the month Göring generally authorized the plan, gave it the highest priority level (*allergrösster Dringlichkeit*) and ordered a monthly output of 50 aircraft by the end of 1944.[185] Final authorization came with the inclusion of the aircraft in production plan 223, published in April.

4. From Technological Expertise to Slave Labor 173

Night fighter. The He 219V-16 prototype, used to test radar and armament installation. It is seen here equipped with the FuG 202 short-range radar and the SN-2 medium range radar. Production plans of this aircraft led to expansion of Heinkel's operations in the east (courtesy National Air and Space Museum, Smithsonian Institution, SI 2005-3453).

Lack of capacity continued to upset Heinkel's newest enterprise. At the end of May 1943 Heinkel decided to turn the Vienna-Schwechat factory into the main He 219 production center on the premise that it would be possible to produce the plane in Vienna mostly with German-speaking inmates. The firm also planned to turn a nearby workshop at Lichtenwört into a concentration camp factory (*KZ-Betrieb*),[186] but Vienna also suffered from a growing workforce shortage. During a meeting with Milch on 22 July 1943, Major Hoffman from the Vienna factory reported a shortage of 336 workers. As a result the factory's planned monthly output was reduced from 10 to 4 or 5 aircraft. Milch replied that the factory must deal independently with the problem because he could offer no extra workers. He also asked if they were already using inmates. Unfortunately nobody bothered to write the answer in the meeting's protocol.[187] Elsewhere Heinkel continued to suffer from skilled manpower shortage. Milch even ordered the allocation of soldiers to its different plants as a sort of temporary replacement workforce.[188]

In summer 1943 the focal point shifted again to Poland, and Heinkel made plans to turn the "Block Budzyn" complex into the main He 219 manufacturing center. In a memo from 29 June, von Pfistermeister, the firm's representative in Berlin, discussed the prospect of dedicating the Polish factories exclusively to He 219 production by terminating their He 111 production. The main advantages offered by the Polish complex were the availability of floor space and manpower. The firm's production manager Hayn also reported that it was possible to find high-quality manpower in Poland.[189] Ernst Heinkel opposed this plan because it could have taken around one year to achieve full production in Poland,[190] but shortly afterwards the RLM authorized the concentration of He 219 production in Poland.[191] As Heinkel feared, relocation of the entire program to the East meant a long delay. Heinkel and the RLM estimated that full series production could begin in Poland only in January 1945, mainly due to difficulties in providing the factories with the necessary tooling. They sought a way to compensate for the loss of production in Marienehe by commencing low-pace production elsewhere.[192] Eventually all parties agreed to start low-pace production in the *Ostwerke* by moving the necessary tooling and personnel from Rostock and Vienna, but part of the production was to stay in Vienna. The main manpower sources for the new scheme were Rostock (including workers from a terminated He 177 engine production plant) and various concentration camps. Karl Hayn sorted out the supply of inmates with *Obersturmbahnführer* Gerhardt Maurer, the WVHA's chief salve dealer. As Heyn reported, the negotiations with Maurer were good, and he found it unnecessary to contact Maurer's boss, Pohl. Hayn made plans to fly on 17 August with Maurer to Vienna and make the practical arrangements on the spot. He also asked Carl Francke of Heinkel's Technical Department to be present in order to discuss the feasibility of establishing a "concentration camp design office" (*KZ-Konstruktionsbüro*) in Vienna.[193] During September the Vienna branch received 850 extra workers—250 of them were concentration camp inmates.[194] By the end of October the number of workers in Vienna reached 3,322, including 701 inmates, 336 inmates still in training and 1,391 foreigners. In order to complete the total number of workers, plant director Schaberger met the commander of Mauthausen, *Obersturmbannführer* Ziereis, on 27 October and received from him an additional 10 skilled inmates and 200 unskilled inmates.[195]

Lack of skilled workers and machine tools in the occupied Polish territories largely frustrated their production program.[196] Although it was not as troubled as the He 177, initial production of the He 219 was painfully slow. Part of it was due to continual development of the aircraft, but another reason was the diffusion of its production. Hayn's optimistic projections considering production in Poland proved largely wrong, as he admitted in a letter to Milch in early September.[197] The fact that main subassemblies were required to travel long distances to reach the final assembly plants complicated the production scheme. The fuselages from Mielec, for example, were usually delivered to Vienna-Schwechat two at a time inside giant Me 323 transport planes—and these planes were very scarce, so delivery was slow and dependent on the availability of the transporters.[198]

The low-pace production of the He 219 almost caused the RLM to terminate its production in December 1943, but eventually it authorized the opening of a second production line in Marienehe, only because the Ta 154, the He 219's competitor, was not ready yet.[199] Milch persisted, however, in his attempts to kill the project early in 1944. His main arguments against it were that it could not do anything else other than night-fighting and that its production demanded large numbers of man-hours, which made it expensive to manufacture compared to other types.[200] In May, Ernst Heinkel tried to convince the RLM to renew its support of the plane by pointing out, among other things, at the dramatically sinking costs of its production. Heinkel promised to reduce the man-hours required for the manufacture of a single plane to 10,000 by the eight-hundredth plane, by the second quarter of 1945, compared to 40,000 man-hours it took to produce the one-hundredth plane in the second quarter of 1944.[201] The bottom line was an acknowledgment that although the He 219 was a highly effective night-fighter, as proved by its successful initial combat deployment, it was extremely expensive and difficult to produce. The RLM and Göring were not convinced by Heinkel's promises, and on 25 May 1944 they decided to terminate the production of the plane. The less advanced but cheaper Ju 388 was selected as the main future night-fighter type.[202]

In mid–June Heinkel tried again by pointing out the advantages of the He 219, both in terms of combat effectiveness and ease of production. He even suggested equipping it with wooden wings and tailplane. When discussing a possible increased output he wrote: "The question, how to obtain increased manpower for such increased production is already solved, because we were promised more allocations of concentration camp inmates."[203] In an act of indecision, RLM officials eventually allowed Heinkel to continue low-rate production as well as some further development, including — again — extra versions and variants of the basic design.[204]

By outsourcing production of main components of the He 219 to the "Block Budzyn" complex, Heinkel sought to compensate for the amount of work required in order to produce the aircraft, by relying on the cheap and large manpower reservoir in Poland. This is an excellent example of the way increased workforce was considered as more capacity, therefore enabling the production of a complicated and work-intensive aircraft. It is striking, indeed, how Heinkel shifted so rapidly to reliance on slave labor for the production of advanced and complicated aircraft. At least up to October 1942 Heinkel strove to use a mostly German workforce in the He 177 and He 219 projects. These planes were at that time still in the development phase, which demanded a high percentage of skilled workers. The firm tried to move as many Germans as possible from the He 111 production and to replace them with foreigners. As a result, 65 percent of the workforce producing the old bomber was composed of foreigners, and output suffered due to their lower skill and low motivation.[205] As Heinkel pointed out in a memo on 8 January 1943, using foreign and mostly unskilled labor was not an ideal solution, because the construction of modern planes still demanded a high percentage of German specialists. However, in the same breath he declared his willingness to use any available workforce.[206] By that time Heinkel suffered deeply from the manpower crisis caused mostly by recent draft,[207] so no wonder he was prepared to accept the disadvantages associated with foreigners and inmates in order to carry out his ambitious aviation projects. By May 1944, Karl Hayn acknowledged that the concept had partially failed with the He 219:

> The production run of the He 219 is still in its initial stage. The prototypes of such a double-engine aircraft cost around 100,000 man-hours. This figure cannot be reduced significantly in short time with the current disproportional workforce of foreigners and unskilled concentration camp inmates. Experience shows that these workers show an initial efficiency factor of 10 percent, increaisng eventually to a maximum of 70 percent. According to this experience the cost can be reduced to around 9,500 man-hours by the manufacture of the five-hundredth aircraft. Aircraft 63 to 73 were delivered in April.[208]

Therefore, Milch was right in his assessment that the He 219 was an expensive and difficult-to-build aircraft. Nevertheless, early business opportunities, combined with Heinkel's ambition to become Germany's greatest aircraft manufacturer, provided an important motivation to the increased demand for slave labor in 1942 and 1943.

The "Block Budzyn" initiative was the climax of Heinkel's pioneer work of introducing slave labor into the aviation industry. It took full advantage of economic opportunities offered by Germany's new status. It was part of a two-pronged experiment carried out while trying to produce two of the most modern aircraft of the Third Reich. One part of the experiment was carried out deep inside the Reich and was based on concentration camp inmates provided by the SS. The other took place far away in the "Wild East" of the General Government, where both Poles and Jews from several ghettos represented a cheap and plentifully available workforce. By late 1942 the RLM and Junkers acknowledged this potential too by deciding to place the giant "Ultra" bomber plant near the Polish border. Only 10–15 percent of around 80,000 workers designated to work in this factory were to be Germans; the rest were composed of various eastern foreigners.[209]

The problems and the solutions were practically the same, both in the heart of the Third Reich and in its provinces. The core problem was how to produce advanced and work-intensive aircraft even as the available manpower reservoir was drying up. The solution was to seek new workers among those population groups persecuted or occupied by the Germans. Internal shifts of power within the Nazi system subsequently made it more convenient for Heinkel — and soon other firms — to rely mainly on those persecuted groups instead of on foreigners and POWs.

Turning Forced Labor into an Industry Standard

Heinkel definitely profited as an aviation business from the early use of slave labor and from the subsequent increased productivity of the Oranienburg complex. Slave labor also solved several problems associated with the German workforce. While in early 1942 the firm lost 15 percent of the normal work hours due to authorized leaves, sick leaves, air-raid alarms and absenteeism, the average lost time in 1943 sunk to only 6.9 percent. Higher payment levels in the area around Berlin also motivated Heinkel to increase the number of its slave workers and thus cut expenses on higher-than-average salaries. Furthermore, since no overtime was paid for slave labor, the amount of budget the firm spent on extra hours also decreased drastically.[210] At the end of 1942 manager Karl Hayn viewed the massive use of concentration camp inmates as a magic solution to most manpower and productivity problems of the aviation industry and as one possible answer to the challenge posed by the enormous production capacity of the American aircraft industry.[211]

As Heinkel made its experiments with slave labor, other firms sought to solve the same problems using different strategies. Some of them tried to find the missing extra capacity outside Germany. Focke-Wulf was the leading firm in this regard, but Junkers, Dornier and Messerschmitt also sought to expand their outsourcing and foreign contracting. Even Heinkel outsourced the development and production of its He 274 heavy bomber — a development of the troubled He 177 — to the French firm Farman.[212] Focke-Wulf initially sought to solve its manpower and capacity shortage by contracting French firms and outsourcing to them some of its parts production, aircraft production and even development work. On the long term this initiative largely failed due to different factors described previously. As a result Focke-Wulf gained little from its 1941–1943 ventures and enterprises in France. Despite relying heavily on outsourcing, Focke-Wulf employed foreigners in increasing numbers in its factories, but it was done in a different way than by Heinkel. By the end of 1942 some 1,395 of the 2,370 workers at the Posen factory were foreigners. They included Italians, Danes, Poles, Frenchmen (including POWs), Spaniards, Russians and Ukrainians. This high percentage of foreigners is easily explainable by the location of this plant on former

Polish territory annexed to the Reich. Of the 1,395 foreigners, 1,175 were Poles—mostly locals. The percentage of foreigners in Focke-Wulf's other plants, located in Germany, was much lower. In its Marienburg plant, for example, only 32 of the 574 workers were foreigners.[213] Therefore the Posen factory can be viewed as sort of an eastwards outsourcing, which provided a solution to the manpower problem by relying on a salaried local Polish workforce.

As we saw, Focke-Wulf also turned to Italy in early 1944. This move was a continuation of its business strategy of outsourcing production tasks to firms outside Germany. Focke-Wulf was therefore persistent with its own way of trying to solve the lack of manpower. Focke-Wulf was one of three firms which had tried repeatedly since late 1943 to recruit Italian workers, particularly skilled, and recruit them for work in its German factories. Focke-Wulf, Arado and Messerschmitt sent their own representatives to Italy in January 1944 to try to solve the problems associated with the recruitment of Italian workers. This initiative formed yet another approach to solving the manpower shortage, which indicates the difficulties encountered with workforce recruitment through other agencies, especially the GBA. Up to that point recruitment of Italian workers had largely failed because the Germans failed to offer local workers an agreed-upon and unified contract, providing them satisfactory payment, housing, and feeding, and regulating other issues affecting their employment conditions. The GBA never bothered to achieve a similar settlement, so the three firms took the initiative and sought to offer their own contracts in order to attract Italian workers.[214] It was too late, and few—if any—Italians were recruited in this way.

The generally positive experience gained during the early employment of inmates at Heinkel's factories soon spilled over to other places. Firstly, the use of inmates was constantly expanded within the Heinkel organization; inmates were soon allocated to the Vienna-Schwechat branch, and then to the main Rostock plant. The number of slave workers employed in these factories never reached the magnitude of Oranienburg and the "Block Budzyn" complexes, but inmates increasingly replaced German workers—even specialists—on the production lines.[215] During 1943 more Heinkel factories, including minor ones, were turned into concentration camp factories. One of them was the Barth factory, not far from Rostock, which manufactured fighter wings under license.[216]

Towards the end of 1943 the ratio of Germans and foreigners on the production lines tipped decisively towards the foreigners. Even though their number within the entire workforce—around 30.5 percent out of 1,852,000—was relatively low, since many Germans filled clerical, administrative and professional positions, the share of foreigners and inmates on the production lines was significant. As a result, in firms like BMW, foreigners of all sorts formed 85 percent of the productive workforce. In October 1943 Milch brought the series production of the Ju 52 transport plane in Bernburg, which was done by 6 German foremen and around 2,000 foreigners, as an extreme example of this trend.[217] Foreigners of 32 different nationalities[218] also rapidly replaced Germans in various other functions within the aviation factories. By 1944 foreigners even served as auxiliary firemen with the factories' fire brigade units. This assignment required the issue of special regulations concerning their status and rewards for this work. The main reward for western foreigners was an exemption from carrying the special badge signifying foreign worker status. The Germans did not trust these foreign firemen and some firm managements recommended that they should not form more than 50 percent of the fire brigade force in any factory. In any case, their supervisors were instructed to closely watch them and to be extremely strict with them.[219]

Increasing numbers of inmates were also allocated to the aero-engine sector. Since the early negotiations between the RLM and the SS, BMW figured as the main potential employer of slave labor. In late 1942 inmates from Dachau were allocated to the Allach plant and worked on the construction site of a new bombproof production hall. These slave construction workers were not employed immediately afterwards in the production of the BMW 801 engine, as was originally intended. Plans to use inmates in the production of the BMW 003 jet engine also came to nothing at this stage, mainly because of severe delays in its development. Therefore, BMW's factory in Allach never produced the 003 engine, and when its series production finally started in summer 1944, most of it was carried out in various dispersal factories. As a result, Allach continued to be the main production center of the BMW 801 engine, which powered the FW 190 fighter as well as other aircraft. Allach was a large factory, covering an area of 235 acres and with 457,200 square meters of floor space.[220] Employment of inmates from Dachau on its production lines started in February 1943. A labor camp was constructed next to the factory to accommodate the inmates — most of them

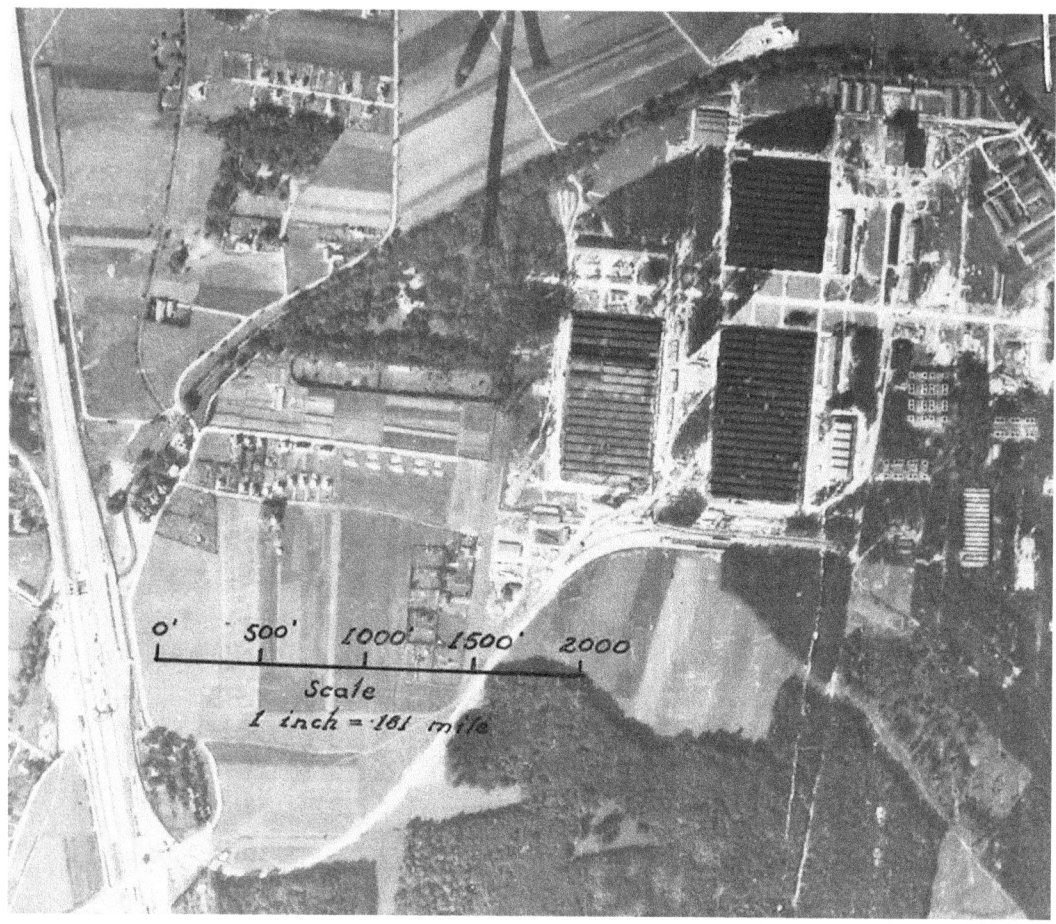

Two aerial photos of BMW's Allach aero-engine factory. The photograph above was taken in early 1943. The photograph on the opposite page was taken on 6 September 1943. Clearly visible on this photograph are the foundations of two new large production halls and a labor camp on left (courtesy U.S. National Archives and Records Administration).

selected specifically because they possessed one or another technical skill. This strategy proved to be quite efficient and BMW noticed improved productivity in Allach.[221] The use of slave labor increased in the following months and by March 1944 the share of non–German workers in Allach reached its peak at 71 percent. This was an exceptionally high percentage when compared to the rest of the aero-engine industry. At that time the foreigners' share within the entire aero-engine sector was 41.4 percent — which in turn was almost four times higher than their share in January 1942 (10.6 percent).[222]

Just as corporate interests motivated Heinkel to use slave workers, Messerschmitt started using slave labor as its scheme of using large numbers of women shattered in late 1942. Absenteeism of women increased as war dragged on and it became obvious that a dramatic increase of production was impossible with this workforce. Foreign workers were viewed as an unsatisfactory replacement for women, and as a result Messerschmitt's board of directors decided to make use of concentration camp inmates. From October 1942 the firm tried to obtain inmates for the production of its large transport gliders at Leipheim. First contingents of inmates from the Dachau concentration camp near Munich arrived, though, only in March 1943 — right at the time when the firm started preparing for the production of the Me 262. Thus most of the inmates were allocated instead to the central Augsburg plant and by July 1943 some 2,299 inmates worked in this factory.[223] In the same way as in Oranienburg, the use of inmates proved to be quite effective and the management noticed an increase of productivity. On 20 July 1943, Willy Messerschmitt wrote directly to the commander of *Sturmbannführer* Martin Gottfried Weiss, and informed him about the significant increase of productivity following the introduction of slave workers. He therefore asked him to provide additional inmates for work at the Augsburg plant and expressed his sincere hope that this initial direct contact will lead to "larger mutual successes in the future."[224] It was the beginning of a close association of Dachau and its inmates with the production of Messerschmitt aircraft. This cooperation ultimately resulted with Dachau's providing work detachments to around eighteen Messerschmitt factories and workshops.[225] In most cases the SS and the firm established sub-camps right next to the factories or close to them. A sub-camp of Dachau was constructed at Haunstetten, just south of the Augsburg factory,[226] and a contingent of inmates with an SS guard detachment was dispatched there. The inmates worked initially at Messerschmitt's Augsburg plant and in component production workshops located at Haunstetten, sometimes referred to as *Werk I*. Messerschmitt not only paid the SS for the workers, but also paid special premiums at irregular intervals to the SS men guarding them.[227] A second concentration camp was later constructed near the 12,800-square-meter Kottern machinery factory, where jigs and other production fixtures were produced. An American air raid heavily damaged this factory on 19 June 1944, and as a result its production was largely dispersed to other locations.[228]

At the same time, Messerschmitt's Regensburg plant started a close association with two more concentration camps. After fully grasping the business potential of slave labor in the war industry, the SS started converting its own DESt quarry industry around the Flossenbürg concentration camp into a manpower reservoir for the armaments industry in Saxony and northern Bavaria. The same DESt business operated in and around the Mauthausen concentration camp in Austria also gradually converted to armaments production. For the SS this conversion was a major shift in the functionality of the camp system and its business orientation. The conversion of Flossenbürg was particularly significant because at that time it was the most profitable DESt operation. Among the most prominent firms expressing interest in using former DESt slave workers was Messerschmitt. It started negotiations with DESt through *Rüstungskommando Regensburg* at the end of 1942. In mid-January 1943 Messerschmitt offered DESt the opportunity to open workshops in the camp and to provide equipment, tooling, raw materials and training personnel at its own costs. The deal was that DESt would manufacture parts and sell them to Messerschmitt for a much reduced price than they would have cost to manufacture in the normal way. DESt keenly accepted the offer and production had started as early as 5 February.[229] This business model soon proved to be profitable for both parties. Messerschmitt significantly reduced its spending on wages and reduced the cost of the parts. The firm paid DESt only 3 RM a day for a skilled worker and 1.5 RM for an unskilled worker. DESt also reduced its operational expenses and functioned from this point more as a manpower agency and less as an industrial enterprise.[230] The number of inmates working in these workshops grew slowly and by July it had reached 300. This cooperation picked up pace following the bombing of the Regensburg factory in August 1943, and more workshops were relocated to the vicinity of the camp. At the same time Messerschmitt concluded a similar arrangement with the Mauthausen camp, and production of Me 109 parts started in the camp during late autumn. By the end of 1943 the number of Flossenbürg inmates working for Messerschmitt increased to 1,375, and in Mauthausen 140 inmates worked for the firm.[231]

Earlier, Erla, another licensed Me 109 producer, also decided to introduce slave labor. In March a sub-camp of the Buchenwald concentration camp code named "Emil" was constructed in Leipzig, and in April its inmates started working in Erla's three factories in and around Leipzig. Later more sub-camps were constructed next to these factories to accommodate

Concentration camp inmates in striped uniforms working at an unidentified damaged Messerschmitt factory, probably in late 1943. Note the improvised repair work of the production hall (courtesy U.S. National Archives and Records Administration).

the increasing number of inmates arriving from Buchenwald.[232] In December Erla relocated workshops to a new sub-camp of Flossenbürg in Johangeorgenstadt, which produced Me 109 tailplanes. In January 1944 a sub-camp was established next to a wings factory in Mülsen–St. Micheln and in March at a dispersal location at Flöha, which manufactured fuselages. A total of around 2,650 inmates worked in these factories in early 1944.[233] All firms using slave labor found out that it helped to cut the large amounts spent on wages. Heinkel paid the SS a monthly sum of 132 RM (4.4 RM a day) per inmate.[234] Erla paid 6 RM a day (10 daily work hours in 1943) for a skilled worker and 4 RM for an unskilled worker. These sums formed only 60 percent and 40 percent, respectively of the normal payment of skilled and unskilled German workers.[235] When taken into account that Erla and other firms saved other expenses related to each employee, the amount of money spared was even higher.

By February 1944 Flossenbürg and its sub-camps became an important parts supplier for the Me 109 production and formed a central component in the dispersal system that saved the fighter's production following "Big Week." Around 2,000 inmates working in Flossenbürg alone produced 900 engine cowlings and radiator fairings for the Me 109, as well as other smaller parts. After "Big Week" 700 of the 1,300 Soviet POWs working in Regensburg were deported to Flossenbürg to continue their production for Messerschmitt there while the Regensburg factory was repaired.[236] Pohl estimated in mid–June 1944 that at that time the former DESt facilities in Flossenbürg and Mauthausen contributed 35 percent of Regensburg's output.[237]

The widespread dispersal of the aviation industry also caused a significant expansion of the Flossenbürg camp system. Among the new partners of this camp were Weser, which constructed an underground factory at Rabstein and Arado. Arado became one of the main "customers" of the Flossenbürg system. Initially male inmates were allocated to its Rathenow factory, which produced parts for the Ar 234 jet bomber. Later more than 1,000 inmates, mostly Jewish women and girls, were transported to Arado's factories in Wittenberg and Freiberg.[238]

Towards the end of 1943 the use of slave labor in the aviation industry became widespread and extensive. It continued to change the composition and character of the workforce of this industry, a trend that had started with the introduction of POWs and foreigners. The statistics of Messerschmitt at the end of December 1943 are particularly revealing. Of all the men and women working at Messerschmitt's factories in Augsburg, Oberammergau, Kottern, and Leipheim, 8,364 were German men and 2,977 were German women. Besides them there were 3,607 foreign men, 1,285 foreign women and 207 POWs. Perhaps most revealing is the fact that of the total of 20,042 workers only 4,411 were listed as skilled workers. The figures for inmates were included in the foreign workers category, but we can learn about the magnitude of their share by the fact that at the end of 1943 a total of 3,882 male inmates worked for Messerschmitt in Augsburg and at Kottern.[239] These statistics portray the transformation of the aviation industry from a high-tech industry demanding highly skilled workers into a branch using mostly unskilled non–Germans. In 1944 this trend became even stronger.

Other projects and firms who drew on Heinkel's experience decided in 1943 to select the slave labor option. The best-documented and arguably the most dramatic example is the early conception of the A-4 ballistic missile's production. As it was looking into the SS proposal to use slave labor for its production, a team led by production expert Arthur Rudolph visited the Oranienburg plant on 12 April 1943 in order to observe how this model worked. The visitors were hosted by Mr. Hänsslein of Heinkel, who showed them around the plant. Soon afterwards Rudolph submitted a detailed report about the visit:

> The Heinkel Works at Oranienburg has largely used foreign workforce made of *Ostarbeiter*, Frenchmen, Dutchmen, etc., in its production. Due to frequent rotations within these groups production has suffered. The Heinkel Works have made contact with *SS Oberstürmführer* Maurer and requested the assignment of prisoners from the concentration camps. The request was granted and the action began as an experiment in a single hall in August 1942. At first 300 men were put into action. The best experience has been obtained.... This system has proved itself, and generally the use of inmates offered considerable advantages in contrast to the earlier employment of foreigners. Especially useful was the take over by the SS of all tasks which have nothing to do with the work itself, and the higher level of security and secrecy offered by the use of inmates.[240]

Rudolph's recommendations were fully implemented and the projected A-4 production was based on a model similar to that of Heinkel at Oranienburg. At that time Milch became assured again that slave labor represented part of the solution to his production capacity problem. One day after Rudolph and his group visited Oranienburg, Milch wrote to the same Maurer, the SS officer responsible for inmate allocation to industrial work, and pointed out the importance of using inmates supplied by the SS in the aviation industry. He also remarked that: "An adequate supply of manpower for the Luftwaffe industry is of crucial importance for the successful outcome of the present war."[241]

Another case where Oranienburg served as an example is related directly to the aviation industry. In mid–August 1943, Staff Engineer Helmuth Schelp, of department GL/C-B 3 of the Technical Office, dealing with experimental engines,[242] and engineer Schaller, who had worked at the Oranienburg plant in 1942 and was now a hangar director at the Rechlin Flight Test Center, the central military flight-testing organization, met Hans Jüttner, head of the Leadership Main Office (SS-FHA) of the SS. They briefed him about the highly confidential development of jet propulsion and about the forthcoming production of jet engines, mainly by BMW. Based on Schaller's first-hand experience with concentration camp inmates while working in Oranienburg, they suggested using this manpower reservoir also in the production of the new engines. Schaller's role in this affair is revealing and exemplifies how some important initiatives in the Third Reich were based on personal contacts and early political affiliations. As Jüttner later reported to Himmler about this meeting, he mentioned that present in the meeting was one of his staff officers, *Standartenführer* Hoffmann, who knew Schaller from the good old "time of struggle" (*Kampfzeit*). This acquaintance was used to support Jüttner's advice to Himmler to approve the enterprise by implying that Schaller was an "old fighter" of the Nazi Party and therefore could be trusted.[243] Jüttner obviously appreciated the importance of the matter and ordered his staff to look into it.

Following this meeting, Franz Mahnke, head of department GL/C-B 3, visited Sachsenhausen and checked the feasibility of producing there the BMW 003 jet engine. He was impressed by the possibilities and suggested to move BMW's personnel to Sachsenhausen in order to initiate production of the pre-series engines and in order to train inmates in preparation for full-scale production. He also asked Milch to intervene with Himmler in order to accelerate the deal.

Milch preferred instead to use the services of an old contact in order to negotiate the deal with Himmler.[244] His contact man was Helmut Graf von Zborowski, director of BMW's rocket engines development group. He was a member of the SS and although his rank was not high, his status in the aviation industry was high enough to allow him direct access to Himmler, whom he updated regularly about the status of new engines, and generally about the newest developments in aviation technologies.[245] On 2 September 1943 he sent a memo to Jüttner, and asked him in the name of Milch to allocate some of the production capacity of the SS itself to the production of the BMW 003 engine.[246] Jüttner again ordered his staff to look into the matter and see what could be done. The SS subsequently contacted the RLM and BMW in order to work out the details of the proposed cooperation.[247]

SS negotiations with the RLM and BMW came to their conclusion when Jüttner's staff suggested establishing the required production facilities inside Sachsenhausen, instead of moving inmates to an existing BMW factory. The main advantage of this scheme was in eliminating the need to move inmates around. Jüttner declared that the 3 main preconditions for carrying out this scheme were:

1. Supply of all the required machinery and toolling by the RLM and BMW.
2. Construction of two new barracks in the location of the new plant by the RLM.
3. Timely arrival of a German supervisory staff.

The partnership model was similar to the one the SS signed earlier with Messerschmitt. The RLM approved the plan and its officials declared that the requested equipment would be made available within days.

In the meantime BMW submitted its production plan for the Sachsenhausen plant. It was supposed to produce 16 main components and 68 other parts for the BMW 003 engine. BMW was ready to deliver the required production tooling and the relevant blueprints, and intended to commence production within 14 days of their delivery.[248] Himmler probably approved the scheme, which was supposed to bring under his influence another high-tech armament project, but at that stage the BMW 003 engine was still far from ready for serial production. Two BMW prototype jet engines were first test-flown as early as 25 March 1942 on the first prototype of the Me 262, but during the flight both engines failed. Following this failure BMW's engineers found several grave problems in the design of the compressor, which required a massive redesign. The redesigned engine went back to the air only in October 1943 and then it took almost 10 additional months until its series production could finally start.[249]

Therefore the production of components for the BMW 003 engine never commenced in Sachsenhausen, although in January 1944 BMW moved its flight test center to the Oranienburg airfield, located next to Heinkel's Annahof factory and in proximity to Sachsenhausen.[250] However, the negotiations between the RLM, BMW and the SS regarding this production scheme demonstrate the dynamics and channels of introducing slave labor into the aviation industry at that stage. Throughout this period Milch kept and nourished his direct and personal contacts with Himmler. In November 1943, for instance, Milch hosted Himmler in the cozy atmosphere of his country estate near Breslau in Silesia. He told him about his difficulties with Göring and Speer in promoting increased fighter production and asked for his help. Himmler promised his support, and as it turned out, this promise became crucial three months later.[251]

4. From Technological Expertise to Slave Labor   185

The stage was now set for a far more extensive use of slave labor in 1944. The sources of foreign workers and POWs largely dried up during 1943. On 4 January 1944, Sauckel informed Hitler for the first time that according to the latest forecasts he could not guarantee the quota of 4,050,000 workers required by the German economy in 1944. He repeated the same message in a meeting of the Central Planning Committee of the Armaments Ministry on 1 March 1944 — the day the Jägerstab was born. On this occasion he also notified Milch and others that "practically there is no more employment possibilities of German labor — neither men nor women; it must be viewed as exhausted."[252] These declarations signified the bankruptcy of German labor recruitment policy both in Germany and in the occupied countries. Shortly afterwards Milch openly acknowledged the fact that around 60 percent of the workers in the aviation industry were foreigners and that this figure was even higher when taking into account that many German women worked only part time and that many Germans were employed in research and development and were not engaged directly in production-related jobs.[253]

Since early 1944 more concentration camps had allocated work detachment to factories in their vicinity. By the end of January 1944, Auschwitz provided 90 inmates to a nearby Siemens-Schuckert factory, manufacturing electronic equipment for night fighters; Buchenwald allocated 1,550 inmates to Erla in Leipzig and 1,310 to Junkers in Schönebeck; Dachau provided 3,434 inmates to BMW's Allach plant, 60 to Dornier's new factory in Neuaubing, 2,695 to Messerschmitt's Augsburg-Haunstetten, 352 to Messerschmitt Gablingen, 192 to Messerschmitt's workshops in Dachau, 341 to Messerschmitt Kottern and 374 to Kempten; Flossenbürg provided 1,911 inmates to Messerschmitt and 576 to Erla's factories in its vicinity. These detachments were only part of the network of cooperation that developed by now between different concentration camps and the aviation industry. However, at that time Dachau and Sachsenhausen appear to be the most significant suppliers of slave labor to the aviation industry.[254] Specific numbers for later periods are unavailable, but it seems that Flossenbürg also became one of the top suppliers. Historian Hans Brenner estimated that around 18,000 men and women from this camp were employed by the aviation industry complex established around it.[255]

At the same time the SS also increased the allocation of female inmates to the aviation industry. First detachment of women was allocated in March 1943 from the main women's concentration camp, Ravensbrück, to work at the Mechanische Werkstätte in Neubrandenburg. The firm produced different parts for the aviation industry, including bomb release mechanisms for Arado. By early 1944 the SS allocated around 2,500 women from Ravensbrück to work detachments at parts manufacturing workshops in Neubrandenburg.[256] Heinkel was again among the first firms to employ female slave workers in airframe production. In July 1943 the first women from Ravensbrück arrived at its Rostock factory, and in August female inmates started to work in its new factory in Barth.[257] These early allocations set the pattern for a massive use of female slave workers in 1944.

Some of the largest projects involving slave labor began with the comprehensive dispersal scheme of 1944. The underground relocation program, ordered by Göring in late January 1944, was the first step towards a much-increased use of slave labor. On 14 February 1944, Göring asked Himmler to allocate the highest possible number of concentration camp inmates to the aviation industry. This workforce was to be used foremost for the underground relocation of the aviation industry and for production of modern aircraft, especially in those new underground plants. Göring referred in his telegram to earlier discussions between the RLM and the SS on this matter, suggesting that general terms of an agreement were already agreed upon and that all parties just waited for a final authorization.[258] Four days later Himmler agreed principally to help, but only on 8 March 1944 did he send a detailed reply to Göring's request. In the meantime, and just before "Big Week," Göring submitted initial requests for inmates' allocation to several underground relocation projects. One request was for 10,000 inmates from Buchenwald to work in the relocation of several Junkers programs, and the other was for additional 10,000 inmates from the Mauthusen concentration camp to underground construction projects in Austria.[259] Himmler's detailed answer to Göring's requests came only after "Big Week" and the establishment of the Jägerstab. Himmler described in his reply the contribution of the SS to the aviation industry so far. According to his statistics, by March 1944 some 36,000 concentration camp inmates worked in aviation-related plants. He intended to increase their number in the near future almost threefold, to more than 90,000 inmates. Himmler also promised to provide around 100,000 inmates required for the underground dispersal — the first time this number is mentioned in official correspondence related to aviation production.[260]

This correspondence took place within a broader framework of increased reliance on concentration camp labor in most branches of the German war industry. The lack of foreign workers and POWs was felt everywhere. The only remaining manpower reservoir was the concentration camps and some still untouched large Jewish communities, especially in Hungary. The employment of these two unfortunate groups now reached an unprecedented magnitude. A striking indication of this tendency was the growth in the numbers of new sub-camps established in early 1944 in order to provide accommodations for the allocated prisoners. In contrast to the earlier trend, most of the inmates sent to these sub-camps from early 1944 were Jews. The increased use of female slave workers — also mostly Jewish — was another typical characteristic of this expansion.[261] The decision to dramatically increase the number of Jewish slave workers in Germany and its annexed territories was in stark contrast to the decision of the Reich's leadership from October 1942 to make the Reich "free of Jews." As a result, throughout 1944 thousands of Jews were transported back *into* the Reich in order to solve Germany's shortage of workforce, particularly in the aviation industry.[262]

The reorganization and reshuffle of the aviation industry in the aftermath of "Big Week" happened within this framework and definitely added its own impetus. The main challenge facing the Reich's construction agencies at this stage was the construction of the underground and bunker factories. These projects were particularly labor-intensive and demanded a large number of construction workers at a time when Germany's construction capacity was stretched to its limits.[263] The Jägerstab initially considered professional miners as the best workforce for the underground relocation. Milch argued in early March 1944 in a Jägerstab meeting that the SS should obtain increased numbers of miners from Slovakia and from Italy to form the main core of the workforce for the underground relocation. Saur commented that getting miners should be no problem because the SS was responsible for the entire miner training program in Germany and operated the best mining schools.[264] By this he referred almost certainly to DESt, the SS-owned giant quarries and earthworks enterprise, and the training schemes it established in different concentration camps. During March the Jägerstab continued considering the type of preferred bombproof factories. Factories in caves and tunnels were easier to construct, but their floor space was usually limited.

Bunker factories offered more floor space, but were difficult to construct and were labor intensive. Early planning suggested that their construction alone demanded an extra 25,000 construction workers. In mid–March the Jägerstab estimated that a total of 100,000 additional workers would be required for its production programs—including the construction of underground and bunker factories.[265] Therefore, it became clear that miners could form only a small fraction of the required workforce.

Additional workers were also required for the construction of underground V-1 factories, especially one planned in a cave system in Tiercelet near Metz in Alsace. It was a key factory because of its proximity to the V-1 launching areas in western France. Professor Ferdinand Porsche, general manger of Volkswagen, had met Oswald Pohl on 14 March 1944 and asked his help in carrying out this project. Pohl promised to allocate 3,500 concentration camp inmates. For the construction of the factory they were to fall under the jurisdiction of *Organisation Todt* in France, but most of them were to stay in the factory after its completion and to be employed in the production. Immediately after the discussion Pohl sent two officials to inspect the location of the factory and make plans for "*Unterbringung und Verpflegung*" (housing and provisions)—in other words, for a labor camp. Porsche announced this agreement in a Jägerstab meeting three days later.[266] The protocol of this meeting, as those of other meetings, clearly proves that Porsche and the Jägerstab members were well aware of the fact that most if not all of the 100,000 workers they required would be concentration camp inmates because of references to *KZ-Häftlinge* and *KZ-Leute* were repeatedly made by participants. Some of them demanded to mix German foremen and supervisors in this army of slave workers in a ratio of at least one German to 10 slave workers as a precondition for the entire deal.[267]

Initially the German leadership considered constructing at least one bunker factory in a border region in order to draw the necessary workforce from across the border and thus solve at least partially the manpower problem. In early April, Hitler formally asked Himmler to allocate 100,000 Hungarian Jews to the bunker factories' construction project in case the border scheme proved impossible to implement.[268] As the SS started collecting these people, the Jägerstab continued its efforts to find at least some of the required personnel somewhere else, even in the remotest corners of the Reich's war economy. On 25 April 1944, for example, Milch remarked during a Jägerstab meeting that 30 to 40 Frenchmen were working with some Germans at a central gas distribution facility between Birkenwerder and Oranienburg and that they were obviously not working too hard. He demanded to pull these men immediately out of this place and employ them in the aviation industry. He remarked: "I believe that if we will look around we will find 100,000 people, who just stroll around exactly as in this place."[269] Milch was, however, too optimistic with this assumption. The gas distribution center and other places offered only minuscule contribution, and the 100,000 inmates promised by the SS quickly became the only viable option. In the following two months the SS gradually fulfilled its promise. Around a quarter of more than 400,000 mainly Hungarian Jews deported to the Auschwitz extermination camp in May–July 1944 were selected to work in the Jägerstab projects. They were selected from among the fittest inmates, loaded onto trains and trucks, and transported to different sub-camps and labor camps located next to or in the vicinity of aviation factories and related construction sites. Other inmates were first transported to one of the main camps, which became in early summer 1944 major hubs in the German slave labor network. In these camps the SS conducted additional selection and allocation of inmates considered fit for work. These inmates were consequently distributed from these main camps to sub-camps and work detachments associated with aviation factories or construction sites. Most of those who stayed in Auschwitz were killed in a matter of hours or days.[270] Transports of Jewish inmates continued to arrive at factories and construction sites throughout the second half of 1944.

Finding manpower for the construction projects was a particularly urgent task. During a Jägerstab meeting on 26 May 1944, Saur estimated that there were three times more construction projects than the available workforce could perform. As a result, most the first transports of Jews that started to arrive in mid–May were allocated to the bunker plants project. Initial allocations failed to fulfill the needs and after being asked by Speer about his requirements, Kammler stated that he needed, at least 50,000 additional inmates for the construction projects under his supervision.[271] Furthermore, both Kammler and Schmelter, the official responsible for labor allocation, were disappointed by the quality of the manpower they received. Schmelter remarked that the first two transports to arrive from Auschwitz were only of kids, women and old men. He warned that if the next transports did not bring enough men fit for work, the whole *Aktion* would fail.[272] Obviously the selection process in Auschwitz was sometimes inefficient and many inmates unfit for difficult construction work slipped through. Another explanation for this failure was that the SS tried to fill the quotas it promised without working too hard. From testimonies of inmates shipped from Auschwitz at this period it seems that the selection process was indeed sometimes slack, and inmates, especially women, were able to get on board outward transports without somebody seriously checking their fitness for the work they were about to perform.[273]

It is difficult to determine how many inmates were allocated to work in the aviation industry after "Big Week." The *Generalluftzeugmeister* department of the RLM estimated in late summer 1944 that by that time around 100,000 inmates were delivered to the aviation industry.[274] More were delivered later, however, so the numbers are definitely higher. The estimated numbers are 130,000 to 160,000. At least 118,000 inmates were allocated to aviation production programs, while at least 30,000 were employed on aviation industry–related construction sites. This last figure however, seems too low.[275]

From any perspective, the number of people shuffled around during these fateful months was enormous. The new bosses of the aviation industry practically became desk-riding slave drivers. As Edith Raim has pointed out, one of the striking aspects of the Jägerstab meeting protocols is the cold, bureaucratic way in which concentration camp inmates were referred to as "packets" (*Paketen*) and "bundled" (*gebündelt*) because they were mostly allocated by the SS in groups of 500 to 1,000. Members of the Jägerstab sometimes discussed the fate of these workers in a most cynical way.[276] On 25 April 1944, Schmelter reported in a Jägerstab meeting that a total of 250,000 construction workers were required for different construction projects of the aviation industry and especially for the bunker factories scheme. Lange, the staff member responsible for production machines and replacing Saur and Milch as chairman during this particular meeting, remarked, "You can get all of them in Hungary. The Jews just hang around in Budapest."[277] In a meeting on 27 June 1944, Schmelter made the following remark about 12,000 Jewish women recently allocated to the aviation industry: "The SS already found out [that it is efficient] to bundle and deliver these Hungarian Jewish women in contingents of 500. This way smaller firms are also able to better use these concentration camp Jewesses. I therefore ask to make requisitions in a size of 500 each."[278] The problem with smaller contingents was that they split the workforce into several small detachments. These small groups proved to be an administrative headache, especially because they required a larger number of guards to watch them than a smaller number of larger groups would have required. Considering the dwindling number of guards available to watch inmates, small contingents and detachments were viewed as wasteful in this regard.[279] The main problem with larger groups was that this system made it difficult to provide inmates to smaller firms, which required fewer than 500 workers. Later the SS agreed to provide smaller groups of inmates in order to solve this problem.[280]

In 1944, especially after "Big Week," more and more "packages" of inmates were allocated to the aviation industry as the extent of the work involved in dispersal and increased production unfolded. But soon other key industrial sectors needed an urgent supply of inmates. A new emergency came following a series of devastating air raids in May 1944 that destroyed large and crucial parts of Germany's oil industry. Suddenly it became necessary to divert large numbers of construction workers to repair bombed oil-producing installations. Soon afterwards the German leadership decided to carry out a huge emergency underground relocation program for the oil industry. Once again the heads of the Armaments Ministry were forced to shift around their meager manpower in order to fulfill all the urgent tasks at hand. Thousands of workers that were supposed to be allocated to the aviation industry were withheld and then reallocated to the relocation of the oil industry. Speer and Dorsch of *Organisation Todt*, and the new head of the Armaments Ministry's Construction Office, agreed to allocate to the aviation industry other available workers, including 1,700 Auschwitz inmates that were supposed to work in the light metal and the petrochemical industries.[281] The underground relocation of the oil industry proved to be much more complicated than the relocation of the aviation industry and it never really got off the ground. Thus, after the initial chaos resulting from the oil emergency, the primacy of the aviation industry was restored and both sectors received a steady stream of new inmates.

In early 1944 the focal point of non–German employment also reached its climax in the aero-engine industry. BMW's Allach factory, as well as the rest of the aero-engine industry, relied heavily on concentration camp inmates for most production tasks. This trend continued and inmates gradually replaced not only Germans, but also foreigners. In June 1944, for instance, 2,500 female inmates were allocated to the aero-engine industry to replace 4,600 "troublesome" Czech foreign workers. While in some factories they were supposed to partially replace them, in Allach the intention was to fully replace the Czechs with these women.[282] We can assume that this replacement was never completed, because by July, 972 new concentration camp inmates started to work at the plant, while only 49 foreigners departed. The July report submitted by the Allach factory to the Jägerstab also mentioned that 115 inmates were taken off the payroll. We can only guess what happened to these inmates.[283] In early July the demand for slave female workers from the aero-engine sector reached a new height: "20,000 female concentration camp inmates and Hungarian Jewesses."[284] In December 1944 the total number of workers at Allach reached 17,313. Some 11,636 of them were foreigners, POWs and concentration camp inmates — including 496 women.[285]

The employment of slave labor in such large proportions demanded the cooperation of different organizations and firms. There were two main modes of workforce allocation at the time. In some cases the Jägerstab "pushed" the workforce to the firms. In other instances, and it seems to be the more common mode, the firms submitted requests for manpower allocations to the central employment office of their region, or directly to the SS. Generally, the Jägerstab demanded that aviation firms and factories use the slave workforce it allocated to them. If a firm or a factory declined the offer it was obliged to provide a satisfactory explanation as to why it chose not to use the offered workforce. A special committee, working directly under the Jägerstab, micromanaged the employment of allocated workforce and reported severe cases of inefficiency. In July 1944 Saur demanded the punishment of the director of Focke-Wulf's Posen factory for declining 500 inmates offered to him from the Wronki concentration camp and for offering an unacceptable excuse. As it came out, it was another case in which changes in the production programs of different aircraft played a role in the employment of slave labor. The factory director, Schwichtenberg, was offered in late May 1944 a list of concentration camps with available manpower in the vicinity of his factory. At that time Posen was still heavily engaged in preparing to produce the Ta 154, but soon afterwards this unsuccessful aircraft was terminated and then the Posen factory was bombed. As a result, work at the plant came to an almost complete standstill and the factory lost around 1,000 of its workers to other factories. After preliminary repairs of the bomb damage, the factory was prepared to restore production of FW 190 fighters. This task demanded skilled workers and the factory could not spare the 20 German foremen required to supervise the 500 inmates from Wronki. Schwichtenberg therefore tried to explain in detail to the chairman of Special Committee F4, responsible for Focke-Wulf production, why he insisted that there was nothing he could do with the Wronki inmates.[286]

It was, however, more common for factory directors to repeatedly demand workforce allocations from their local employment offices in order to get the slave workers they needed. Even high-priority production programs, like the Ta 152 high-altitude fighter development and production in Focke-Wulf's Adelheide factory, suffered from this problem. The plant director repeatedly failed to obtain the required workforce from the regional employment office in Bremen. Initially 100 Polish POWs were allocated to him but none arrived. Then he was promised 200 Slovakian workers, who also never showed up. Simultaneously, he lost an increasing number of his German workforce to other factories, as the firm tried to move around its dwindling workforce among its different factories.[287] Some factories were never allocated inmates or received only negligible numbers of them. On the last day of January 1945, Junkers' Leopoldshall and Stassfurt factories reported a total number of 2,589 workers. Although 1,089 of them were foreigners, the only workers falling somehow into the category of slave workers were 4 Belgian POWs. The box in the reporting form reserved for inmates and Jews (*Häftlinge und Juden*) remained empty. Interestingly, three of the POWs were registered as skilled workers and the other as a trainee.[288]

The underground dispersal was especially challenging because it involved various firms and contractors — including mining and construction firms hired for the task by aviation firms — and various state and local authorities.

The "Bergkristal" (aka B8 and "Esche 2") project near St. Georgen in Austria is a good example of a large underground relocation project producing high-tech aerial weaponry using slave workers. It also exemplifies the complex network of cooperation required for the execution of such complicated projects. While Messerschmitt's Regensburg plant provided machinery, skilled workers and the know-how, the SS — through its DESt enterprise — provided the facility and the main workforce from nearby concentration camps. The Luftwaffe

Female slave workers at the construction site of the "Weingut II" bombproof factory. The earth and vegetation cover applied to the cement dome as camouflage is visible in this photograph (courtesy Yad Vashem Archives).

provided extra guards to watch the large number of inmates collected at the Gusen sub-camps of Mauthausen for this project. At that time Mauthausen became the hub of a network of 31 sub-camps, 7 of them dedicated to various underground relocation projects. At least 60,000 inmates incarcerated in these camps worked in 1944 in these projects.[289] The two Gusen camps alone employed in July 1944 around 6,000 inmates in aviation-related production for Messerschmitt. A monthly average of 77 trainloads filled with parts for Me 109 and Me 262 fighters left the workshops in these camps towards different component and final assembly factories. The nearby underground factory formed a massive expansion of this aviation production, aimed at main components manufacture and eventually at final assembly of complete aircraft.[290] Right from the beginning this underground factory relied completely on inmates not only for its construction, but also to perform most manufacturing tasks. After it was completed and production of some Me 262 components started in early 1945, most of the manufacture of this advanced aircraft, including quality assurance of completed aircraft, was carried out by inmates.[291]

Another example of the close cooperation between state, industry and the SS was the "Ringeltaube" bunker factories project. OT prepared the blueprints and supervised on-site the work of different construction firms. The SS provided a large proportion of the workforce from its camp manpower reservoir. Even more illuminating in this regard was the planned operation of the completed factories. From the 90,000 workers assigned to this factories complex, 30,000 were supposed to be inmates provided by the SS. The primacy of inmates in daily production was so dominant that the managing firm, Weingut-Betriebsgesellschaft, hired the ex-commander of Dachau, *Obersturmbannführer* Martin Gottfried Weiss, to deal with all the organizational aspects of their employment, including the construction of several concentration camps near the factories. The firm also established within its management an "inmate matters" (*Häftlingswesen*) department to deal with different administrative tasks related to the employment of the inmates. No wonder, therefore, that one scholar of slave labor referred to the Weingut-Betriebsgesellschaft as the climax of cooperation between the SS and German industry.[292]

Production of advanced aircraft and engines by slave workers proved to be problematic from several points of view. The nature of the workforce and the poor and brutal treatment it received was the source of the problems. Poor motivation and lack of training were reflected in doubled monthly wastage of material following the arrival of 2,500 Soviet workers at Daimler-Benz's aero-engine factory at Genshagen in 1942. This wastage and other problems with the POWs led K.C. Müller, the director of production, to claim in July 1942 that "You can't build engines with Russians."[293] The situation was somewhat alleviated later as the firm improved both training and the integration of the Soviets into the production process. These measures failed, however, to fully solve the problems involved in producing modern aero-engines with foreign and slave workers. At the end of October 1943, Müller voiced his desperation over the difficulties of producing aero-engines when 65 percent of the workforce were foreign workers "with no inner interest" in their work. He complained of the "great and difficult, if not impossible task" of producing such a complex item as an aero-engine under these conditions.[294]

Even more disturbing were the problems associated with the Me 262 production. At the beginning of March 1945 it was reported that between 6,000 and 6,500 concentration camp inmates worked for Messerschmitt, which became the most important aircraft producer with its entire capacity dedicated to the jet's production.[295] After some highly critical reports about bottlenecks in the production and flight-testing of the aircraft were submitted, the RLM decided to look closely into the matter. The RLM appointed Major Wilhelm Herget to inspect the four main flight-test centers of the aircraft. Soon he found out that the real bottlenecks were in the industry. A particularly troublesome bottleneck was located in the supply of wing sets from the Leonberg "Leo" factory located in a double highway tunnel. It produced only 150 sets each month instead of its planned monthly output of 750 sets. As he visiting Leonberg, Herget was supposedly horrified by the poor condition of the inmates working there. The SS officer in charge of the inmates told him that he asked for increased food rations for the inmates but never got them. Herget reported his finding to his boss, General Kurt Kleinrath, who bore the pompous title "Reichsmarshall's Plenipotentiary for the Control of Acceptance Flight Testing." Kleinrath reported to Willy Messerschmitt and to Messerschmitt's chairman Friedrich Seiler. According to Herget, they were both shocked by his findings and immediately fired Leonberg's manager. Since this move failed to improve the situation at Leonberg, Kleinrath took the matter to Göring. The only recorded result of this intervention was that Herget was reprimanded for acting beyond his authority.[296] Although some parts of Herget's story should be viewed with suspicion, it is believable that production pace and quality suffered because of the poor conditions of inmates in Leonberg and in other places. The growing deterioration in the manufacture quality of the Me 262s can be related partially to the increase use of inmates in their production.

Some of the production problems were caused by intentional sabotage performed by inmates — a problem that will be detailed in the next chapter.

The problems associated with the employment of slave workers did not stop the increased reliance on this workforce during 1944. Towards the end of this year slave labor became an industry-wide standard in the aviation industry. According to statistics found in one USSBS report, Heinkel's Oranienburg plant, the first important employer of slave labor, become the largest employer of slave and foreign workers within the aviation industry. It is estimated that in October 1944, a couple of months after production of the He 177 was terminated and the factory started to produce fighters, 53 percent of its workers were inmates, POWs and Jews, and 10 percent were foreigners. Heinkel's Wittenberg plant was in second place, with 41 percent inmates and slave workers, and 16 percent foreigners. Number three on the list was Messerschmitt's Augsburg factory and its branches, with 35 percent slave workers and 28 percent foreigners.[297] These statistics were based, however, on German documents available to U.S. official researchers in 1945–1946 and they covered only the most important factories. Unfortunately, late-war major production centers, like the dispersed and bunker Me 262 factories, were not dealt with in the German reports read by the Allies. Therefore, we can assume that the actual share of slave workers was higher.

Karl Frydag made the following revealing observation when asked after the war about the proportion of foreign labor in the aviation industry: "If you look over the whole company — perhaps 25 to 30 percent foreign labor. If you take the product — the labor on the whole — it will come to about 70 to 80 percent foreign labor."[298] It is another indication of the well-acknowledged fact that the foreigners' share on the production lines was much higher than their nominal share within the workforce of specific factories. This observation underlines the heavy reliance of German aviation production on slave and foreign labor when it came to series production. On the one hand, the massive use of inmates and foreigners enabled the Germans to achieve the "production miracle" of 1944. It was done basically by filling the manufacture halls with cheap slave workers and forcing them to work long shifts — exactly the mode of operation which could not have implemented earlier with German workers. On the other hand, this mode of operation brought several disadvantages. The main drawback was the generally deteriorating quality of the end products, but it also resulted in generally inefficient employment of human resources. As an American intelligence report summarized the matter in a postwar study about the German aviation industry: "Evidently it was believed that the use of a large number of slave laborers made unnecessary the attention to the production problems relating to the efficient use of labor that characterizes the production in the United States."[299]

Most firms viewed the use of slave labor purely as a temporary measure and planned to rely again on highly skilled workers once the crisis was over.[300] Nevertheless, slave labor formed a solution to a major structural problem of the German aviation industry. This solution, combined with the introduction of new production methods, new organization and new management structure, resulted in a significant increase of output in 1944. The massive use of slave labor in the aviation industry since 1943 fully supports Rainer Fröbe's argument, that at least in 1944–1945 the National-Socialist policy of extermination of certain human groups according to racial ideology became more theoretical and less central in light of wartime contingencies. Accordingly, the Germans tried to preserve a large proportion of the workforce available to them in the concentration camp reservoir and to fully exploit it in the interests of their wartime economy.[301] This cold-headed conclusion largely ignores the harsh daily reality in those production halls and construction sites. The next chapter will deal with this unhappy aspect of the "Production Miracle."

5

On the Production Lines — Daily Life in the Factories

So far we have dealt with various technical, organizational and political aspects of the German aviation industry and the "Production Miracle." At the same time millions experienced the war from below while working in this prosperous industry. Whereas decision-making and production data are quite well documented and readily available to the researcher, it is not as easy to find details about daily life in the factories of the aviation industry in the official files. Nevertheless, through the use of factory documents, brochures, newspapers and oral testimonies, it is possible to sketch a general picture of daily life in the German aviation industry during the late phases of World War II.

The United States Strategic Bombing Survey teams made a clear division in their postwar reports about the aviation industry between German workers and foreign workers. It was clear to them that these were completely different workforce categories, which worked under completely different terms and conditions. This division was based on a distinction made in captured reports and by former German managers during their postwar interrogation. This division was important, however, not only on the level of statistics, work allocation and management; these two groups also experienced the war and daily work routine in a different way. Writing a history of the German aviation industry without looking at the daily experiences of the millions who worked in its factories would be therefore incomplete. Furthermore, by looking at daily life we can learn more about the social environment in wartime Germany, about work routines, about economic aspects and about the treatment of foreigners and inmates in individual locations and in general.

General Working Conditions

General conditions in and around the factories of the aviation industry crucially influenced their daily routines and the wartime daily life of their workers. Some of them influenced Germany nationwide and therefore also had an impact on the aviation industry. For instance, night bombing of cities by the RAF, food rationing and chaotic wartime transportation, especially after September 1944, were general conditions that also influenced the aviation industry. Other conditions, like U.S. strategic attacks on aircraft factories, on Luftwaffe installations, and on underground relocation sites, affected the aviation industry in a more direct way.

Allied strategic bombing was one of the main factors influencing the German aviation industry from 1942. The Allies never repeated the concentrated effort of "Big Week," but statistics reveal that in its aftermath the total number of bombs dropped on aviation plants actually increased. A record number of 5,155 tons of bombs was dropped on aviation-related factories in April 1944, surpassing the 4,732 tons dropped in February. In May a new record was set with 5,642 tons. This increase reflected the fear of the Allies that German jet fighters were expected to enter widespread service soon. In response they attacked plants thought to be involved in the production of jet planes. In June and July the numbers dropped significantly as Allied airpower was busy supporting the Normandy invasion. Late in the summer the focal point shifted again to the aviation industry, alongside increased attacks on the oil industry. In August an all-time record number of bombs was dropped on the aviation industry: a total of 6,573 tons of explosives. Afterwards the numbers decreased considerably as the fighting power of the Luftwaffe dropped dramatically and as the focal point of Allied strategic bombing shifted increasingly to attacks on oil and transportation targets.[1] The American Eighth Air Force alone dropped during World War II a total of 132,805 tons of bombs on the aviation industry and on other aviation-related installations. This total formed 19.3 percent of the total tonnage it dropped during the war. Only transportation targets surpassed this tonnage of explosives with 235,312.5 tons (34.1 percent of the total tonnage).[2] These statistics clearly reflect the importance the Allies assigned to the destruction of the Luftwaffe and its supporting industry.

The Germans closely monitored the development of the bombing campaign and its changing trends by recording the number of sorties flown against each target category and against each part of the Reich and occupied Europe. Monthly reports by the Intelligence Division of the Luftwaffe's high-command (*Ic Abteilung*) regularly produced tables that graphically illustrated the ever-growing crisis of strategic bombing. It must have been a demoralizing experience to compose and read these reports — especially as the number of recorded sorties against targets in Germany increased dramatically after the Normandy invasion. Although since mid–May 1944 the focal point of attacks shifted to the petrochemical industry, according to the Luftwaffe's intelligence reports the number of sorties against the aviation industry formed 35 percent of the total number of sorties in mid–July and 41 percent in mid–August.[3] The slackening onslaught after the August climax is also reflected in these reports. Only 15 percent of the sorties recorded in September were against the aviation industry, and in October only 6 percent.[4] Therefore, the Germans were aware of the priorities set for the Allies' strategic bombing campaign and of the dire air-war situation. This detailed knowledge was restricted to the higher level of the military and political leadership. The men and women on the factory floors, however, received similar impressions by experiencing daytime strategic bombing firsthand and by hearing or reading about air raids all over the Reich. The population of big cities also experienced firsthand British night bombing aimed at "de-housing" German civilians and undermining their morale. This type of bombing sometimes heavily interrupted aviation-related production by either accidentally hitting a factory, or more likely, by affecting workers and causing massive short-term absenteeism.

Direct attacks on factories affected people and equipment in several ways. The foremost effect of such attacks was the destruction of production facilities and machinery. In some cases specific attacks caused more damage than others. On 29 May 1944, for instance, the USAAF attacked three of Focke-Wulf's old dispersed factories in the Eastern part of the Reich. The attack heavily damaged the Sorau, Posen and Cottbus factories, which produced

Air raid. Smoke billowing from fires burning at Junkers' Dessau factory drifts across the factory airfield following a bombing raid by the U.S. Eighth Air Force on 16 August 1944. This post-strike photograph was taken by an F-5 photo-recon plane of the 7th Photographic Reconnaissance Group (courtesy U.S. National Archives and Records Administration).

fuselages for the FW 190 fighter. Besides the damage caused to the factories, a large number of jigs and rigs were destroyed. The loss of this valuable tooling caused a sudden and severe bottleneck in the production of fuselages for the FW 190. Production was resumed later in the bombed factories, and in the meantime the production rate of other firms licensed to produce the plane was increased in order to compensate for the loss, but output was affected for a relatively long period.[5]

In numerous cases a byproduct of these attacks was the destruction of completed or semi-completed aircraft in the factories and in their adjacent airfields. In a daylight raid by the USAAF on Erla's Leipzig factory on 4 December 1943, for example, 19 completed aircraft and 12 fuselages were destroyed. An additional 63 Me 109 fighters and 40 fuselages were damaged. A similar attack on 20 July 1944 destroyed 48 aircraft.[6] Attacks carried out towards the end of the month proved to be especially effective in destroying aircraft. As Albert Speer told his interrogators during his postwar interrogation, experience proved that monthly production was concentrated mainly in the later part of the month. It meant that daily output increased towards the end of the month because plant managements pressed their workers to fulfill the monthly target output. As a result, late-month bombings often hit crowded production lines and hangars and destroyed more aircraft. The first few days of the month were also busy in aircraft factories because it took some time to test-flight and deliver the larger number of aircraft produced in the last part of the previous month.[7] A particularly effective end-of-the-month attack was the bombing of the Neuburg Me 262 dispersal factory and its airfield at the end of March 1945. The airfield was badly damaged and around 60 brand-new Me 262s were destroyed on the airfield premises. A further 20 aircraft were destroyed in the nearby production line.[8]

Air raids were not limited to strategic bombing of production centers and population centers. From the last months of 1944 tactical western air power became dominant over Germany in the shape of long-range fighters returning from escort missions, or fighter formations on independent sweeps deep into enemy territory. Later in the war low-level attacks became an increasing menace, especially to transportation targets and airfields. Prowling Allied fighters also increasingly attacked factories. The threat posed by such attacks caused several firms to issue a special system of warning against low-level attacks. A special siren was used to warn factory workers and the civilian population in the vicinity of the factories of the approach of low-flying fighters.[9]

Such attacks could not cause considerable direct damage to production facilities because of the light armament carried by fighters, but their indirect damage could be considerable, and they were effective against personnel and vehicles caught in the open. The most important indirect damage caused by tactical air power was to the transportation system, as described later. Further damage was often caused by attacks on factory airfields, where new aircraft were parked on the tarmac, waiting for the completion of their test flight or for their delivery flight to the Luftwaffe. These attacks not only destroyed valuable aircraft on the ground, but also disrupted airfield routines.

Sometimes these strikes were especially destructive. On 16 January 1945 a low-level attack on Focke-Wulf's Neuhausen factory airfield destroyed 14 brand-new Ta 152 fighters, which represented almost the entire initial production run of this new fighter. The aircraft were kept at the airfield because quality assurance checks discovered several defects, which had to be repaired before ferrying the aircraft to operational units. According to eyewitness reports the American fighters located the airfield after they spotted two landing trainers. Although the brand new fighters were parked on the edge of a wood under camouflage, the American fighters found them and strafed them individually.[10] Allied fighter activity generally disrupted delivery flights of new aircraft from factory airfields to operational units. As a result of low-level attacks, like the one described above, the Luftwaffe forbade ferry pilots to take off whenever low-flying Allied fighters were in the area. Whenever local ground controllers broadcast the codeword "Mayo," pilots were ordered to stay on the ground.[11]

An American P-51 Mustang (lower left) banks away after strafing Me 262s parked in the open next to the Leipheim forest factory on 18 November 1944. Three of the parked jet fighters are burning (courtesy U.S. National Archives and Records Administration via Richard Eger).

In other cases factory-fresh aircraft were stranded at factory airfields simply because there were no means to fly them out. Fuel shortage was a problem, which affected every type of German military activity towards the end of the war — including all types of flying. This problem mostly affected operational units, but also all kinds of factory activity that required engine starting. Another problem peculiar to the Luftwaffe was a severe shortage of pilots due to the collapse of the training organization and due to the high attrition rate of operational pilots. Thus, it happened often that there were no ferry pilots available to fly new aircraft to the operational units.

The problem of flying out completed aircraft from factory airfields was aggravated by the extremely harsh weather that dominated the winter of 1944–1945. The Germans encountered similar weather-related difficulties in the previous winter, but this time the general situation was far more serious.[12]

Aviation production also suffered indirectly from air attacks on other industrial sectors and on different infrastructure targets. One of the main problems created by late-war Allied air campaign and badly affecting war production was the ever-worsening transportation situation. Most of the German transportation system relied on traditional means of transport — railroads and inland waterways — which relied heavily on large and complicated infrastructure. Around three fourths of all freight traffic of the German industry went by rail. The Allies heavily bombed transportation-related targets in West Germany during the pre–D-Day air campaign. The accumulative effect of these attacks caused a meaningful decrease of rail traffic in the western regions of the Reich from mid–August. Air Marshall Lord Tedder, Eisenhower's British deputy, consequently urged the concentration of air attacks on the transportation system, but was opposed by advocates of the anti-oil campaign. In mid–September Tedder got his way and the transportation network became the second-highest priority target after oil. It was agreed that Allied strategic air power would join tactical air power in concentrated attacks against the transportation system inside Germany from the last week of September. The attacks first aimed at the inland waterway system, and from early October numerous rail targets were attacked.[13]

The offensive against the railway system reached its climax in January 1945, when around 30,000 tons of bombs were dropped on rail targets. This bombing effort was directed against a large and widely spread target complex, and caused a gradual deterioration of the transportation network. It was, however, not enough to bring German transportation to a complete standstill. In December 1944, American and British planning staffs began discussing a plan for a single-day all-out attack on numerous railway and waterway targets throughout Germany. The Allies also hoped to strike a major blow on German popular morale with this simultaneous massive demonstration of air power over parts of Germany that were never seriously attacked before. Internal prioritization debate and bad weather throughout most of winter 1944/1945 delayed the execution of this plan.[14] The attack, code named "Operation Clarion" was finally launched on 22 February 1945. For 24 hours around 9,000 Allied aircraft attacked hundreds of transportation-related targets. These included at least 150 marshalling yards. In addition, around 500 railway cuts were affected and around 300 locomotives were destroyed. "Clarion" was a crippling blow to the German transportation system, although the damage was easily repaired in many places.[15] Furthermore, Allied planes indiscriminately attacked towns and smaller cities during the operation and badly hit their civilian population.

Karl Frydag, the chairman of the Airframe Main Committee, argued somewhat simplistically immediately after the war that attacks on the transportation system were the main reason for the sharp drop in output from September 1944 onwards. According to him, particularly destructive were attacks on rail hubs and strafing of running trains by low-flying tactical fighters, which caused the Germans to abandon rail and road traffic in fair weather and during daytime.[16] Attacks on rail targets were especially effective, considering the fact that approximately 90–95 percent of the freight of most aviation firms was shipped on trains.[17] The wide dispersal of the aviation industry made it greatly dependent on reliable transportation and therefore extremely vulnerable to transportation disruptions. Smooth-running transportation was essential in order to provide factories with raw materials, parts, and components. Shipping out end products was also highly dependent on functioning transportation, because in many cases the end products — aircraft components or engines — were sent to their final destination by rail or by road.

The gradual degradation of the transportation system had therefore a dire effect on the aviation industry. Towards the end of January 1945, for example, production of the Me 262 came almost to a standstill because fuselages and engines failed to arrive at final assembly factories.[18] BMW's aero-engine production was also badly affected In October 1944, when its output, peaking in May with 1,450 engines, dropped to only 499. Until 1944 BMW experienced difficulties in fulfilling its production plans due to organizational and managerial deficiencies. This time the drop was related directly to the difficulties in supplying parts and components to the firm's main factories. Towards the end of 1944 it took freight trains up to 14 days to travel from middle Germany to the Munich area, where BMW's main engine factories were located. Sometimes the train never arrived.[19]

Junkers' Leopoldshall factory was one of those components manufacturers that suffered heavily from the deteriorating transportation situation. It produced control surfaces and various parts for different final assembly plants. Factory officials reported that transportation was the worst bottleneck in the late-war operation of this plant and that the factory suffered from frequent shortages of parts and components. This situation frequently forced the management to dispatch couriers with cars in order to bring some urgently needed parts.[20] Other factories of the Junkers complex were forced to rely on a strict priority marking system, couriers carrying vital parts as baggage on passenger trains, and using all the factory's available trucks. Towards the end of the war, as gasoline became scarce, electric powered vehicles were used almost exclusively to move parts and materials between the Junkers factories.[21] Furthermore, the use of couriers carrying parts in bags between factories and using whatever transportation means available became widespread in the aviation industry. This mode of operation received the unofficial name *Rucksack Aktion* (rucksack operation).[22]

Focke-Wulf was also forced to rely more and more on couriers just in order to bring the required raw materials and parts to its factories. Department directors demanded in December 1944 "draconian measures" in order to obtain more couriers for its Cottbus factory, because since early November production schedules became increasingly dependent on their efficiency.[23]

The deterioration of the German transportation system influenced the aviation industry also indirectly. During the winter of 1944 coal supply to power stations declined by 20 percent.[24] The deterioration of coal supply contributed to an increasing number of power cutoffs, which brought some ill-equipped factories to a standstill and caused production

A low-level attack by American fighters on a railway station somewhere in Germany (courtesy U.S. National Archives and Records Administration).

breaks. Even BMW's well-developed factories suffered from the escalating coal shortage starting in December 1944. During the first half of January 1945 most of the firm's factories stood idle due to lack of electricity and heating materials.[25]

The disruption of the transportation system was aggravated by earlier concentrated attacks on another key sector of the wartime industry: Oil. Germany always lacked its own oil resources and relied almost exclusively on oil import. Its oil situation became especially critical in summer 1944 after the Rumanian oil fields were destroyed and the Allies began systematically bombing the German oil industry. As mentioned above, oil scarcity had an immediate impact on the operations of the Luftwaffe and especially on its training activities. It also had an impact on the aviation industry. On 19 June 1944 the Jägerstab discussed at length ways to save aviation fuel. Colonel Ulrich Diesing, at that time head of the RLM's Planning Office, accepted the responsibility for finding ways to save fuel wherever possible. Among other measures, the Jägerstab ordered engine firms to cut back any unnecessary operation of aero-engines. It also ordered aircraft firms to cut acceptance flights of new aircraft to less than 20 minutes.[26] These fuel-saving measures contributed to the slackening of quality assurance, which required test running of engines and careful test flying of completed aircraft before delivering them to the Luftwaffe. Fuel shortage badly affected ground service vehicles in factories and airfields. Towards the end of October 1944, ground personnel at

Messerschmitt's Obertraubling factory airfield moved aircraft around primarily with horses and oxen because there was no gas for the tow tractors.[27] Fuel scarcity also affected motor vehicle road traffic, which became crucial later in the war, after Allied air attacks largely paralyzed rail transportation.

The most important measure to deal with air raids was through interception of the enemy formations — as demonstrated by the RAF during the Battle of Britain. It was basically the job of the Luftwaffe, but as Allied attacks intensified, some firms organized their own factory protection flights. They were composed of test and delivery pilots based in the factory's airfield, flying fighters set aside from normal delivery batches. Messerschmitt's Regensburg flight was one of the largest factory protection flights. It was established in mid–January 1943 and consisted of civilian firm pilots and Luftwaffe pilots detached temporarily to the plant as acceptance and delivery pilots. The flight kept two fighters at constant readiness and four more aircraft in reserve during daytime. The fighters were normally tasked to operate within a 30 km radius around the factory and were scrambled upon order from the local Luftwaffe ground control unit.[28] The military value of these flights was minuscule and they were unable to provide efficient protection for their factories. They could also be costly in terms of lives and aircraft. Several firm pilots and aircraft were lost during these "private" fighter operations. On 6 and 8 March 1944, for example, Focke-Wulf lost a test pilot and a delivery pilot from the Sorau flight when they tried to engage bomber formations heading towards their factory. Another pilot was shot down by escorting fighters, but managed to belly-land his aircraft and run to a wood before a fighter strafed his aircraft and destroyed it.[29]

Under these circumstances it is quite surprising that aviation factories were able to produce anything at all during the early months of 1945. In fact, at least until March 1945 output was surprisingly high. The aviation industry produced a total of 3,185 aircraft in January 1945 and 2,479 in February.[30] Total documented output for 1945 was 7,539 aircraft, among them 4,935 conventional fighters and 947 jet and rocket fighters.[31] Specific firms were more productive than others in this period. Junkers, for example, produced in January–March 1945 a total of 422 aircraft, most of them night-fighter versions of the old twin-engine Ju 88.[32] A former Focke-Wulf executive estimated immediately after the war that Focke-Wulf and some of its licensees produced around 4,420 FW 190 fighters in 1945 — 70 of them as late as April.[33] Although these figures seem to be too high, it is clear that the aviation industry produced several thousand planes in the last months of the war as everything around it was falling apart.

The aero-engine industry also continued to produce a substantial number of engines as Nazi Germany collapsed. The aero-engine division of Junkers produced in 1945 a total of 2,388 Jumo 004 jet engines and 2,955 Jumo 213 piston engines.[34] The MIMO factory at Taucha and its dispersals delivered 400 Jumo 213 piston engines in January 1945, 340 in February and 267 in March. The firm also delivered several hundred compressors for the Jumo 004 jet engine during these months. MIMO's output dropped meaningfully in March 1945 because of difficulties in receiving components from other suppliers, and because, at the beginning of that month, it began moving its Jumo 004 production facility into an extension of the Mittelwerk underground complex.[35] Under an ever-worsening-supply situation, some factories were able to continue production at a reasonable pace thanks to large inventories of raw materials and parts kept in their stores.[36] Firms that kept small or no inventories at all at their own facilities experienced a large drop of output due to lack of materials. These materials piled up at different depots as the transportation situation became desperate and then hopeless. As a result of lack of materials, some factories were forced to cancel complete shifts and send the workers home.[37]

The advance of Allied armies forced the Germans to evacuate some of their factories in 1944, but these were in most cases relatively minor production centers. Massive evacuation of factories began only once Allied armies, especially the Soviet Army, began to penetrate the heart of the Third Reich. Timely evacuation of a factory required early planning and a massive logistical effort in order to ensure complete evacuation of most important items and quick commencement of production at the new location. In many regards this wave of evacuations was a repeat of the earlier relocation scheme, but now time was much shorter and untimely evacuation meant the loss of a factory with its entire production capacity.

The experiences of Focke-Wulf in January–February 1945 demonstrate well the difficulties involved in the orderly evacuation of aviation factories under late-war circumstances. Focke-Wulf was the first firm to disperse its factories eastwards to East Prussia and Silesia, well beyond the range of British bombers in 1941. As a result, these factories were among the first factories that had to be evacuated in 1945 due to the advance of the Soviet Army. Otto Lange, an Armaments Ministry official and longtime member of the Jägerstab and the Rüstungsstab, supervised the firm's evacuations. Focke-Wulf tried to continue production as long as possible and to evacuate machinery, parts and semi-finished products to facilities located in advance. This exercise ended mostly with the loss of most of the production machines, much of the other equipment and most partially finished products.

The main reason for these losses was the catastrophic situation of rail transportation during this phase of the war. Even the move of medium-sized factories, like Cottbus and Sorau, required 300 to 350 train cars, and these could be hardly assembled in time.[38] When the relatively small Gassen tooling factory was evacuated at the beginning of February 1945, it was impossible to find the 60 train cars required for its move. The fact that Saur ordered in January the evacuation of Focke-Wulf's eastern factories—all of them involved in the production of the high-priority Ta 152 fighter—to central Germany, helped little when it came to train car allocation.[39] Speer's personal appeal to the local *Gauleiter* at the end of January to mobilize all the resources of the Nazi Party in this region in support of this important evacuation also failed to solve the transportation problem.[40]

As a result, each factory had to rely largely on road traffic. Air power in the East posed no serious threat to road traffic as in the West, but chaotic conditions caused much disruption and confusion. When the Posen factory was evacuated at the end of January 1945, the truck columns transporting a large proportion of the equipment traveled along roads clogged by retreating columns of military and civilian vehicles. Nobody briefed military commanders about the evacuation, and they refused to clear the roads for the loaded trucks.[41] Sometimes the logistical nightmare did not end upon arrival at the new location. When trains carrying equipment from Gassen reached one of its new facilities in a former porcelain factory at Altrohlau in the Sudeten, it was found out that there were no cranes available to offload the heavy equipment. As a result 120 much-needed train cars became stranded in the place for several days until a mobile crane was located and brought to the factory.[42]

These hasty evacuations required a high degree of managerial skills and leadership in order to succeed. As World War II drew to its end and the Third Reich was squeezed between the vises of Allied armies from East and West, there was not much point in evacuating factories.

In any case, late in the war many factories came to a complete standstill due to lack of parts, raw materials and power. Workers and machinery stood idle and waited for evacuation. During the last phases of the war evacuation of the machinery and equipment was impossible, and only the personnel were evacuated, if possible, before the plant was overrun. In some cases, factory managers lost their nerves and fled long before the arrival of Allied ground troops, leaving their workers behind. Dr. Anselm Franz, the Austrian director of the Junkers aero-engine division and chief developer of the Jumo 004 jet engine, fled from the main factory at Dessau two weeks prior to the arrival of U.S. troops, leaving behind workers extremely bitter over this action.[43] Some factory managers received orders to destroy their facilities as far as they could. Mr. Bock, the technical manager of Junkers' Tarthun underground factory, received from the local military authorities 24 aerial torpedoes and was ordered to use them to demolish his plant. He refused because he was afraid of collateral damage and the U.S. Army captured his factory intact.[44]

Shortly before the end of the war some destroyed or damaged smaller factories began converting to civilian production in preparation for the postwar era. A factory in Huchting, manufacturing parts for several firms, including parts for the Jumo 004 jet engine, started manufacturing metal furniture and continued doing so even under British occupation. A Focke-Wulf branch factory in Hemelingen was badly damaged in bombings, and after it was partially repaired it started manufacturing kitchenware, using sheets of aluminum and aluminum processing machinery formerly used to manufacture aircraft parts.[45] Some parts manufacturers simply returned to their prewar production. A U.S. intelligence team found out that the Hugo Danger metalware factory in Hamburg, which worked as a subcontractor to Junkers, Heinkel and Blohm & Voss, simply resumed its prewar production of bakery equipment.[46]

At the end of the war Germany's powerful aviation industry was largely gone. The Allies dismantled what was left and took away a large portion of the intact equipment and data as war booty. For several years after World War II no aircraft were produced in either part of divided Germany.

German Workers

In contrast to the foreign workers employed in the aviation industry since 1940, German workers were always there. Their daily life, however, changed considerably during the war. Besides rationing, bombing, and other issues effecting civilian life on the wartime home front, their working environment and working conditions changed dramatically throughout World War II.

One of the most important features of daily life in this period was the time workers spent at work. If earlier in the war working hours increased only insignificantly, in 1944 people spent more time at work than ever. One of the first measures taken by the newly-established Jägerstab was to increase the work-week of every firm involved in fighter production from 60 hours to 72, and to establish a mandatory two or even three shifts a day.[47] In some cases, when tight schedules had to be adhered to, German workers spent even more time at work. For instance, when Daimler-Benz's Genshagen aero-engine factory ran far behind schedule in October 1944, its management confined the German workers to the

A damaged Me 109G-6 at the Messerschmitt Augsburg factory, May 1945. Scattered incomplete fuselages of Me 410s two-engine fighters are also visible (courtesy U.S. National Archives and Records Administration).

floors of the factory for the last two weeks of the month. The workers slept on improvised straw mattresses laid on the floors of the factory and in offices, and spent all their free time in and around the factory until the job was done.[48]

Although there are indications that the general increase of work hours was not implemented globally and uniformly,[49] it nevertheless increased dramatically the time workers spent at work. It also increased their income. Naturally in wartime this increase was superfluous, because there were not too many opportunities to spend hard-earned money. Factories worked seven days a week from 1944, but workers were usually allowed one day off a week. The firms generally allowed this day off and the Jägerstab approved this practice in 1944 in order to enable workers to do their weekly shopping and take care of their personal matters.[50] Surprisingly, even towards the end of 1944, at least one firm — Focke-Wulf— allowed most of its workers three days off during Christmas and two days off on New Year.[51] Messerschmitt's Augsburg factory was forced to allow only half of its German workforce to have a Christmas leave, because its slave workers worked normally during the holidays and Germans were required to supervise them.[52] The way Focke-Wulf supervised its slave workers during the holidays is not on record.

German workers usually worked around a ten-hour shift each day, including a one-hour break. Shifts in underground factories were usually shorter. In some places, like in BMW's

Springen factory, located in a salt mine, Germans as well as foreigners and inmates worked daily in three shifts of eight hours each. As with other firms, BMW's executives also considered eight hours as the practical daily limit for human work underground.[53] Those working underground were usually allowed to leave the plant and breathe some fresh air during their breaks. All underground workers received somewhat increased daily food rations, amounting to 5 grams of fat, 50 grams of bread and 30 grams of meat.[54]

Basic gross wage rate for a skilled worker in 1941–1942 averaged somewhere between 0.99 and 1.39 RM per hour in most aviation factories. Salaries depended largely on the geographical location of the workplace. In big cities or near them, workers usually earned somewhat higher wages. As the war dragged on, wages rate sunk somewhat and practically reached their prewar rates. In December 1943 the normal gross wage went down to between 0.82 and 0.98 RM per hour, although there are some indications that in some factories the wages stayed higher at 1–1.85 RM. Women, who were always viewed generally as an unskilled workforce, earned between 0.30 and 0.56 RM per hour.[55] Income also depended on skills and position. Professional office workers averaged a monthly wage of around 800 RM, and engineers averaged 1,200 RM[56]; professional production technicians earned 277–307 RM and women 251–271 RM.[57] These wages were not so bad considering the fact that an unmarried fighter pilot earned at the same time only around 220 RM,[58] but these wages were only one-quarter of what an equivalent American worker earned. This income rate, however, enabled the average worker to decently support a family in wartime Germany.[59]

Overtime hours were paid according to the prewar regulations — 25 percent extra for over 72 hours a week (over 48 hours before the war), 50 percent for Sunday work and 100 percent for working on holidays.[60] Extra hours were only rarely paid following the increases of the workweek because most workers adhered to their prescribed shift plan, which was usually within the weekly working hours framework set for the industry by the RLM or the Armaments Ministry.[61]

Firms normally offered premiums to specialists or as a reward for excellence. Particularly important in this regard were the premiums offered to workers submitting improvement suggestions. This scheme was a general practice in the entire German aviation industry, although firms like Heinkel particularly encouraged their workers to submit suggestions.[62] The premiums for a serious submission were around 20 RM, and if the suggestion was found useful and was implemented, the worker could expect a premium of 100–150 RM. Suggestions which brought a particularly useful improvement of the product were rewarded with a 150–200 RM premium. The improvement suggestion scheme was generally useful and effective. As Focke-Wulf noticed, such suggestions led to the streamlining of the mass-produced FW 190 fighter production and reduced its cost. Focke-Wulf also remarked, though, that the premiums for improvement suggestions were kept relatively low, because the management disliked the idea of workers spending their time at work pondering new improvements as a way to increase their income.[63]

It seems that as the situation became more desperate, the Germans tended to rely even more on improvement suggestions submitted by workers and rewarding the successful ones generous premiums. At the end November 1944 the Armaments Ministry announced a general improvement suggestions drive in the aviation industry titled "Wir sparen Werkstoff" (We are saving raw material), to be carried out from 1 December 1944 to 15 January 1945. During this period the Ministry and aviation firms encouraged workers to submit improvement suggestions on how to save raw materials and improve productivity. The Armaments Ministry appointed Hermann Tödter of the Junkers firm to lead this drive, therefore putting at the helm someone from within the industry.[64] Tödter ordered an industry-wide stop of all other improvement suggestions in order to find new ways to reduce the use of raw materials in aviation production and to shorten production time. Each factory appointed its own director for the drive. These directors spread the news about the drive in different ways — including by spreading rumors. Tödter's staff prepared publicity material and disseminated it to the factories before the drive began. It even prepared recorded humorous commercials for playing on the factories' public announcement system. Each factory sent Tödter weekly reports about the incoming suggestions and about their usefulness. The Ministry offered rewards of up to 100 RM, extra cigarette rations and different commodities, like books, bottles of alcoholic beverages, shavers, women's purses, etc. These rewards were presented as the main motivation for more suggestions.[65] Although there is no summary report about the drive and its results, it is hard to believe that it brought a significant improvement so late in the war. In a later drive in February 1945, Tödter in cooperation with Heinkel launched "Aktion 50 Km schneller" (Operation 50 km/h faster). In its framework workers in factories involved in the production of the He 162 jet fighter were asked to submit ideas on how to add 50 km/h to the plane's speed.[66]

The relocation of large portions of the aviation industry to underground and dispersed locations added further difficulties to the daily life of German workers. Until after "Big Week" firms were reluctant to send workers from bombed factories to work in other factories of the same firm, because they did not want to separate them from their families. Instead, they usually let them clear the rubble and repair the damage — a task that could have been preformed by a workforce of completely different training.[67] Even when long-distance relocation took place, firms tried to relocate only those workers who were interested in doing so, as happened when Heinkel relocated large parts of its Rostock operation to Poland in late 1942.[68] This approach had changed once the general relocation had begun. When a factory was relocated, some of its workers were forced to move with the machinery. These cadres were composed mostly of specialists and other skilled workers. The regional pattern of the relocation supposedly eased somewhat the burden imposed on the workers and their managers, but in several cases the new factories were located many kilometers away from their original location. Housing and transportation solutions had to be found for the workers, but this was not an easy task, as some firms found out.[69]

Arranging housing for the moving workforce was perhaps the second most difficult challenge associated with the dispersal scheme, after moving the factories. It seems that housing arrangements for workers of the underground factories posed a particularly difficult organizational challenge. One of the main concerns was not to compromise the location of these factories by constructing anything new above the ground near them. The Jägerstab basically ordered that workers should sleep and live in the tunnels of their factories. Firms tried, however, to arrange normal living quarters for their workers in spite of this order. Werner Peres, the manager of the Askania underground instrument and optics factory constructed in a potash mine at Bartensleben, was not allowed to construct huts for his workers near the mine entrance in order to make it less conspicuous to aerial reconnaissance. He was therefore forced to arrange housing in surrounding villages, some of them more then 32 km away from the factory. This arrangement necessitated a special bus service to transport the workers to and from work each day.⁷⁰ Daimler-Benz arranged quarters for around 1,800 Germans working in its "Goldfisch" underground factory in private residences and hotels in the vicinity. The local Nazi Party authorities and the SS helped the firm with the required arrangements.⁷¹

It is interesting to note that the French firm SNCASO, which had earlier relocated large parts of its production to an underground factory near St. Astier, faced similar difficulties. SNCASO constructed a barracks complex at Perigeux for the approximately 2,500-strong workforce of this factory. Since the barracks complex was located away from the factory, the firm was forced to change the work schedule and to make special transportation arrangements—at the firm's expense—in order to allow workers to travel back to their home in Bordeaux each alternate weekend and have 3 days off.⁷² In 1944 the deterioration of the rail and road transportation in France made it difficult to bring the workers to work and back. In early July 1944 this difficulty became one of the main obstacles facing Focke-Wulf and the Jägerstab when they sought to add a second shift in this factory. The main cause of the problem was a massive increase in French guerrilla activity following the Normandy landings. Guerrilla fighters often sabotaged and attacked the rail stretch between Perigeux and the factory. German security forces in this region were weak and ill equipped. Mr. Elten, the Jägerstab representative in the factory, even tried to obtain an armored car in order to give the trains on this stretch some protection. He hoped to acquire an American vehicle from the "booty depots in Normandy"—a completely illusionary notion. Even when the trains were able to run to nearby St. Astier and back, delays of 2 to 3 hours often occurred.⁷³

Guerrilla warfare was nonexistent in Germany, but here relocation of workers within the industry also posed a great difficulty for the workers. Such relocations became common in the wake of "Big Week," when workers were transferred from bombed factories to the new dispersed factories or to other undamaged factories. Furthermore, reduction of types and shifting priorities under a tight centralized control left the workers little choice but to adjust to their new assignments. Thousands of workers became surplus due to these changes and were transferred to other factories within the same firm, or to another firm in need of extra manpower. In some cases factory managers failed to make surplus manpower available to other factories upon demand, and they were usually severely reprimanded for that.⁷⁴ When workers moved to other factories it often happened that they found the new location totally unprepared for their arrival. They often found out that no living quarters or feeding facilities were available in the new place. On-location mangers were also sometime slow in integrating the new arrivals into the production process, and the new workers sat idle for long periods of time.⁷⁵

If the move to a new factory away from home was not difficult enough, the harsher working regime in 1944 made it even harder. In numerous cases workers far away from home could not go home at the end of their shift and they were allowed fewer and fewer weekend leaves. From late August 1944 such workers were allowed to have a weekend leave only once in 4 weeks. In October 1944 such a leave was allowed only to workers who had to travel less than 100 km to their home. Those who had to travel longer distances were required to submit a special application to the personnel department of their factory and to obtain special permission.⁷⁶ These restrictions were imposed in the framework of a general leave suspension directive the Armaments Ministry issued in early September 1944. The leave suspension directive was also widely published in the press. Exceptions were made only in the case of older workers (from 65 years old for men and from 50 years old for women), of women whose husbands were on leave from their military service, in case of family emergency and in cases of illness.[77]

The working environment in the aviation factories deteriorated considerably in the last years of the war. It happened not only as part of the proliferation of the underground factories, but also in most conventional factories. The formerly modern aviation industry turned in 1943–1945 into the same bleak industry it strove to replace. Aviation factories became depressive workplaces. Hygiene and nourishment deteriorated gradually in most factories. The chief physician of Heinkel's Vienna factory resigned in August 1943 in protest against the unbearable conditions at that factory, caused mainly by bad nutrition and poor hygiene at the cantina. He pointed to a gradual decline in the workers' health and at the inability of the factory doctors to do their job.[78] Work environment considerations became largely irrelevant, as more production capacity was crammed into smaller spaces—especially following the dispersal of 1944. In stark contrast to the prewar trends, providing a comfortable working environment became largely irrelevant in factory planning and layout in 1944. As Saur has put it: "Especially in the underground factories I guarantee that we have to come to a different use of space, particulalrly in relation to height, width, productive spaces, etc. We must free ourselves from the 'Strength through Joy' mentality and the like. We must be very tough...."[79] Saur also hinted that the Germans had become spoiled and suggested they take as an example German productivity under more primitive conditions in World War I, and Soviet productivity in the current war.[80]

Among the depressive elements in many factories, including above-ground factories, was poor lighting. American intelligence teams noted after the war that lighting was very poor by American standards. Fluorescent lighting was found only in some offices, and light level in factory spaces was estimated to be less than half of what was considered essential in American plants.[81]

The introduction of conveyor-belt production lines and the increased mechanization of the production process also influenced the work environment. The level of noise became higher and in some cases almost unbearable. The predetermined pace at which the conveyer belt moved left many workers hardly time to breathe. Otto Liebl worked on the production line of Me 109 fighters at Messerschmitt's Regensburg plant and experienced firsthand the changing conditions: "The *Takt* method of production which allowed for a certain pause for rest was eventually replaced by a conveyor-belt method of production. As an electrician it was no simple matter to lay the masses of cabling in the restricted fuselages, and work safety hardly existed at the time. During the pre-assembly and metal construction phases, an indescribable noise pervaded...."[82]

The deteriorating working conditions made it particularly difficult for women working in the aviation industry. In some factories the number of women increased steadily and significantly. In other places the numbers of women decreased as the war dragged on and the situation at the home front became more desperate. In Heinkel's Oranienburg factory, the number of female workers declined as soon as an alternative workforce—namely foreigners and concentration camp inmates—was found. While in early 1942 women formed around 25 percent of the workforce there, all through 1943 their share stayed at around 15 percent.[83] In spring 1944, after the Oranienburg complex was bombed, an increasing number of female workers asked the local employment office for permission to resign from work in the factory, probably fearing more attacks. The employment office reportedly declined all these requests, but they point to a problem characterizing this workforce.[84]

It can be assumed that the main reason for the decline in the number of female workers cannot be attributed only to gender ideology, but to a more general tendency of women not to go to work in wartime Germany. Increasing bombing and other hardships of civilian life under the conditions of a total war aggravated this situation. In some places the authorities noted women's increased absenteeism due to illness.[85] Moreover, women tended to spend more time at home while their husbands, fathers and brothers were away with the army or at work. The number of female workers in the BMW factories in Munich peaked in December 1941 with 1,456 and then decreased steadily until in March 1945 only 848 women were left on the payroll.[86] This decrease seems to be directly connected to the fact that Munich was repeatedly bombed during the later stages of World War II. Even when women were employed, many of them decided to work only on a part-time basis. The Reich's Ministry of Labor was willing to tolerate such employment contracts in order to obtain more workers in those desperate times. On 17 April 1943 it ordered factories to organize half-day shifts in order to effectively employ women unable to work full days.[87] Several aviation firms soon implemented this guideline. Of the 100 women allocated to BMW Munich in April 1943, only 8 worked a full day. All the rest worked a half day, and it can be assumed that the main reason for that were different domestic obligations.[88]

The authorities and firms contributed to the low effectiveness of this potentially formidable workforce. They recognized, among other considerations, the need of female workers to spend time with their husbands or sons while on leave from the Wehrmacht. Focke-Wulf, for example, offered its female workers up to 18 days of paid leave a year for such happy occasions.[89] Such gestures were a clear sign of the willingness of the German war industry to compromise and improve the employment conditions of women just in order to fill the ranks.

This trend continued as late as 1944. The Armaments Ministry instructed factories allocated with extra female workers to arrange special training courses for them and to allocate special social workers in the factories to take care of their special needs.[90]

Increased absenteeism was definitely not peculiar to female workers. It proved to be a major problem in wartime German industry and it also affected the aviation industry. Absenteeism can be attributed to low morale, daily hardships on the home front, emergencies created by bombing, or a mix of these elements. Allied bombing raids formed a central factor contributing to absenteeism by influencing several domains of private life. Even early RAF night attacks on German cities caused surges of absenteeism, "illness" reports, lateness, leaving work early and what was generally termed as "laziness" (*Bummelei*).[91] This tendency increased as the Allied bombing campaign against German intensified in 1943–44. In 1944 many firms seriously viewed the slackening discipline of their workers, and directors repeatedly issued circulars and announcements aimed at fighting this problem. Focke-Wulf, for example, issued an announcement on 3 September 1944, in which it reminded its workers the sort of behavior expected from them. The management allowed the firm doctors to overrule doctors from outside, and workers were warned to stick to work and break times. They were also urged to take into account train delays due to air attacks and to leave home earlier on their way to work. In case of delays, they were expected to bring a special confirmation note provided by the *Reichsbahn*. Violators were threatened with disciplinary action. The announcement also made it clear that: "The gravity of the situation leaves no more room for recurrent reprimands, but requires the carrying out of the [above] measures in a form similar to what the soldier serving at the front takes for granted."[92]

After the war it was noticed that successful Allied attacks on aircraft factories caused not only short- and long-term material damage; they also increased absenteeism. After Focke-Wulf's Marienburg plant was effectively attacked on 9 October 1943, there was a significant increase in the number of absentees in the following days. On the day after the raid, only some 100 workers of the plant's 669-strong workforce came to work. On the second day, out of 497 male German workers, only 208 came to work, and out of 172 female employees only 59 appeared. It took an entire week to bring all the workers back to work.[93] It was also reported in several cases late in the war that workers failed to return to work after an air-raid "all clear" alarm was sounded.[94]

Absenteeism peaked in the BMW aero-engine production and development centers in Munich in August 1944. It reached almost 20 percent and can be attributed to several bombing raids on BMW's factories and on Munich in July. The annual average of absenteeism rates in these factories in 1944 reached almost 15 percent.[95] It was estimated that absenteeism due to all reasons averaged 10–13 percent at Junkers Aschersleben factory.[96] It was also estimated that absenteeism due to different reasons caused a loss of 8 percent of the monthly work hours at Fieseler's factories in 1944–45.[97] Ironically, plant managements of several firms indirectly encouraged and approved absenteeism by allowing up to 14 days special paid leave to workers whose homes were damaged in the bombings.[98] This so-called "bombing leave" (*Bombenurlaub*) was strictly regulated and was given only to workers whose own homes were damaged, but sometimes workers misused the leave. As Focke-Wulf clearly instructed its workers, "After taking care of the family and home the interest of keeping production in the factory stands in the foreground, so excuses like helping relatives or friends can not be accepted, even if the need for mutual help must be fully recognized."[99] The amount of productivity lost due to such absences can only be guessed.

Until early 1944, firms usually punished absenteeism and other misconduct related to presence at work with fines of up to one week's salary, depending on the severity of the offense. From March 1944 absenteeism and idleness were punished by withdrawal of the extra rations provided to workers in the aviation industry. In some firms workers who failed to turn up on Saturday afternoon or Sunday night — two shifts marked traditionally by increased absenteeism — were punished by withholding their extra rations for up to one week. In some exemplary cases absentees were sent to a Labor Education Camp, as happened to one German female typist working at Daimler-Benz's Genshagen factory in November 1944.[100] Such camps had been established since late 1942 next to large factories and came under SS control. Their main aim was to combat "idleness" (*Bummelantentum*). People sent to them continued to work in their factories but lived in the camp and could be forced at any time to perform hard physical labor.[101]

The health-care departments established earlier in most factories continued to work during the war. The aviation industry found it difficult, however, to continue its prewar medical care standards, since doctors and other staff members were conscripted while the number of workers steadily increased.[102] However, the prewar system was put to an additional good use, since it helped to prevent absences from work due to medical treatment. Speer issued in late 1944 a decree which ordered workers of the armaments industry to use only the medical facilities of their factories. This decree excluded dental treatment, but some factories took the initiative and urged their workers to visit only the factory dentist.[103] In some factories health-care centers became a "watchdog," looking for any malingering or other troublesome behavior. Junkers' doctors examined all sick employees and looked specifically for fakers. If the employee was thought to be faking, and if it was the first time, he was warned. On a second time the employee was threatened with disciplinary measures, and on the third time he was either fined, or in more serious cases, reported to the *Gestapo*. Junkers reported that only a few such cases were ever reported,[104] but the activities of Focke-Wulf's firm doctor suggests that reporting troublesome employees to the *Gestapo* was a regular practice.

In Focke-Wulf's Bremen main factory, chief doctor Herbert Warning turned his 60-some-worker-strong department into a policing organization, which monitored mainly foreign workers, but also Germans. Warning was obviously an ardent National-Socialist. He wrote an article about the racial struggle of the Reich for the firm's newspaper in early 1943.[105] He regularly recommended general director Kurt Tank to take more draconian measures against misbehaving workers, and he carried out his monitoring energetically.[106] In 1941, for example, he reported that his department investigated 1,120 cases of laziness at work (*Arbeitsbumelei*). According to his report, 16 workers were verbally reprimanded and 1,104 received a written reprimand. From those, 298 were reported to the *Gestapo*, which sent 119 of them to labor camps.[107] It is obvious that Warning's brutal approach was not only based on his National-Socialist conviction, but also upon a practical need to deal with counterproductive tendencies among the workers. As he pointed out, these problems stemmed from low morale due to the long duration of the war, the evacuation of children from Bremen to the countryside, air attacks and rationing.[108]

By summer 1942 the disciplinary problems with the German workforce were serious enough to reach Milch's desk. Milch was shocked by the number of reports about constant lateness, absenteeism, etc. On 26 August 1942 he demanded extreme measures to be taken against problematic workers. He urged to call in the SS to deal with the problem because "[the workers] have not been condemned and in no way violate the existing laws, but only act against their country, which certainly has not yet got rid of the old legal nonsense. That is why Himmler should get these people into his clutches, because he can treat them outside the law."[109] Milch suggested issuing offending workers a special numbered ID card, in which their offenses would be registered. They were to be put subsequently under close surveillance of the Reich's Security Service. There are no indications that this measure was implemented, and troubles continued.

In October 1943, after several months of heavy bombing raids, Hitler expressed his dislike of workers' not showing up to work for several days after an air raid. This misconduct was especially unacceptable in cities like Wiener-Neustadt and Regensburg, where only the factories were hit and where collateral damage to residential areas was minimal.[110] After "Big Week," Milch repeated his brutal threats to execute anyone caught sabotaging the war effort. He mentioned the story of a commander of a construction battalion sent to Regensburg to clear the ruins at the Messerschmitt plant. Because he was unsatisfied with the accommodations offered to his group, he refused to start working. Milch said that such people must be brought before a court-martial and summarily hanged.[111]

As far as is known, workers were never executed, even after the security of the aviation industry was entrusted to the *Gestapo* in late 1944. However, late in the war the *Werkschutz* (factory security) in different factories was entrusted with over-seeing discipline and dealing with minor offenses as lateness and sloppiness. Entrusting these tasks to the *Werkschutz* was supposed to signal to the workers the seriousness of the situation.[112]

In several factories security was partially entrusted to a nearby *Gestapo* station, or alternately, a small *Gestapo* team was detached permanently to the factory. In Daimler-Benz's "Goldfisch" factory a small *Gestapo* staff occupied a barracks next to the entrance to the tunnel system. Its main task was to keep a watchful eye on the foreigners and the concentration camp inmates working there, but it also monitored German workers. Among other matters, in September 1944 it advised the firm's management on how to deal with widespread theft of light bulbs from its factories. While the firm considered using new light bulbs that would be more difficult to dismount, the SS suggested dealing with the matter more actively through "Kontrollen der Gefolgschaft" (controlling the workforce) and "Bestrafung der Glühlampendiebe" (punishing the light-bulb thieves). The directorate declined these suggestions and pointed out that most theft was probably done by inmates, and therefore stricter control of them should solve the problem.[113]

There were also genuine efforts to counter disciplinary difficulties with more positive measures. Absenteeism reflected to some extent low morale due to the general war situation, and in many factories morale-boosting programs and events were organized for the workers. Leading Luftwaffe aces were often invited to give morale-boosting talks in factories. Usually special efforts were made to bring them to the factories producing the aircraft they were flying. Top ace Major Walter Nowotny, for example, who flew FW 190 fighters, visited Focke-Wulf's Bremen factory in December 1943. The firm's general director Kurt Tank hosted him as he gave several morale-boosting talks to the workers. It was reported that his speeches were received with great enthusiasm.[114] In 1944 high-ranking Luftwaffe officers and famous war heroes, like Adolf Galland and Hannes Trautloff, visited mainly bombed factories and gave morale-boosting talks to their workers.[115] The propaganda machinery highly publicized these aces, and in wartime Germany they became famous celebrities. Their tours in factories formed a continuation of their propagandistic appearances in the press, newsreels and radio broadcasts. Propagandistic articles published in internal firm newspapers also sought to boost morale by repeating the old community-building rhetoric used before the war. Other articles published the combat accomplishments of the aircraft manufactured by the firm.[116]

The prewar social welfare activity of the factories and of the *Deutsche Arbeitsfront* (German Labor Front) continued during the war and was also used to reward workers. Social welfare helped to keep workers at work by taking care of some of their problems when possible. Sometimes individual factory managements expanded their individual prewar welfare plans in order to cope with the new challenges. The Arado factory in Babelsberg — located on the southwestern edges of the Berlin metropolis — sought to improve the social and medical services it offered to its workers. The factory offered free of charge clock and wristwatch repair services in a special workshop in the factory, and the factory's cantina served hot meals twice during a single shift. A special welfare center was established in order to help workers affected by bombing to deal with their difficulties. Furthermore, the health-care package offered within the factory was enlarged to include massages, X-ray, lab services and dental treatment.[117]

In most factories workers obliged to work more hours each week were offered extra rations of cigarettes and food as compensation. Soon after the establishment of the Jägerstab, Hitler approved the allocation of extra food rations and clothing to workers involved in its production programs as a form of social welfare.[118] Dr. Birkenholz was responsible within the Jägerstab for social welfare. He dealt with providing an extra supply of luxury items to the hard-pressed workers of the aviation industry in order to motivate them. As Birkenholz put it: "The additional social walfare has proved itself also here as a way to increase the joy of work, because humans work happily when they receive gratitutde and appreciation for their work."[119]

Generally the Jägerstab approved the allocation of extra rations to every male worker engaged weekly in 69 to 72 hours' fighter production-related work. The required quota for women was 66 weekly hours. Theoretically the weekly extra allocation included 50 grams of fruits, 30 grams meat, 15 grams fat and 150 grams bread or flour. Birkenholz estimated in late May 1944 that around 350,000 workers regularly received these extra rations.[120] The cantina's kitchen of each factory received the rations through the regional food supply centers and was responsible for distributing them the eligible workers. The Jägerstab approved and provided the extra rations according to weekly reports submitted by the firm.[121] In some cases extra rations of cigarettes were distributed to the workers of bombed factories — in some cases immediately after the attack — in order to improve their morale during the ensuing rubble clearing and repair work.[122] In some instances, adequate supply of clothing proved to be a problem. After visiting Junkers' Bernburg factory in early March 1944, the Jägerstab found out that although low-rate production resumed in this factory only four days after it was bombed, workers working in roofless structures in frosty cold lacked warm clothes.

"Strength through Joy." Heinkel employees during a social evening in Mielec. Slave workers of the same factory spent their evenings in a different way (courtesy Yad Vashem Archives).

In a matter of several hours Birkenholz organized a supply of warm clothing for this factory.[123] The special benefits offered to workers of factories involved in fighter production obviously aroused complaints from other sectors of the armaments industry, who did not receive similar benefits despite their hard work. Hitler therefore reapproved the preferential status of the fighter industry towards the end of May 1944 and pointed out that a similar motivation scheme would be impossible to implement industry-wide.[124] It should be stressed again that the benefits were also used to discipline workers. Their withdrawal largely replaced fines as a punitive measure available to the factory management. This change reflected the importance of wares and food in 1944 over money, as at a time of shortage and rationing, money could not buy much food and luxury goods.

Besides the increasing daily hardship, working in the aviation industry in the last years of the war could be quite dangerous. Strategic bombing brought the war to the home front in its utmost brutality and working in an aviation-related factory meant working at a potential target for quite precise USAAF daylight strikes. Since RAF bombings happened almost only at night and were generally not aimed at specific targets, they posed a danger to the workers mainly while at home. USAAF bombing raids usually caught the workers during the morning shift, towards midday. Many factories were attacked repeatedly, especially in 1944. MIMO's aero-engine factory in Tauche, for instance, was attacked nine times in 1943 and 1944, and around 850 bombs fell on its premises. These attacks destroyed 50 percent of its production space and killed 37 Germans and 25 foreigners.[125] In 1943 most factories were ill prepared to deal with air raids. Messerschmitt's Regensburg complex was thought to be beyond the range of Allied air power, but some air-raid precautions were taken. A siren system was installed and air-raid shelters were prepared, but workers were not allowed to leave the factory's grounds in case of an air-raid alarm. As a result, the 17 August 1943 bombing of this important plant killed 395 workers and wounded 754 in one of the deadliest raids on an aviation factory.[126]

The Eighth Air Force's raid on Focke-Wulf's Marienburg factory on 9 October 1943 also struck an ill-prepared plant, in which no alarm was sounded before the bombs started falling. The attack caused heavy damage, and since no warning was given, most of the workers were caught in unprotected areas. Consequently 114 people were killed and 76 injured.[127] Even after the Germans became aware of the seriousness of the threat, some raids caused spectacular loss of life. Some 450 workers were killed — mostly in their slit trenches — by an accurate USAAF bombing of the Erla factory complex in Leipzig on the first day of "Big Week." Tens of shocked survivors of the raid fled the area and refused to return to work. Milch, who visited the site a couple of days later, ordered: "In the future as soon as air-raid alarm sounded the entire factory staff was to form up in column of threes and 'march singing' out of the works to a distance a kilometer away.... They can dig their slit trenches there and watch their factory, and then if necessary return for rescue, salvage and fire-fighting operations."[128] It seems that Milch's orders saved many lives, because when the USAAF attacked Messerschmitt's Regensburg factory on 25 February 1944, the entire staff was evacuated more than 2 kilometers away, and as a result only 5 workers were hurt.[129] Timely warning and evacuation also saved most of the workers on 21 July 1944, as the plant was bombed again by 134 Flying Fortresses. The plant was heavily damaged and 350 people became homeless — including 300 Italian POWs, whose barracks camp was destroyed — but only two persons were wounded.[130]

After the initial wave of attacks in 1943, air-raid shelters were hurriedly constructed in most factories and other measures were taken in order to ensure sufficient early warning and efficient evacuation of the workers to the shelter in case of an air raid. Low-level attacks by fighters became widespread during the last six months of the war and added to the hazards. Late in the war the air-raid warning system sometimes malfunctioned and in numerous cases the alarm sounded too late. A late alarm could mean that a large number of workers would be caught in the open, especially by low-flying fighters. As a result it was decided not to sound the alarm at all in case the warning came too late and workers were ordered to stay in the factory buildings.[131] Strafing of groups or individuals by low-flying fighters became such a hazard that the air-defense director of Focke-Wulf issued on 9 November 1944 a general order to all workers not to wear light-colored work garments outside factory buildings in order not to attract the attention of low-flying Allied pilots. Violators of this order were threatened with disciplinary action through the police.[132]

The continued bombing and Allied air activity over the Reich was an obvious sign that something was wrong with Germany's battlefield performance. This reality, together with news and rumors about other defeats, heavily influenced the German public mood towards the end of the war. Just like other Germans during World War II, workers listened to BBC news broadcasts from London and were thus aware of the grim military situation. Some of them even told foreigners and inmates working with them what they heard.[133] Because of their special position within the German war industry, workers of the aviation industry were sometimes able to obtain particularly sober "insider" information about the military situation. In early November 1944, for example, it was reported in a Rüstungsstab meeting that special teams sent by firms to operational units to perform field modifications on different types of aircraft were amazed to find most aircraft standing idle in camouflaged wood dispersals because there was no fuel to fly them. These teams returned to the factories and spread wild rumors about the dire situation of the Luftwaffe. Workers were especially depressed to hear that the aircraft they produced with so much effort were sitting idle on the ground.[134] It was also reported that workers involved in the production of the Me 262 fighter were especially annoyed by the fact that it apparently made no impact on the air war and that only small numbers of these aircraft actually reached operational units. Furthermore, very few Germans actually saw this aircraft and therefore regarded it largely as a creation of the propaganda machinery.[135] Messerschmitt technical personnel supporting units that operated the Me 262 encountered much criticism regarding the wartime leadership of the Luftwaffe. They also learned about the largely failed operation of the Me 262.[136] The German leadership was aware of such signs of low morale in one of its key war industries. In mid-January 1945 it was noted in a Rüstungsstab meeting that twenty Ta 152 fighters produced in December still stood idle at their factory airfield. Besides being concerned by the failure to deliver these valuable planes to operational units, members of the staff were also concerned by the effect this failure was having on the morale of the factory's workers.[137]

Defeatism was by no means restricted to lowly workers. Eberhardt Schmidt, a Messerschmitt executive allocated to work at Sauckel's Kahla factory staff, took his brother's and his own family in a firm car across the Swiss border one day in late December 1944. He explained to the Swiss authorities that he had no problem with the regime, but that he was afraid for the life of his family in case of a German defeat.[138] Directors fleeing their factories late in the war also reflected this sort of low spirit among some executives. There were cases when the unsuccessful execution of a production program had dire effect not only on the workers but also on the local population in the vicinity of the factories. The termination of the Ta 154 program in June 1944 badly affected the workers of Focke-Wulf's Detmold factory, one of the main production centers designated for this aircraft. As the factory manager reported, the workers of Detmold became very depressed following the cancellation since nobody could understand why this fine aircraft was terminated. The consequent transfer of workers to other factories involved in priority production also did little to improve moral. The story of the failure and its consequences became widely known in Detmold and its vicinity, and local people used to ask each other sarcastically, "Arbeitest Du oder bist Du bei Focke-Wulf?" ("Are you working, or are you with Focke-Wulf?"). Some civilians also openly expressed the wish that the factory would be moved somewhere else because it could attract air attacks and put their community in harm's way.[139]

Towards the end of the war, advancing Allied land forces threatened most factories. In order to protect the factories from ground attack the armaments industry and the local military authorities ordered each of them to established its own *Volkssturm* unit from fit German workers still there. Big factories were able to establish quite large *Volkssturm* units. Messerschmitt's Augsburg factory became the proud owner of a full *Volkssturm* battalion comprising four companies.[140] The military value of these units, however, was negligible, like the entire *Volkssturm* scheme, and there was no way these ragbag outfits could seriously resist Allied ground troops.

During the last weeks of the war, aviation factories ceased production one after the other due to lack of raw materials and due to evacuations. Most of the staff left them, as there was nothing else to do. As a result, by the time the Allies captured these factories they found in them only skeleton staffs and some representatives left behind to destroy different pieces of equipment. Other German workers simply went home or were recruited by one or another military organization to put up a last resistance.

Foreign and Slave Workers

The composition of the workforce of the aviation industry changed dramatically during World War II. In 1944 most aviation factories became a modern-day version of the Tower of Babel. Their workforce was composed of groups of different nationality and different ethnic identity. In 1943 BMW, for example, employed in its Allach factory — where Germans formed a minority right from the start — widely diverse groups of workers: concentration camp inmates, SS stockade prisoners, Soviet POWs, Eastern workers and foreigners from around 30 nations, mostly west Europeans.[141] Eva Bohcher, a Slovak Jew, described the workforce she encountered upon arrival at the Arado aircraft factory in Freiberg in autumn 1944: "There were around 3,000 foreign workers there at that time, among others from Belgium, Italy, France and Russia. The Russians were POWs recruited by the Germans. There were 1,000 Jewish women: 500 from Hungary and 500 from Slovakia."[142]

While usually sharing the same workplace with German workers, daily life and routines of these foreigners and prisoners were quite different. Looking into the daily life of the foreign and slave laborer enables us to answer several important questions. Perhaps the most important question is the prospect of survival for individuals forced to work at the aviation industry. This question is closely related to another question: how did the Germans treat these groups? The recruitment of slave labor for work in the German aviation industry cannot be viewed as a direct continuation of the Nazi "annihilation through work" scheme used in 1941–1942 mostly in order to get rid of the Jews. In 1943 and 1944 the inmates working in the aviation industry were under brutal SS custody, and theoretically still lived within the camp system, but the long shifts on the factory floors offered them some refuge into normality and outside the bubble the SS created around them. Therefore, it is reasonable to assume that working in the aviation industry offered those inmates somewhat better living conditions and therefore better chances of survival.

Furthermore, looking at daily life patterns of inmates and foreigners offers us an insider view into the operation of the German aviation industry during the later phases of World War II.

Although this chapter deals with the daily life of foreigners and slave workers, there was usually a clear hierarchy not only between these two general categories, but also inside them. Foreigners were placed generally higher in the hierarchy, especially if they worked willingly for the Germans. Among the foreigners there was further hierarchy. Workers coming from Western Europe usually received a better treatment and better conditions then their eastern colleagues. More hierarchy existed among inmates. German criminals were usually at the top, while Jews and Gypsies were at the bottom of the hierarchy as viewed both by the SS and by other inmates.

Initial ventures of individual firms into using slave labor were largely motivated by business considerations. The primary benefit of using this type of workforce was its cost-effectiveness. Compared to early foreigners, who received a salary on almost the same level as contemporary German workers, slave workers were extremely cheap. According to a Jew who worked in the accounting department of the Mielec factory, Heinkel paid the SS 5 RM a day for each worker. This payment was the normal payment the SS demanded at that time from firms employing inmates from its camps. It was much lower than the payment of the Polish workers working in the same complex and served as an incentive to gradually replace Polish workers with Jews.[143] Thus in 1942, while most factories operated one shift a day, slave workers working at Heinkel's factories, including Mielec, worked in three shifts of 8 hours each.[144] The labor camps in the General Government in 1942 were not yet part of the established concentration camp network, but the harsh living conditions prevailing in them were equivalent to those in "mainstream" concentration camps.[145] Sidney Birnbaum, a former inmate in Mielec, described the day's routine in this camp-factory:

> Regular working hours were, I used to get up at six and I used to get on about 7, 7:30. I used to work to 18:00, 19:00. Then I used to come back to the camp. Came back to camp, you used to get your soup or whatever they gave you at the time. If they had any extra work in the camp, they used to catch a group of men, boys and they used to work in camp.... Every day we had roll calls.... At night also ... we used to eat only when we came back. In the morning we used to have coffee and a piece of bread."[146]

From 1943 two shifts a day become a standard in the aviation industry. Inmates usually worked on a 12-hour-shift basis, seven days a week, which increased their monthly work hours. While in March 1943 each worker worked an average of 175 hours each month, during the first quarter of 1944 the figure for slave workers increased to an average of 235 hours, while the average figure for German and foreign workers stayed at around 183 hours.[147]

In some places, where production was behind schedule, inmates worked much longer. In Daimler-Benz's Genshagen aero-engine factory, around 1,000 female inmates from the Ravensbrück concentration camp worked in winter 1944 up to 14.5-hour shifts and sometime 100 hours per week. This intensified work schedule was expected to cover the firm's backlog in supplying 800 new engines by the end of October 1944, but it failed and eventually Genshagen delivered only 410 engines.[148]

Such schedules meant long miserable hours spent by slave workers in the production halls of the German aviation industry. If they were lucky, their camp was located right next to the factory and therefore they were spared long trips — sometimes on foot — to their workplace. In other cases, however, the inmates were accommodated in camps far from the factory and spent hours on the way to and from their workplace. Generally, the inmates' day was long and tedious. At Arado's Wittenberg factory (which formed a sub-camp of Sachsenhausen) in late 1944, inmates woke up each morning at 4 o'clock, went to the roll call, and stood there in the bitter cold while SS guards counted them. Then they marched for around an hour to the factory. In the factory they worked for 12 hours, performing mostly exhausting physical work. In the evening they marched back to their camp, where they received their meager daily meal.[149] At the "Bergkristall-Esche 2" underground factory, inmates from the nearby Gusen II camp also spent in early 1945 a total of 14 hours at work or on the way to their workplace and back. Inmates worked one week in the day shift and every other week in the night shift. The tunnels were very cold and some inmates stuffed old bits of paper into their clothes as an insulation layer. Peter Erben, a young Jewish Czech student, worked in the factory as a quality inspector, doing quality assurance on completed Me 262 fuselages. He was given an inspection lamp that generated some heat and helped him keep himself warm. At the end of the day he and his fellow inmates were taken back to the camp in an industrial train, received their daily food rations and went to sleep.[150]

On the production lines and workshops a single German foreman or supervisor usually supervised a group of inmates. At Heinkel's initial operation in Oranienburg each German supervised around 10 inmates. In April 1943 there were around 600 Germans and around 4,000 inmates working at Oranienburg. In parts production, 20 percent of the workers were Germans, while the general average in the plant was only 10 percent. Although it was mentioned that some inmates, especially from the criminal category, became foremen and supervised their own groups of workers, it is clear that the Germans also closely supervised them.[151] This form of employment — an inmates/foreigners group led by a German — was used whenever inmates and foreigners were employed.

Inmates worked in practically every part of the production process. Here, again, Oranienburg set the benchmark and the role model for the rest of the industry. Inmates worked at the parts manufacturing departments, at parts assembly, component assembly and system installation. Reliable inmates were even allowed to work as quality inspectors in the quality assurance department. Even positions that demanded specific training, like welding, blueprint drawing and jig-making, were soon occupied by inmates.[152] Later in the war foreigners and even non–Jewish inmates trusted by the Germans became supervisors.[153] In situations, where more than 80 percent of the workers in a single workshop were foreigners, like in Daimler-Benz's "Goldfisch" underground factory, there was no choice but to let them perform most of the production tasks at hand.[154]

While the allocation of forced labor in 1942–1943 was rather well organized and within the normal (and extremely brutal) concentration camp routine in most cases, the intensification of slave labor in 1944 somewhat changed the picture. As early as 1942 Heinkel tried to screen its foreign workers in order to find the most appropriate job for them. They were asked to fill different questionnaires and take a psycho-technical exam in order to establish their professional abilities. Following the test the new workers were divided into four groups, with the highest cleared for complicated tasks and the lowest allocated to jobs that required no skills at all.[155]

In 1944 the criteria for the choice of workers was lowered in order to fill the quota. In some concentration camps the Germans tried to locate inmates with some kind of technical skill or experience. In autumn 1944 Daimler-Benz experienced a severe shortage of technical personnel and asked the SS for help. The demand went through Kammler's staff to different concentration camp commanders, which subsequently tried to locate technically skilled workers among their inmates.[156] Sometimes firm representatives appeared in camps and sought skilled inmates, as happened to Josef Pinsker in Buchenwald in October 1944, when a Junkers representative sought skilled inmates for a factory in Niederorschel. Pinsker claimed that he knew how to operate a lathe — which he did not — and was picked right away.[157]

This practice was quite common and sometimes the representatives came looking for inmates possessing specific professions. Sometimes firm representatives evaluated candidates from among the inmates with an exam and an interview, or even asked them to demonstrate their ability to perform different technical tasks.[158] Volkswagen sent one of its engineers, Arthur Schmiele, to Auschwitz in late May 1944 to locate skilled workers for the V-1 production. Schmiele interviewed inmates that responded to the call for skilled metal workers and also asked them to perform different technical tasks in order to assess their abilities. According to the results of the interview he divided those who passed into two groups. The first, with around 300 of the most successful interviewees, was sent in June to Fallersleben, where its members received additional training. They were then transported to the new underground factory in Tiercelet, where they formed the initial cadre of slave workers. The others were sent directly to Tiercelet, where the former group later integrated them into the work.[159]

In other instances in 1944, inmates were picked out of the crowd just to fill quotas. SS men pulled women, elderly, youths and even children out of transports, roll calls and work detachments and hurriedly transported them to their new workplace. In several cases the Jägerstab noted that these workers came barely clothed and barefoot to the factories.[160] The integration of such inmates reduced immensely the efficiency of the slave labor system. This situation motivated some firms involved in construction projects for the Jägerstab to measure the efficiency of different groups of workers using the efficiency of the average German worker as a yardstick. While considering German workers as 100 percent efficient, it was determined that foreigners were generally 83 percent efficient, Italians 71 percent and Jews only 33 percent. In other words, it took 3 Jewish inmates to accomplish the work done by a single German. Since from mid–1944 an overwhelming number of Jews were employed on the construction sites of the aviation industry, the Germans needed an ever-growing number of them to accomplish construction schedules and tasks associated with underground and bunker factories.[161] A similar efficiency level was noted also on normal production lines. Daimler-Benz found out after it started employing 1,000 female inmates in its Genshagen factory that it took these women three times as long to accomplish different tasks performed by German workers.[162]

Since most inmates and foreigners preformed technical tasks that required, some skills, some sort of retraining was required not only for the unskilled workers. Those inmates who possessed one or another technical skill, but never worked in aviation production, also required some retraining. In Mielec and probably also in Oranienburg inmates received their training while working. A former inmate in Mielec gave the following description of the way Heinkel trained its inmates:

> Not a single person in our group was a craftsman.... We were told on the first day that they would train us for work in the aircraft factory and that those who fell behind would be "trained" by the guards. Each morning we stood at the roll call and were then divided into work detachments. Some went to the wing department, some worked on fuselages and others worked on engines. Small air-hammers were used for riveting. Some of us learned the profession; others had it the hard way.[163]

Later it became customary to give foreigners and inmates brief training before starting to work. In Messerschmitt's Regensburg plant even female workers from the occupied East received training in drilling, filling, riveting, etc. from one of the German female employees.[164] BMW's Allach factory trained its inmates from the early phases of its operation. Newly received inmates from Dachau — most of them skilled or half-skilled — were first sent to training workshops allocated especially for this purpose. They spent between 4 and 10 weeks learning different metalworking skills under the watchful eyes of SS guards. Around 2,000 inmates received this training before the war's end.[165] Female inmates arriving in 1944 to Junkers' aero-engine factory in Markkleeberg were trained by firm technicians, who were sent from one of the main factories to train some of them to take micrometric measurements. Only after finishing their training did the inmates start working on a production line of parts for the Jumo 004 jet engine. It is important to note that most of the female inmates engaged in the production of this highly modern product in Markkleeberg were Jewish housewives and schoolgirls from Hungary — arguably the most unskilled workforce one can imagine.[166]

The skilled inmates collected in Buchenwald in October 1944 were sent to work in a Junkers dispersal factory at Niederorschel, also known as Langenwerke AG. This factory produced wing sets for Ju 88 night-fighters and FW 190 fighters. The management sent the supposedly skilled inmates immediately to work. However, the Germans soon found out that most of the inmates lacked the special skills required for work with aluminum, and the management was forced to pull the inmates out of work and train them.[167] Even in late 1944, inmates were sometimes trained before being put to work. Several hundred Jewish women arriving at the Rochlitz labor camp, a sub-camp of the Flossenbürg concentration camp, received technical training at the facilities of the Mechanik GmbH firm. They learned how to use production machines before being allocated in January 1945 to work in a factory producing propellers.[168]

Initial intentions to separate slave workers from Germans mostly failed due to practical reasons. As a result, Germans, inmates and foreigners normally worked next to each other on a daily basis. Theoretically, contacts between Germans and inmates were supposed to take place only through their guards or *Kapos*. In some places factory management showed racist and brutal attitudes towards workers belonging to "inferior" races and demanded their workers not to contact them directly and not to fraternize with them. Heinkel choose not to employ German female workers in production halls, where inmates worked. As a result of this policy, inmates working in such halls preformed easy tasks usually preformed by women.[169]

Sometimes German workers were disciplined for being too friendly towards inmates working with them. In Henschel's Rzeszow aero-engine factory the manager sacked a German foreman without any pressure from outside "because he lowered himself to greeting the Jews with a handshake and had swapped postage stamps with them."[170] In some factories the onsite security force or the *Gestapo* staff maintaining security and surveillance were also responsible for overseeing and enforcing the contact prohibition.[171]

In practice it was impossible to adhere to these contact restrictions. Many former inmates working in aviation factories reported constant direct contacts with their German supervisors and with other German workers while inside the factories. In some instances Germans and inmates even socialized while at work. A Messerschmitt employee working in the firm's workshops at Flossenbürg summed up the problem: "Contacts with prisoners were not allowed to be made and were forbidden under threat of punishment. But because of the daily nature of the work, these interim human contacts could not be avoided, insofar as most of the Russians spoke a little German. Thus, in all secrecy, private conversations did take place." This particular worker also brought some food to the Soviet POWs working with him.[172]

Even Jewish inmates sometimes received fair treatment from German workers. Tzipora Shwiatowitz's German foreman treated her well after he saw the quality of her welding. He regularly brought her a sandwich and protected her from SS guards.[173] Other inmates also reported after the war friendly contacts with German coworkers. Oddly, in at least one factory—Junkers' engine factory at Markkleeberg–inmates were allowed to eat in the employees' cantina.[174]

Germans and foreigners lived, however, almost always in completely separated quarters and under different living conditions. In most places foreigners were housed in separate quarters according to their nationality or ethnic identity. Also here there was a hierarchy, insofar as the Germans viewed workers from Eastern Europe as the most inferior, and therefore usually allocated them the worst quarters.[175] Different German inmates, including SS convicts, were locked in their own camps or in separate compounds.

The use of thousands of inmates and foreigners in 1944 demanded an expansion of the security force used to guard them in their camps and at their workplace. Earlier, SS guards were almost always present at the different workplaces, ready to solve any disciplinary problem or to respond to a call from a German foreman. At Heinkel's factory complex in Poland, Ukrainian auxiliaries were used as guards. They were supervised by a small group of SS men.[176] The massive expansion of the slave labor scheme in 1944 made it difficult to find enough SS guards to watch the growing numbers of slave workers and foreigners dispersed among diverse factories. During April the Jägerstab established that an additional 11,900 men were required to guard the increased number of inmates about to be engaged in its production and construction programs. The Luftwaffe was asked to provide most of them, but by the beginning of May only 7,850 extra guards were authorized.[177] The SS was especially worried by the possibility of prisoners attempting mass escape if their factory were bombed. Therefore, Himmler demanded a relatively high guards-to-inmates ratio in every inmate detachment.[178] Because of the fear of mass escapes, usually more guards were required to escort inmates to air-raid shelters and trenches during air raids. A particularly difficult situation developed in Heinkel's Barth factory, where employment of inmates from the Ravensbrück women's concentration camp started in August 1943.[179] By May 1944 some

3,000 of the 3,300 workers in Barth were concentration camp inmates. Only around 20 guards were available to guard them, and this number was much too low for air-raid evacuation security. As a result the factory management armed some of its German workers and allocated them to guard the inmates in case of air raids.[180] Therefore, although in most factories inmates were evacuated to air-raid shelters in case of air-raid alarm, they were closely guarded throughout the process.

In order to perform general security and fire-protection tasks, most firms established earlier in the war *Werkschutz* (factory security) detachments. The importance of these units increased during the last years of the war, as factories swarmed with foreign and slave workers and as the fear of ground or airborne attacks on factories increased. The function of the *Werkschutz* was constantly broadened during the war and they also dealt with counterintelligence. Its members were armed and fulfilled more and more security tasks in and around the factories. In most factories members of the *Werkschutz* and their auxiliaries supervised and monitored the foreigners. Among others, these security guards kept foreigners away from where they were not supposed to be, prevented them from taking photographs, escorted them during emergency evacuations, watched them during air raids, and generally monitored their discipline. The guards were instructed specifically to report immediately any act of passive resistance and any sign of mutiny.[181] The SS principally rejected such amateur security solutions and demanded that only its own men or trained Wehrmacht soldiers be allowed to watch inmates. There was, however, little choice, and the SS sought to improve the efficiency of the *Werkschutz* by attempting to exercise more control over these units. The *Werkschutz* units first came under *Gestapo* jurisdiction on 1 October 1943.[182] In November 1944 the Armaments Ministry handed over to the *Gestapo* overall responsibilities for security and protection of the aviation industry. From now on the *Gestapo* fully supported professionally and administratively the factory protection units. This support included training their staffs, providing them with unified regulations, supervising their professional skills and managing their manpower. Their financing, though, still came from the RLM's budget.[183]

Although *Werkschutz* members fulfilled some guard duties, especially on the factory premises, they formed only a fraction of the force required to watch the inmates in 1944. In early summer 1944 the Jägerstab sought Wehrmacht soldiers to watch hundreds of work detachments in factories and on construction sites. Usually the Wehrmacht allocated old men from reserve units to guard duties, but younger Luftwaffe ground and administrative personnel were also assigned. The Army and the Luftwaffe were able initially to allocate soldiers as guards, but soon their limited reservoir was exhausted. In desperation, the Jägerstab even tried to convince the Navy to allocate sailors from idle vessels to watch inmates.[184] The Navy could not allocate the required number of guards and the desperate Jägerstab considered as a last resort the use of wounded soldiers from convalescence centers as guards.[185] Other creative solutions were sought on the lower level by individual firms. In July 1944 Junkers requested auxiliary Flemish SS men as guards for the large number of inmates and POWs (including British) employed in its Aschersleben factory.[186]

Basically the presence of SS guards was counterproductive. The industry required healthy and motivated workers, able to perform their work satisfactorily. Brutalizing and mistreating workers lowered their performance and that was precisely what SS guards were mostly doing. Generally, treatment in the factories was tolerable, but the camps, controlled by the SS, were a different story. A Messerschmitt employee, who was sent in late summer 1944 to Flossenbürg to install production tooling, reported about the harsh regime in this camp:

> It was here that I saw concentration camp inmates for the first time in their gray striped suits, shaven heads and wearing enforced caps. As soon as any guard crossed their path, the camp inmates tore off their caps and saluted them. If the guard felt they had not saluted quickly enough, they were kicked, punched on the face or beaten with their rifles.... The numerous construction jigs that were delivered stood out in the open and on the exposed parts rust had already set in. In order that these be cleaned and made free of rust, Russian POWs were assigned to us.... Blows were constantly showered here too, on the prisoners by the guard personnel.[187]

As mentioned earlier, although some of the labor camps were initially outside the jurisdiction of the SS Concentration Camps Inspectorate, conditions in most of them fully resembled the conditions and regime of the concentration camps. At Erla's Johangeorgenstadt plant, producing tailplanes for Me 109 fighters, general living conditions in the adjacent labor camp were extremely harsh and resulted frequently in large numbers of deaths due to illness. Brutal treatment by SS guards only aggravated the situation. Medical treatment was almost nonexistent, and death rates were so high that the SS regularly brought in new groups of inmates to keep a nominal number of 1,000 inmates in the camp. Brutal conditions prevailed in all sub-camps of Flossenbürg involved in Erla production.[188]

Furthermore, in many cases prisoners coming from some of the worst concentration camps were in bad shape when they started working in the factories. Junkers' aero-engine underground factory at Venusberg (code named "Knurrhahn") received in late 1944 several hundred Jewish women, who were already in bad shape as a result of the time they spent at the concentration camps of Ravensbrück and Bergen-Belsen. They were immediately incorporated into the 12-hours-shift cycle in the underground workshops located in a former lime mine. Within a short time 69 of them died.[189] This case exemplifies how sometimes a work regime dictated by factory managements disregarded the need to preserve the workforce and contributed to the death of inmates. In this absurd situation the German system basically defeated its own cause.

Violence was not restricted to the camps. Corporal punishment was a common feature even on the factory premises. The SS officer in charge of the labor camp next to Arado's Freiberg factory used to beat inmates publicly in the main production hall.[190] Inmates at Heinkel's Barth factory were punished for the slightest disciplinary offense by beating or by food ration deprivation.[191] An inmate at Heinkel's Heidfeld factory near Vienna reported daily physical maltreatment by the SS *Kommandoführer* (detachment leader) and by the capos, usually veteran Polish or Russian inmates.[192] Sometimes factory staff also mistreated foreign workers and inmates. Frau Mirter, who trained Eastern female workers at Messerschmitt's Regensburg plant, witnessed the abuse of foreign female workers:

> One day they [the women] came to me with tears in their eyes and asked me to obtain for them some bandages and wraps. When I asked what these were needed for, they showed me their blood-covered heels. The women wore wooden shoes that were open at the back, and each time that they did something wrong, their *Meister* or supervisor kicked them in their heels with his shoe until they turned bloody. I naturally reported that to my supervisor, but whether the guilty ones were ever punished, is something I do not know.[193]

A German Jew testified after the war in a West German investigation against a former foreman working at Arado's Freiberg factory: "Zimmermann had a group of around twenty prisoners to supervise. He repeatedly abused me physically. He threw shop tools, which I was required to bring him, at my back, or he tore the tool from my hand and beat me with it."[194]

In several cases firms pointed out at the futility of coercion methods exercised in industrial sectors manufacturing complicated items, which usually required not only higher technical skills, but also some motivation. The Dornier factory in Munich explained the core problem of brutalizing workers: "Nobody can seriously believe that difficulties which arise in this way can be solved using police methods — quite aside from the fact that police resources on the scale required are simply not available. Moreover, the deployment of manpower in a highly sensitive production process like aircraft construction can never be effectively handled solely by the use of police coercion."[195]

Besides further degrading the already low motivation of these workers, poor conditions also caused higher illness and death rates, which forced the Germans to constantly seek replacements. The mentality of viewing inmates as an easily replaceable commodity was well embedded in the SS camp system, but it ignored the profitability of preserving qualified inmates for work demanding high skills. Thus, manpower employed in the aviation industry acquired different skills through training or while working, and therefore was more difficult to replace. The management of MIMO articulated the problem in a particularly clear way in a report it submitted in April 1942: "If 2,000 Russians are deployed on road construction in the Eastern territories and, within three months, a few hundred Russians fall by the wayside due to the meager rations, then the missing laborers can be replaced by new Russians. But in an armaments plant, it is impossible to replace a man operating a specialized piece of equipment at a moment's notice."[196] It seems, however, that such cold-headed calculations hardly impressed the SS and others responsible for running labor camps and managing their manpower.

It seems that foreigners and inmates employed in the construction of underground and bunker factories fared worse than those working in production. Although the tunneling and cement casting in the construction sites were somewhat mechanized, large portions of the work involved backbreaking menial toil. Inmates were usually allocated to construction sites in large detachments and were accommodated in labor camps constructed nearby. These camps usually formed a sub-camp of one or another concentration camp. Sometimes extra inmates were required for the construction of the sub-camp, as happened in the case of the Neckargerach camp, constructed near Daimler-Benz's "Goldfisch" factory. In mid-May 1944 the local management demanded 500 to 700 inmates to accomplish this construction project, which in turn was needed for the completion of the underground factory.[197] In 1944 Kammler's staff decided to restrict the daily work hours of inmates working in underground factories to eight hours, thereby equalizing in most cases the length of their shifts with those of German workers. Consequently, the workday in most tunneling projects under Kammler's supervision was divided into three 8-hour-shifts. The significant time it often took the inmates to get to work and back was excluded, of course. This time could take up to 6–8 additional hours.[198]

In some underground factories the inmates lived in the tunnels they dug. While working underground the inmates were exposed to all the dangerous and unhealthy elements prevailing in these facilities. Inmates working in underground factories lost their lives in different accidents associated with this special type of facility. Most of the people killed in a series of deadly accidents in the "Goldfisch" and *Nordwerk* factories in August–September 1944 were foreigners and inmates.[199] Even without accidents, the unhealthy living environment underground proved to be fatal for many already weakened inmates. The arrival of winter in 1944 was marked by a significant increase in cases of illness in "Goldfisch." Cold air was the main cause and at a certain point at the end of 1944, around 50 percent of the workers in most departments became ill. The situation became so desperate that Daimler-Benz's management asked the local SS construction engineer to urgently find ways to improve insulation at the entrances to the tunnels.[200]

Working outdoors on a construction site meant no significant improvement for the inmates. They worked outside regardless of weather conditions and other factors. Being exposed to the elements, combined with extremely exhausting work and malnutrition, caused higher than usual death rates. In 1944 the death rate in the Mauthausen camp complex of inmates employed on construction sites, most of them related to the aviation industry, was over 30 percent.[201]

A former Polish inmate working in the "Malachit" project told an American intelligence team about his experiences after the U.S. Army captured the place. These experiences exemplify the conditions prevailing in other construction sites. Project "Malachit" was an underground factory intended for production of the Jumo 004 jet engines, and later as an emergency production facility of He 162 fighter components.[202] It was constructed in a wood-covered sandstone hill near the Halberstadt airfield and was supposed to provide almost 246,000 square meters of floor space. Slave workers from a nearby concentration camp performed most of the construction. Initially the food rations were an improvement over what most workers received in their former concentration camps. Rations were issued, however, on a daily basis only if the worker had satisfied his foremen with the day's work. This distribution system led to internal rivalry among the inmates, which sometimes tried to claim extra daily rations by sabotaging the work done by other work groups. The foremen, many of them inmates, were also rewarded and paid according to the progress of their allocated tasks. This system turned them into brutal slave drivers. The work on "Malachit" progressed slowly because the hill in which it was dug consisted of several layers of soft sandstone and this geological composition demanded a large number of support structures. There was a shortage of support girders and of steel bolts, cement and coal (consumed by the small-gauge rail engines)—all ascribed to the bombing of railways and neighboring towns. After February 1945 conditions became worse. Food rations dwindled and the death rate among the inmates rose rapidly. The factory was never completed, although some machine tools were moved in during February 1945 and production commenced for a short time.[203]

Like the German workers, foreign and slave workers also bore the brunt of air raids and lost their lives in their course. A "Big Week" daylight raid by the USAAF on Messerschmitt's Augsburg factory on 25 February 1944 claimed the lives of around 250 slave workers.[204] After the USAAF bombed the same factory again on 6 March 1944, the management reported that 15 Germans and two Italians were killed. The slave workers, however, suffered worst: 70 of them were killed.[205] It is not clear if this heavy loss of life was caused by direct hits on the factory's concentration camp at Haunstetten or by untimely evacuation of the slave workers. A bombing raid by the USAAF on Heinkel's Germendorf factory on 18 April

1944 killed 260 workers, 230 whom were prisoners. Scores of SS guards were also killed during this attack.[206] Generally the Germans evacuated their inmates to air-raid shelters or slit trenches in case of air-attack warning. The reasons for this care for their lives were quite practical. First, they did not want to leave inmates unattended for any length of time. Factory managements clearly ordered that in case of air-raid alarm all workers must leave their workplace and find shelter. Nobody was allowed to stay behind for fear that foreigners or slave workers would use this opportunity to sabotage or to escape.[207] Second, it made no sense to go through all the troubles of obtaining inmates only to lose them in an air raid because they were left exposed. The inmates were required not only for daily production, but also for repair work in case the factory was damaged. It was practical to take the inmates out immediately after a bombing raid and assign them to rubble clearing without the need to wait for the arrival of special repair teams.[208] Not much is known about the fate of inmates injured during air raids. In at least one case, in the aftermath of the bombing of Messerschmitt's Regensburg factory, a local Nazi leader ordered improper treatment for wounded Soviet POWs. According to eyewitnesses, he said, "Paper bandages are good enough for them!"[209]

Using undermotivated and mostly hostile workers to manufacture highly sophisticated machines bore with it several risks. One of the prime risks was sabotage. Slave workers often carried out different acts of sabotage, and aircraft were particularly prone to be severely damaged even by minor sabotage. The main method of sabotage was to damage parts or components on purpose. Mark Stern, a young Polish Jew who worked at aircraft factories in Mielec and later in Flossenbürg, described how he and other inmates sabotaged products:

> The only way that we knew of how to fight back was by sabotage. I was part of it. I remember being told by somebody that the proper way to sabotage is to weaken the wings of the planes that we were building. So, we riveted; we drilled out rivets and replaced them with other ones; replaced them with weaker rivets. I am sure that some of it worked. We didn't know exactly how to sabotage but we tried our best.[210]

At Oranienburg, German political inmates used the professional knowledge of an inmate in order to weaken the aluminum rivets used in the production of the He 177 bomber. Instead of dipping them in a salt solution of 600 degrees Celsius, they heated the solution only to 400 degrees. Large numbers of these weakened rivets were discovered during quality assurance and were never used to assemble aircraft, but the delays and wastage caused by the need to replace them also disrupted production.[211]

The main damage was caused, however, by damaging products and by not adhering to manufacturing standards. The Americans discovered after the war that the *Nordwerk* plant of Mittelwerk experienced severe difficulties in producing reliable Jumo engines. The rejection rate of engines on test sometimes reached 100 percent.[212] This extremely poor performance may be attributed to the large number of slave workers from the nearby Dora concentration camp employed in this factory and to the generally primitive conditions prevailing in this underground facility.

Acts of sabotage by slave workers were well known to the German leadership and to lower-ranking Luftwaffe officers and soldiers.[213] Adolf Dilg, who served as a factory test pilot at Arado's Warnemünde factory, producing FW 190 fighters, told aviation historian Alfred Price about the effects of sabotage and about a peculiar flight emergency he experienced:

Aircraft were sabotaged in all sorts of ways. Sometimes we would find metal swarf in electrical junction boxes, or sand in oil systems. On two or three occasions brand new Focke-Wulfs took off for their maiden flights and as they lifted off the ground one of the wheels fell off; the pin holding the wheel retaining ring had "accidentally" come adrift. Once, when I was delivering an aircraft, the engine suddenly burst into flames. I bailed out and the aircraft crashed into a marshy area where the water rapidly extinguished the flames. When the wreckage was examined, it was found that somebody had jammed a couple of pyrotechnic flares between cylinders nos. 7 and 9, the two at the bottom of the rear row which became hottest when the engine was running. During the delivery flight the cylinders had duly heated up, "cooked off" the flares and up went the engine. Every time we had such an incident the *Gestapo* would make a lot of fuss, but although they would take the odd scapegoat the problem of sabotage was one we had to live with.[214]

Slowing up work and general sloppiness were other forms of resistance. In some cases foreign workers and inmates slowed up work, even coordinating such acts while attending factory meetings. In several Focke-Wulf factories supervision became more lax during the last months of the war and foreign workers took advantage of it. Some workers slowed up work by hammering their pedestal instead of the part they were supposed to work.[215]

Reports about sabotage and general misbehavior of foreigners reached the higher levels, and as early as May 1942, Milch demanded severe punishment of foreigners refusing to work. After receiving reports about Frenchmen refusing to perform the work given to them or simply being lazy, he demanded putting offenders against the wall and shooting them, or alternately transfering them to a concentration camp. He also demanded to discuss the problem with Himmler, whom he obviously viewed as an expert in dealing with troublemakers.[216] Later in the war, as active sabotage became widespread, he repeated his demand to severely punish any foreigner, POW or inmate caught sabotaging or refusing to work: "If he performed an act of sabotage or refused to work, let him be hanged right on the place where he works. I am convinced that this will not miss its effect."[217]

Saboteurs and other offenders were indeed punished, although not necessarily by execution. It seems that especially during 1944 supervisors tended to physically punish offending inmates if caught for the first time. Beating was the most common form of corporal punishment, but there were cases where inmates were forced to stand outside their barracks at night in the winter cold. This "milder" policy was borne out of the necessity to preserve every possible worker. Mark Stern told about his experiences while working for Messerschmitt in Flossenbürg:

> I was caught, twice. The first time we got 100 lashes for sabotage. A hundred times they hit me with a broomstick; for a week I couldn't sit. The second time they were going to shoot me. I was lucky. There was a German civilian who worked there. He took a liking to me. He said that he told me to do it that way, so I got only 100 lashes again. A hundred times I was hit again, but I survived.[218]

In 1942–1943, while he worked at Mielec, several inmates were executed on the spot because the Germans suspected they were involved in acts of sabotage. In Junkers' aero-engine factory in Markkleeberg, French and Polish women did most acts of sabotage. If caught, they were usually punished by being forced to stand outside in the snow for long hours. According to a survivor of this camp, this form of punishment was quite effective, because this way the Germans kept their workers alive and at the same time it deterred Jewish inmates locked in the same camp from performing acts of resistance.[219] In order to prevent sabotage it was strictly forbidden to let foreigners work on their own without German supervision. If a German supervisor took a break he had to find a replacement, or halt the work of his foreign workers and send them out until he returned. It happened repeatedly, however, that foreigners were left to work alone while the German supervisors and foremen went to their lunch break.[220]

Escapes from labor camps and work detachments of the aviation industry were quite widespread, even within the Reich. Negligence by guards and supervisors, lax security and desperation motivated escapes, even though it was an extremely risky matter. Negligence on behalf by guards and foremen resulted in one of the most spectacular attempts to escape. In mid–February 1944 two Soviet officers, one of them from the Soviet Air Force, working in the factory airfield next to Messerschmitt's Regensburg-Obertraubling plant, sneaked into an Me 109 fighter. One of them sat in the cockpit while the other stuffed himself in the fuselage behind him. They attempted to take off, probably intending to fly to Switzerland. Lack of training combined with a difficult-to-fly aircraft resulted in a stall and crash immediately after take off. The two POWs were found injured but alive on the crash site and were severely mishandled before and after being handed over to the police. It was argued that Hitler heard about the incident soon after it happened and that he personally ordered the execution of the escapees. They were both shot by a police firing squad in the airfield's gun range on 14 February. Consequent inquiry showed that slack supervision by army reservist guards and factory foremen enabled this escape attempt. According to some sources, a similar escape was attempted toward the end of 1944.[221]

Most foreigners and inmates escaped in less dramatic ways. Foreigners from Western Europe usually simply failed to return from authorized leaves and went underground. Eastern Europeans were in a more difficult position, and those who fled normally stayed within the Reich's boundary looking for better jobs and better living conditions. Allied air attacks destroyed several important registration offices where files of foreigners were kept, thereby incidentally helping some escapees to stay at large. No comprehensive statistics about escaping foreigners and inmates from work at the aviation industry have been found. The magnitude of the phenomenon can be deduced from the fact that between January and September 1943 MIMO alone lost 860 foreigners, who escaped in one way or another.[222]

In some rare cases inmates openly rebelled. In the night of 1 May 1944, Polish and Russian inmates working at an Erla factory at Mülsen–St. Michelm in Saxony (a sub-camp of Flossenbürg) started a brawl, during which they set the factory on fire. The resulting fire heavily damaged the factory and 100 to 200 wing sets were destroyed. The German police rushed in and shot around 200 inmates to death on the spot. Some 80 other inmates were severely wounded and 20 escaped. During a *Jägersatb* meeting on the following day, Kammler commented that this kind of thing happens whenever inmates notice lax German supervision. He reported that he ordered to hang 30 inmates as a "special treatment" in order to deter further unrests.[223]

Around 300 American POWs sent to work at Dornier's Oberpfaffenhofen factory in June 1944 instigated a different kind of rebellion. The Americans simply sat down and refused to work, even after they were threatened with a firing squad. The Germans feared mostly that such acts would spread to other places and to other groups and were eager to take harsh deterrent measures. It was reported that a proposal to shoot some of the rebels was declined by "higher authority" (probably by Hitler himself) for fear of reprisals, and that 500 Soviet POWs eventually replaced the Americans. Less severe punitive measures, like depriving food and suspending mail and Red Cross packages, were also considered in this case.[224]

Fighting sabotage, sloppiness and indiscipline among foreign workers and inmates was usually the task of different police organizations established within most factories. Sometimes factory managers simply called in assistance from the outside. There were cases when managers denounced and handed over foreigners or inmates to the *Gestapo* for slight offenses. Director Josef Sommer of the technical department of Daimler-Benz's aero-engine factory at Genshagen handed over several young Russian workers to the Gestapo after they were caught stealing potatoes during an air raid. They were summarily executed.[225] Dr. Warning's medical police department in Focke-Wulf's Bremen factory was mentioned before as a control organ within the firms. Warning also organized regular raids on foreigners' camps under the disguise of medical inspection. His so-called *Sprechstunden* (office hours) there ended in numerous cases in beatings of his "patients."[226]

Getting sick was bad news for foreign workers, and especially for inmates. They could not rely even on the help of a doctor like Warning. In most camps there was a nominal sick bay, operated usually by foreigners or inmates with one type or another of medical knowledge. In cases of foreign workers the factory doctor and his staff supervised these sick bays. This supervision was practically an extension of the normal health care offered to the German employees.[227] However, in some cases this supervision did not mean better treatment. Dr. Warning, for example, used to terrorize his foreign patients while treating them. He kicked many patients asking for his help and in at least one case he let a sick woman die just because she dared ask him for a couple of days off.[228] However, foreigners normally received much better medical treatment than prisoners. At Mielec, for example, there were no medical facilities whatsoever, and inmates tried to help each other as far as they could. If an inmate became too weak and couldn't work anymore, he was taken to an isolation barracks and was left there to die. Typhus was a widespread disease in the Polish camps. It weakened hundreds of inmates, who were subsequently sent to the death barracks.[229]

In light of the generally ill treatment of foreigners and inmates, it should be noted that in some places their employers offered them more positive incentives in order to increase their motivation. Foreign workers from Western Europe usually received somewhat lower wages than German workers.[230] Skilled foreigners — excluding those from the East — in Focke-Wulf's Cottbus and Sorau factories earned in late 1943 between 0.56 and 0.72 RM (depending on their age and productivity) per hour. Female workers of the same group (none of them considered skilled) earned between 0.23 and 0.43 RM.[231] In order to improve the motivation of French skilled workers and in order to improve the prospects of recruiting more of them, the RLM agreed in November 1943 to equal their payment to the payment scale of skilled German workers.[232] It was done after both Messerschmitt and Focke-Wulf encountered difficulties with newly received French workers. In some cases workers that were picked out specifically from the French aviation industry because of their skill, refused to travel to Germany. Some of those who agreed to go declined signing their contracts after they learned on the spot about the living conditions and treatment awaiting them.[233] Besides improving the salaries, firms initiated other measures aimed at awarding their foreign workers. Focke-Wulf even included some foreigners in its special Christmas allowance scheme at the end of 1944. All Italian workers received the full allowance of 25 RM, and 20 percent of the Soviet workers received this allowance in recognition of their good work. Poles and POWs, however, were excluded from this scheme.[234] Messerschmitt's foreign workers were also allowed to take leaves on special occasions and holidays. Even as late as December 1944 foreign workers were included in the limited holiday leave arrangements advised by the management despite the pressing production needs. The slave workers were excluded, of course, and were forced to work throughout the holidays.[235]

Foreigners engaged in fighter production also sometimes received the extra rations and clothing allocated by the Jägerstab. Extra cigarettes were considered as a highly effective motivation booster — especially of workers from the East.[236] Firm managements allowed foreign workers to take part in different entertainment events organized by the firm or by other organizations in the vicinity of their workplace. Junkers, for example, organized for its Flemish workers a special *Kameradschaftsabend* (social evening) in Dessau, which was reported on a propaganda radio broadcast in Flemish.[237] Eastern workers at Genshagen had their own movie theater and were allowed to organize musical events, as reported by Heinkel's Staff Engineer Baatz, who toured the factory in mid–1942 as part of his firm's cooperation with Daimler-Benz in the He 177 project. On the same occasion Baatz also reported about poor food supply in this factory.[238]

Surprisingly, inmates were initially supposedly paid for their work, at least while working at Heinkel's factories. The payment was delivered on behalf of Heinkel by the SS and was viewed as part of a successful scheme aimed at giving the inmates a motivation to work harder and better. Five levels of payment were established and inmates received a weekly payment of between 1 and 4 RM according to their skill.[239] It is almost certain that this arrangement never became a general practice, and that from 1943 the SS kept all the payment it received from the firms for the workers it delivered. It is certain that the masses of prisoners allocated to the aviation industry in 1944 received no payment for their work.

Perhaps the most bizarre method used to increase the motivation of prisoners was the improvement suggestions scheme in the Heinkel firm. In 1943 Heinkel encouraged inmates at the Oranienburg factory to submit their own suggestions to the factory management. The firm agreed with the SS that if a suggestion submitted by an inmate were found useful, he would be rewarded with extra rations. This scheme led to an increase in the number of submissions, and in July 1943 some 42 from the total of 107 improvement suggestions received by the plant management were submitted by inmates.[240] Both Oranienburg's manager Hayn and Himmler viewed the large number of improvement suggestions submitted by inmates as a positive sign of the successful integration of inmates into the aviation industry. Himmler boasted in a report he sent to Göring on 9 March 1944 that 200 suggestions by inmates were already received and implemented in Oranienburg and that the concerned inmates were rewarded.[241]

It is extremely difficult to make general observations about the treatment of slave workers in the aviation industry. U.S. investigators determined after the war: "In the larger government controlled plans, such as the Mittledeutsch (sic) Motoren Werke and the Nordhausen V-plants, the slave laborers were badly mistreated; in the small privately controlled plants their lot was comparatively good and efforts were made to properly house and feed them."[242] It is difficult to grasp the way people lived in these camps and factories without experiencing it firsthand. Good conditions reported by former inmates in several places would look unbearable to people who never went through similar experiences. Therefore, the yardstick used here is based on comparisons made by survivors when referring to different places they experienced. The general picture created by these comparisons is that sub-camps and factories, where inmates worked in production, were always an improvement when compared to main camps, like Buchenwald, Dachau, Mauthausen and Auschwitz.

Conditions also changed according to location and time. Conditions in the early phase of slave labor in the aviation industry were just an extension of the harsh conditions in other concentration camps. At Oranienburg-Germendorf, inmates lived in half-buried barracks located right next to their assigned production halls. These barracks were overcrowded and the prisoners slept on four-story bunks, sometimes sharing the same bunk. An SS detachment from the Sachsenhausen concentration camp was responsible for supplying the inmates with clothes, food and medical treatment, therefore sparing the firm the need to take care of the workers and keeping them under strict SS camp regime.[243] In Heinkel's other main early enterprise at the Mielec and Budzyn factories, conditions were also rough. Ukrainian auxiliaries guarding the inmates treated them brutally, especially if they were Jewish. They often beat the prisoners whenever they were in the camp.[244] The camp was located some distance from the factory and the inmates were forced to march each day to work and back. This routine resulted in countless hours spent outside, and as a result the inmates were able to sleep only few hours each night.[245]

Labor camps constructed later next to factories could also be a rough place. A German worker at Messerschmitt's Regensburg factory testified that living conditions of the more than 1,000 Soviet POW officers working in his factory were extremely miserable. They received meager rations of "unidentifiable soup and *Ersatzbrot*, dark brown and moist bread mass." Furthermore, the hygiene in their barracks camp, located north of the Obertraubling airfield, was "indescribable," and people could smell its stench from a great distance.[246] Jewish workers at Henschel's aero-engine factory in Rzeszow, Poland, lived in five 30-man barracks. Around 100 Jews lived in each of these barracks equipped with extremely rudimentary provisions. Only two transportable toilets were provided for the 500 workers living in this camp. The barracks were surrounded by a barbed wire fence that was patrolled by armed and eager-to-shoot guards.[247]

Many of the 100,000 Jews allocated to the Jägerstab in mid–1944 generally viewed the conditions in their new locations as an improvement. This is not surprising, considering the fact that most of them had spent some time in Auschwitz. Furthermore, being picked out of an Auschwitz transport usually meant rescue from certain death, especially with hindsight. In some places, especially where the firms were involved in making arrangements for the new arrivals, conditions were indeed an improvement. A former inmate, aged 16 years old by the time she arrived at Arado's plant in Freiberg, described the lodging she and other newly arrived inmates found upon arrival: "It appeared to be a good change from Auschwitz. We slept only two to a bed, had pillows and a type of blanket."[248] Another woman survivor described the conditions in Junkers' aero-engine factory in Zittau as "paradise" when compared to Auschwitz. This plant was one of the main production centers of the Jumo 004 jet engine. Her group of women, mostly teenaged Jewish girls from Hungary and Yugoslavia, was billeted in a three-story building with bathrooms and showers. Each of them received her own bed with straw mattress and pillow. They were also given spoons and plates. Their living conditions deteriorated, however, as an SS detachment took control of them during the last months of the war and as supply became scarce.[249]

Food rations can hardly be viewed as a motivating factor, although foreign workers were relatively well-fed with daily meals of 3000 calories.[250] It seems that the food rations given to slave workers depended largely on where they were accommodated. If they lived in a normal labor camp they received meager concentration camp rations. Those could amount to a daily ration of two slices of bread and a bowl of watery soup.[251] Those who lived in the factories usually fared better. In Junkers' Markkleeberg factory, for example, women inmates received, up to late 1944, three meals a day in the factory's cantina.[252] It is noteworthy, however, that firms used to include their slave workers in the listings sent to the Jägerstab in order to get the extra rations allocated to workers engaged in the 72-hours-work week. Messerschmitt, for example, reported on 9 May 1944 a total of 1,552 Eastern workers, 1,394 foreigners, 213 POWs and 2,717 concentration camp inmates as eligible for extra rations due to their extra working hours.[253] As the experience of the inmates in Markkleeberg may suggest, sometimes these extra rations actually found their way to the inmates.

There are indications that some factories rewarded at least some of their prisoners in different ways. Josef Pinsker was a Rumanian Jew who worked in 1944 at a Junkers factory in Niederorschel. He was trained to work on a lathe and manufactured fine parts. The Germans rewarded him for his good work with an extra monthly ration of 5 cigarettes, which he traded for food.[254] Underlining the difference between the camps of the aviation industry and other camps, in Junkers' Zittau factory the guards threatened inmates with a transfer to the notorious nearby Gross-Rosen camp if they misbehaved.[255]

As Allied armies advanced into Germany, factories were evacuated and their slave workers were taken eastward or westward into the shrinking heart of the Reich. Usually firms tried to keep their workers and moved them to another factory. When Daimler-Benz evacuated its plant in Rzeszow in late summer 1944, it moved all its Jewish workers to a new underground factory near Wesserling-Urbès in Alsace. The firm put experienced Jewish workers to work again on the production lines, so their expertise was not wasted. Inmates still working on the construction of this factory at the same time under SS supervision even complained that the Jews received better treatment from their Daimler-Benz supervisors, who obviously appreciated their value.[256] When Volkswagen evacuated its Tiercelet V-1 underground factory in Lorraine in late 1944, most of its workers, selected earlier in Auschwitz specifically for this place, were evacuated along the plant's machinery to the "Rebstock" factory located in a tunnel near Dernau in western Germany. After working for a short time in the new factory they were evacuated again with the same machinery to Mittelwerk, where they stayed until the U.S. Army finally liberated them.[257]

During the chaotic last months of the war and in the framework of the evacuations there were inconsistencies in the allocation of inmates. In some cases inmates who never worked before in the aviation industry were evacuated to an aviation-related factory. Some of the inmates of the Glöwen sub-camp of Sachsenhausen, who worked in an explosives factory, were evacuated to Arado's Rathenow factory, where they riveted wings for the Ar 234 jet bomber.[258] In other cases the Germans evacuated inmates working in an aviation factory and sent them to perform a completely unrelated type of work, thus wasting their training and experience. When Heinkel's "Block Budzyn" complex was evacuated in May 1944, the SS transported a large number of its slave workers to the Plaszow concentration camp, where most of them were employed in non-aviation-related work, like garbage disposal.[259]

Lack of raw material and the fear of being overrun were the main reasons for the evacuation of factories during this phase. In the chaos prevailing in these days the Germans had little idea of what to do with the evacuated inmates. In March and April 1945, work in most surviving factories came slowly to a halt, and inmates evacuated to them sat idle and waited for the liberation. In other cases sub-camps simply returned their inmates to the main camp, sometimes forcing them to march long distances under extremely cruel conditions. A Czech woman working at Arado's Freiberg plant told about the evacuation of the camp, after work had stopped on 14 April 1945:

> On 14 April 1945, there was a sudden departure. We were loaded into open cars at the train station and traveled westward into the Protectorate, passing train station signs with familiar city names. The nights were cold and sometimes it snowed or rained. Only sometimes did we receive food. En route we encountered similar transports to ours almost daily. Then we had a long stop in Horní Bríza and were transferred into closed cars. The people of the town brought us something to eat. We were supposed to be brought back to our original camp, Flossenbürg. We owe our thanks to a brave station manager who, despite threats, held up our train. We traveled back in the direction of Budweis.... Once a day the car was opened and someone shouted "Out with the dead." We noticed that the train changed direction. On 29 April we stood in the train station at Mauthausen. Half starved, we dragged ourselves through the town. At a fountain we wanted to at least drink something, but the locals chased us away and threw stones at us.... On 5 May we were liberated by the U.S. Army.[260]

Of the 1,240 inmates working in Junker's Langsalze factory the SS marched 1,177 back to Buchenwald at the end of March 1945. Those who survived the death march and the harsh conditions in the main camp were finally liberated there by the U.S. Army.[261] Himmler personally ordered the evacuation of Flossenbürg and most of its sub-camps on 14 April 1945. He also ordered that no inmates be left behind, which resulted in a series of death marches towards Dachau. Flossenbürg itself was liberated by the U.S. Army on 23 April 1945, but some of its eastern-lying sub-camps were liberated by the Soviet Army.[262]

Sidney Birnbaum's experience during the last months of World War II was typical. His account also reads like a tour of some central locations and names of the late-war German aviation industry. Sydney started his "career" with the aviation industry in 1942 Mielec, living in a labor camp under harsh conditions and working as a riveter at the nearby Heinkel factory. He survived 18 months at Mielec, gaining in the process much experience as an aircraft riveter. As the Soviet Army approached the area in 1944, he and other inmates were evacuated to Wieliczka, where Heinkel constructed an underground factory intended to take over the capacity of the lost factories in the East. After a couple of weeks the Germans gave up this project due to the high humidity in the tunnels. Sidney and some of his fellow inmates were evacuated to Flossenbürg. After a short term in this difficult camp he was transferred to Leitmeritz, where the Germans constructed the "Richard" (also marked as B5 in Kammler's staff documents) underground tank engines factory. He and some other Mielec veterans were allocated to dig ditches on the construction site. At the end of 1944 the group was evacuated to Dachau. After spending a week there, Sydney and other inmates were transported to the underground factory "Leo" in Leonberg. In this factory he returned to his old profession and became a riveter once again. This time he was "promoted" and worked on the wings of modern Me 262 jets. The inmates lived in a barracks camp near the tunnel entrance. There were frequent air-raid alarms and every time the siren sounded, the Germans drove them into the tunnel. At the same time many inmates died due to malnutrition, hard labor and the general rough treatment. In early spring 1945 the Germans evacuated the factory. Sidney and his fellow inmates were marched to Augsburg. The survivors were soon taken on trucks to Mühldorf, one of the camps associated with the "Ringeltaube" bunker factories project. They never worked there, but Sidney was injured from a bomb splinter during an Allied air attack. He and other inmates were soon evacuated again, this time in a train, to Sternberg in Czechoslovakia. On the way they experienced first hand the Allies' campaign against transportation targets as their train was attacked from the air and the German escort personnel fled. Soon afterwards the U.S. Army liberated them.[263]

6

The "People's Fighter" as Case Study of a Late-War Program

Most of the German aircraft produced during the late war period covered in this research were designed long before that. Even the jet aircraft had their roots in the early 1940s. In this regard the He 162 jet fighter — also widely known as the Volksjäger ("People's Fighter") — was an exception. It was conceived, designed, produced and entered operational service within the last seven months of World War II. It was the only completely new design to enter high-rate production in the newly organized industry. Its history therefore represents well the way aircraft were produced in Germany towards the end of World War II. From the way its production was managed to the way it was manufactured, the production story of this aircraft reflects the conditions and mentality prevailing in the German aviation industry at that time. The He 162 epitomized the reorganization and changes of the German aviation industry up to late 1944. Its story also demonstrates the relations between the new bosses, the old bosses, the customer and the contractors in Nazi Germany's last major aviation project.[1]

Idea and Conception

The beginnings of the lightweight "popular" jet fighter came with the realization by the Luftwaffe and the RLM that the Me 262 was too expensive and too late to have a meaningful impact on the air war. Although on paper it looked like an extremely formidable warplane, its combat performance proved to be quite disappointing. Some people in the RLM and Luftwaffe leadership thought that one of the reasons for this failure was that it entered service in relatively small numbers and piecemeal, therefore never reaching the "critical mass" required in order to effectively engage the massive air armadas of the Allies. The first Me 262 fighter unit became operational only in late September 1944 and the Luftwaffe's high command withdrew it from combat after only six weeks, during which it suffered heavy losses and achieved little. From this disappointing experience it became clear that training to fly this sophisticated aircraft was a long and difficult process even for experienced pilots. Introducing this revolutionary fighter into operational service was indeed painfully slow, and at the end of December 1944 only 112 Me 262s were in service with the Luftwaffe.

6. The "People's Fighter" as Case Study of a Late-War Program 237

Volksjäger. An almost completed series production He 162 found by U.S. troops at Junkers' Bernburg factory at the end of the war (courtesy U.S. National Archives and Records Administration).

Only 70 of them served with combat units — mostly as bombers — while an additional 42 were used for operational training.[2] Furthermore, the twin-engine, full-metal construction Me 262 was extremely wasteful in terms of raw materials used to build it and fuel required to fly it.

These were the main problems that motivated Lieutenant-Colonel Siegfried Knemayer, up to August 1944 head of the Aircraft Development Department in the RLM, now subordinated to TLR, to order his experts on 5 September 1944 to study the notion of developing a small and inexpensive jet fighter. Despite the nominal subordination of TLR to the Luftwaffe's high command, two days later the RLM submitted the required specifications to several leading firms and asked them to bid their proposals as soon as possible. The tender specified a light fighter propelled by a single BMW 003 jet engine, enabling a top speed of 750 km/h and an endurance of 20 minutes. Pre-production series of this engine started coming of the production lines in August 1944 after a long and painful development process that started in 1940. It was expected to be available in increasing numbers in the following months. The RLM also demanded a short take-off and landing run (less than 600 m) in order to enable the new fighter to operate from small dispersal airfields. The most interesting parts of the specification, however, were related to manufacture and production of the aircraft. The basic demands were not to interrupt production of other fighters and to come up with a simple easy-to-manufacture design. The specifications and restrictions dictated a compact and light aircraft (weight of no more than 2 tons), constructed partially of wood.[3]

The idea behind the light jet-fighter project was simple: it was supposed to be manufactured in large numbers, destroy one or two much more expensive enemy aircraft, and be scrapped after several missions if it survived them. This concept, together with an extremely tight schedule, made this project extraordinary from the start. Usually aircraft were produced and introduced into service after several consecutive phases of development: conception, design, scale-model testing, prototype manufacture, flight testing, pre-production run, testing of these early aircraft, operational testing, and finally series production. Here it was clear that most of these phases must run in parallel in order to initiate mass production of the aircraft at the beginning of 1945. It was an "all or nothing" project typical of the German military, political and industrial thinking of that time. The whole concept was far removed from the military reality of that time and such a makeshift fighter had little chance to survive encounters with overwhelmingly superior Allied air power. It made sense from the economic-industrial point of view, but the decision-makers largely ignored the extremely short endurance of the aircraft (only 30 minutes) and the inability of the Luftwaffe to train enough pilots to fly the projected numbers of the aircraft.

Five firms submitted proposals to the tender. Blohm & Voss and Heinkel, two former producers of large aircraft, that were left at this stage without much work and therefore with free design and production capacity, submitted the most serious proposals. Ernst Heinkel was first informed about the contract on 8 September by Gerhard Giese, a designer working in his technical department. Giese pointed out immediately that the firm's P 1073 light fighter study seemed to fit almost perfectly the RLM's specifications.[4] From this point Heinkel picked up the project most energetically. Earlier the RLM had pushed this pioneer of jet flight out of jet aircraft development, and his firm had produced mainly bombers during the war. Since bomber production was terminated, it was clear that Heinkel possessed a large amount of free capacity. It was also clear to Heinkel that if he could not come with a design of his own, he would be ordered to produce someone else's design under license. Perhaps more importantly, this tender was his chance to reenter the fast-aircraft arena, as he suggested earlier in letters to Göring and Knemeyer in July 1944, in which he listed his previous achievements in high-speed plane development. At that time his He 177 was already terminated and the jet-propelled He 343 bomber had hung in the balance since late May, pending cancellation (as finally happened in November).[5] Therefore, the new fighter was Heinkel's only chance to upkeep his status as a leading aircraft designer and producer.

Like so many other German late-war aviation projects, Heinkel's proposal was based on earlier development work. The termination of all the firm's planes by early summer 1944 made Heinkel initiate studies of new fighters. Heinkel's designers studied in summer 1944 different light jet-fighter designs, and on 10 July they proposed a design titled P 1073. The P 1073 was another "fantasy" project initiated independently by an aviation firm, and was loosely based on an aerodynamic concept study code named P 1063, which DFS started in its main testing facility at Ainring. Its design included two engines mounted above and below the fuselage, a swept-back wing and a V-tail — all highly modern and largely untried design features.[6] In contrast to its eventual development, the P 1073 was to be a cheap metal plane. On 12 July, two days after receiving the blueprints from his designers, Ernst Heinkel gave his go-ahead and ordered the commencement of initial wind tunnel tests and further studies.[7] On the same day Knemeyer also expressed his interest in the project and authorized Heinkel to continue research and development of the design.[8]

Some progress was made in the following months, but the project stayed largely in the study phase. Chief designer Siegfried Günter's proposal for the RLM's specification was a simplified version of the P 1073 with a single engine mounted on its back, straight wing and a conventional double-fin tailplane. In a meeting on 11 September, Heinkel decided to go ahead with the construction of three P 1073 prototypes without waiting for an official approval. Development and production were to take place at the firm's Vienna complex, with Rostock playing only a supervisory role. The timetable set for the project was extremely tight in order to fit the RLM's schedule: completion of initial design by 15 September, handing over of detailed blueprints to factories on 1 November, first flight on 10 December and initial series production in January 1945. Amazingly, Heinkel's highly skilled and experienced experts expected to deliver 2,000 to 4,000 planes by the end of April 1945.[9] On the following day Heinkel's Technical Director Carl Francke presented to Knemeyer the initial blueprints. Knemeyer was generally satisfied, but demanded some design changes, like increasing endurance from 20 minutes to 30 minutes and the inclusion of a steerable nosewheel in order to enable the plane to have better taxi performance on small airfields. In the afternoon Francke briefed Major-General Ulrich Diesing, chief of the TLR, who was also happy with the design and criticized only the limited rear view from the cockpit due to the dorsal mounted engine.[10]

Now the discussion moved to the next level. On 13 September, Diesing and Francke briefed the Main Development Committee under Roluf Lucht. Diesing emphasized in his opening remarks the importance of an inexpensive fighter that could operate from small airfields. After hearing Francke's briefing, the committee focused their discussion on the capacity of the cabinetmaking industry and on the timetable. Hayn warned not to give any specific promises regarding production figures for April 1945. Colonel Geist, Saur's representative, informed the participants that his boss was already briefed about the project and that he would fully support it if it were approved. Heinkel was allowed to continue working on his project while the decision-making process was still in progress.[11] Saur's early support was of key value to the future of the project and signaled the approval of the industrial technocrats. After the meeting Francke informed Heinkel that Saur also gave an initial approval to the P 1073 project and asked all the relevant Reich's authorities in Vienna to support it.[12]

A key Main Development Committee meeting took place on 14 September, when Lucht and his coworkers reviewed the proposals submitted by the different firms. Arado sent no representative but submitted a blueprint similar to Heinkel's under the designation E 580; Messerschmitt presented an old light fighter project taken hurriedly out of the files, and only Blohm & Voss came forward with a serious contender, designated P 211.[13]

In the meantime Heinkel's engineers constructed a mock-up of the proposed plane. By basing its design on an existing project, Heinkel was able to submit a preliminary design within days. In order to fulfill the weight restriction and the ease of manufacture demands, Günter's design was made largely out of wood and used several parts and components from existing aircraft, like main landing gear and wheels from the Me 109. Heinkel estimated that one P 1073 would use only half of the raw materials required to manufacture a single Me 262.[14] As Günter responded to critical remarks by Messerschmitt, his plane was better suited to the prevailing war situation of an ever-worsening raw materials and fuel shortage.[15]

Contract and Priority

On 19 September 1944 the Main Aircraft Development Committee formally decided to pick Heinkel's design. On the 21st, Knemeyer, Lucht and Major-General Diesing briefed Göring about the new aircraft. The three experts' presentation included a crude animation film prepared overnight by TLR and based on draft design, which showed the fighter in different flight situations.[16] Although General Kreipe, chief of staff of the Luftwaffe, and General Galland, leader of the day-fighters, voiced their misgivings regarding the concept, Göring approved the project and acknowledged that he was ready to accept the risks and limitations associated with the aircraft.[17] On the next day the mock-up was unveiled and presented as the He 500, signifying its transformation from a paper project into a real plane.[18] On 23 September, Hitler also approved the plan and stressed its special character as a highly concentrated effort by numerous firms and organizations (termed *Gewaltaktion*) aimed at making the fighter ready for operational use as quickly as possible. He ordered a monthly output of 1,000 aircraft and also declared his willingness to accept the risks involved in authorizing mass production of an unproven design directly from the drawing board.[19] He also appointed General Director Phillip Kessler, chairman of the Armaments Advisory Board in the Armaments Ministry, to lead the project. Kessler was the former general director of the Bergmann Electricity Firm and was responsible for the reorganization of the ball-bearing industry after the bombing of the Schweinfurt ball-bearing factory in 1943. He was known for his abilities to manage difficult projects under tight schedules.[20] Later on the same day Heinkel presented Lucht and members of the Main Aircraft Committee the crude early mock-up of the aircraft.[21] On 27 September Francke informed all the people involved in the project that its designation was changed to He 162.[22] It was included in production program 226/2 under the designation 8–162.

Even though the Blohm & Voss contender lost, the firm's chief designer, Dr. Richard Voigt, tried to reapply on 2 October with a refined project plan. Saur rejected the new submission and explained in his rejection letter that Heinkel's plane was chosen because of the free capacity his firm could offer and because his design was more mature. These remarks make it clear that Heinkel's plane was the only rational choice right from the start.[23] While Blohm & Voss also possessed some free capacity, its size was smaller than Heinkel's, besides, this firm had never designed and produced aircraft in meaningful numbers.

It was especially important for the Reich's leadership to spread the word about the high priority status given to the He 162 project. It constantly sought ways to support the project through different decrees and nominations. These measures were typical of the haphazard management of the Reich's war industry at that stage. On 12 October 1944, Hitler issued a special decree that allocated special commissioners to each of the most important production projects of the aviation industry. Hitler empowered the commissioners with supposedly unrestricted authority to do everything in order to complete their tasks. Among other previsions, the decree upgraded Kessler's status to a Special Commissioner of the He 162 program.[24] It is interesting to note, though, that existing documentation suggests that Karl Frydag performed most of the daily management of the so-called *Aktion 162*. Frydag's position was perfect for this task because at that time he chaired the Aircraft Main Committee (formerly the Airframe Main Committee) and was deputy chairman of the Development Main Committee.

Special efforts were made to win the support of different Nazi Party and state organizations active in the Vienna area. On 28 October, Kessler met Vienna's *Gauleiter* Baldor von Schirach, representatives of the armaments authorities and other local functionaries, and briefed them about the project and its high priority. His talk was titled "Brechung der Luftherrschaft des Gegners durch Gewaltaktion 162" ("Breaking the air superiority of the enemy through massive operation 162") and emphasized the military importance of such a plane. More importantly, he mentioned that the project was ordered personally by Hitler. Kessler finished his speech with the words: "The Heinkel firm developed the aircraft that will wipe out the terror flyers from our skies."[25]

Kessler emphasized in his speech also the need for stringent secrecy, but by that time at least one security glitch had already happened. A roll of He 162 blueprints was found on the street in the town of Randstein, near Vienna, and was brought to the local police station. Keeping diagrams and plans in Heinkel's Vienna factory complex, where many workers were foreigners or inmates, was also problematic and required special care.[26]

Despite the decrees regarding the urgency of the He 162 project, its status remained unclear for a couple of months and well into 1945. For an unexplained reason the high-priority status of the He 162 got lost somehow in the chaos of this period. In early February 1945, Heinkel executives complained that a lack of a written order, giving the He 162 the same status as other high-priority aircraft, was making it difficult to acquire materials and services necessary for the program. One of the difficulties caused by this misunderstanding was securing adequate power supply to factories involved in the program. As part of a wider power-saving measure, the local authorities in Rostock ordered the suspension of the power supply to Heinkel's Marienehe factory for eight days, beginning on 29 January. Following an urgent inquiry the director found out that the order came from the Rüstungsstab and that no one was aware of the high priority assigned to the fighter produced in this factory.[27] The Vienna factory complex experienced similar problems with power supply disruption at the same time.

On 6 February Francke sent an urgent telegram to Kessler and Lucht (head of the Development Main Commission), and asked them to immediately include the He 162 in the emergency production plan and to issue a written document to prove that.[28] It took some time, however, to set the wheels in motion. As the aircraft was finally ready for series production towards the end of February 1945, Hitler formally included the He 162 in the *Führernotprogramm* (Führer's emergency program) during an armaments conference with Speer, Göring and Messerschmitt that took place on 26 February. This program gave the top priority to the newest aircraft types: Me 262, Ta 152, Ar 234 and Do 335. It also foresaw an output of 1,000 He 162s in April 1945 and 2,000 in May, making it the most numerous type of aircraft.[29]

The official timetable set for the project was without precedent and adhered closely to Heinkel's earlier timetable: full-scale mock-up by 1 October 1944; first flight by 10 December 1944; and series production begin on 1 January 1945.[30] According to production program 227, monthly output was expected to increase to 1,150 and continue on this pace at least until the first quarter of 1946. Later the more realistic production programs 228/1 and 228/2 projected a monthly output of 530 aircraft from the end of 1945. This production rate was supposed to continue until March 1946.[31]

Although approved by the highest Reich authorities, the decision to go ahead with the He 162 encountered some stiff resistance. Its main opponents were Willy Messerschmitt and General Adolf Galland, chief of the Luftwaffe's day-fighters. Although Galland's criticism allegedly became milder after he viewed the mock-up of the aircraft on 7 October,[32] both he and Messerschmitt basically argued that developing and producing a completely new fighter at that time would be possible only at the expense of the proven and more advanced Me 262. They were also pessimistic regarding the date projected for the plane's service entry.[33] Galland also criticized the plane's tactical limitations: light armament, poor view from the cockpit, and potentially dangerous flight characteristics. Messerschmitt and others pointed out the contradiction between the high-performance and the "popular" character of the aircraft; high-performance aircraft demanded well-trained pilots and complicated construction. The He 162 in its proposed shape was definitely not the aerial equivalent of the Volkswagen car, as some officials suggested.[34] General Kreipe had pointed out already during the 21 September meeting that the training organization of the Luftwaffe would not be able to train the number of pilots required to fly the plane. It was an important and realistic argument, which most people involved in the light-fighter program failed to address.[35] The debate around the He 162 was one element within a larger political struggle within the Luftwaffe's high command, and especially among its fighter leaders' fraternity. As a result of this struggle Göring dismissed Galland in late December 1944 and replaced him with his main rival, Colonel Gordon Gollob. The dramatic reshuffle at the top of the Luftwaffe resulted in what Göring called "a mutiny without a parallel in history" of several senior Luftwaffe officers, who openly clashed with Göring in January 1945 and demanded that he reinstitute Galland.[36]

As the German leadership was approving and reapproving the He 162's status and arguing its logic, Heinkel continued to work energetically on the plane. In order to stick to the timetable Heinkel employed almost its entire design staff—370 men—in designing the He 162. It was stated that under normal circumstances only 150 workers would be employed on such a project.[37] It is therefore a little surprising that in accordance with the generally chaotic German system, at the same time Heinkel worked on another light-fighter development project. It was a tiny rocket interceptor code named "Julia" or P 1077, that was supposed to be extremely easy to fly because it took off automatically from a catapult and either landed on skids or the pilot would bail out after the fuel ran out. The aim was to train new pilots to fly the tiny fighter in four weeks after some basic flight training.[38] Its development started in late summer 1944 as part of an RLM program to design a replacement for the unsatisfactory Me 163 rocket fighter. On 8 September 1944 the RLM contracted Heinkel to manufacture 20 prototypes, and two weeks later it issued a preliminary plan for a monthly production of 300 "Julia" fighters. The Main Aircraft Development Committee suspended the "Julia" project towards the end of 1944 after it decided in favor of an improved Me 163 design developed by Junkers.[39] Heinkel continued, however, to develop "Julia," including a new version powered by a pulse-jet engine. A small staff at the Vienna design bureau and staffs in several small cabinetmaking firms working for Heinkel in Austria continued working on the project on a part-time basis.[40]

Development of the He 162

The development work went ahead at the highest tempo. Most of the people involved in the design work worked 90 to 100 hours a week and even slept sometimes in their offices. Air-attack alarms in Vienna and transportation difficulties frequently disrupted their work.[41] Relatively little testing was done with aerodynamic models during the development phase, mainly in the wind tunnels of the LFA in Braunschweig and of the AVA in Göttingen, but some of these tests started only after manufacture of the prototypes was underway.[42]

From early on Heinkel planned to produce two basic variants of the aircraft: a single-seat fighter and a double-seat trainer. Experience with the Me 262 clearly showed that training or converting pilots to fly jets was not simple at all, and a trainer was viewed as indispensable. Galland demanded that 3 percent of the production be trainers. Two type trainers were to be constructed: a wooden glider similar in form to the real aircraft, intended for basic handling training, and a powered version with a limited fuel supply for advanced training.[43]

At about that time some people in the Reich's leadership started to refer to the He 162 as the Volksjäger, or "People's Fighter."[44] The origin of this nickname is allegedly in an idea Colonel-General Alfred Keller, the head of the National Socialist Flying Corp (NSFK), brought up in late summer 1944. Keller conceived a small jet fighter which could be flown in large numbers by hastily trained young pilots.[45] There is no evidence that this concept in any way influenced the decision makers to initiate the light-fighter program in early September, and the term had never been used in the correspondence of the firms involved in the program, nor in Luftwaffe's documents dealing with the aircraft. However, the name soon became unofficially associated with the plane.[46] Such a name was in line with a trend existing at that time to use the prefix *Volk* in relation to different military organizations and an assortment of weapons — usually very basic and cheap. It should also be remembered that on 25 September — two days after authorizing the aircraft — Hitler announced the establishment of the German National Guard — the *Volkssturm*. Just like the name of this popular militia, Volksjäger was a purely propagandistic term supposed to represent the totality of the war for Germany.

The popular nature of the plane was further enhanced when Saur dovetailed Keller's original idea and suggested to train Hitler Youth (HJ) boys to fly the new fighter. He suggested that they be trained on the wooden two-seat glider version of the He 162 and then proceed directly to fly the fighter.[47] This idea was also not completely new, as several suggestions were made earlier to establish a Luftwaffe Hitler Youth outfit on the same lines of what the Waffen SS did in mid–1943 with its HJ Division.[48] The Rüstungsstab discussed the idea on 5 October, and Keller agreed to take the responsibility for pilot training, using training gliders manufactured by NSFK workshops to train cadets from the HJ manpower reservoir. By that time the NSFK had already gained some experience in high-speed pilot training, as it performed basic training of Me 163 rocket fighter pilots, which required glider flying skills.[49] According to at least two sources the NSFK established an HJ training detachment titled *Jagdfliegernachwuchs für Sonderzwecke* (Fighter pilot cadets for special purposes) at Trebbin, south of Berlin, and some engineless He 162s were flown there by instructors from late March 1945. One of the gliders was test-flown at Trebbin on 4 April by well-known test pilot Hanna Reitsch. However, no HJ pilot ever finished the training course, let alone flew an He 162.[50] It seems, though, that although Göring received Saur's memo, the idea was never officially adopted. The mere fact that the grand boss of Germany's aircraft production came up with such an absurd idea reflects the irrational thinking prevailing in the German administrative and technical leadership in these desperate days.

In late October the He 162 project was well underway. Initial design work was completed and manufacture of the first two prototypes progressed despite problems with delivery of components from outside suppliers — especially from the cabinetmakers.[51] Manufacture of the prototypes started in Vienna and in its "Languste" underground dispersal plant on 25 October. By mid–November 1944, some 116 skilled German workers and 257 inmates worked on the first planes.[52] While the prototypes were nearing completion, Heinkel began to manufacture the initial batch of 30 pre-production aircraft. The first prototype (designated He 162V1) was rolled out on 2 December 1944. Test pilot Gotthold Peter first flew the prototype from the Schwechat airfield on 6 December 1944 — only three months after the conception of the project and four days ahead of schedule. The flight had to be cut short after 12 minutes because a wooden main undercarriage door broke away in mid-flight. After the landing, Heinkel technicians found out that the mishap was the result of defective bonding of the glued joint.[53] This minor structural failure was a bad omen, reminiscent of the problems that had doomed the Ta 154. Four days later the prototype crashed in Schwechat during its second flight in front of numerous important guests, including Kessler and his staff, and local Nazi leaders. One of the wings disintegrated during a high-speed low-level pass and the aircraft plunged into the ground, killing Peter. Francke, who was responsible for the flight, was later criticized by a coworker for turning the plane's second flight into a flight demonstration and for allowing it to take place in bad weather conditions.[54] Subsequent investigation, which was aided enormously by a film of the accident shot by military cameraman Lieutenant Helmut Kudicke,[55] revealed that the cause was another and more serious case of defective glue bonding of the wooden component which disintegrated under the stress of maneuvering at high speed.[56] This faulty bonding looked like a symptom of a wider problem in the wood part manufacture. Further investigations suggested that particularly the lamination of the wood parts in "Languste" suffered from the high humidity in the underground galleries.[57] Consequently Kessler ordered to strengthen the wooden wing and to exercise stricter quality control at the cabinetmaking firms "so that even a crazy guy will not be able to cause some damage."[58] Heinkel also imposed severe flight restrictions until the problem could be solved.[59] Problems with the wooden components persisted, and in early January 1945 Heinkel even asked his design team to consider replacing the wooden wing with a metal wing.[60]

Despite the early misfortune, the project continued. Carl Francke, Heinkel's technical director and an experienced test pilot, flew the second prototype (V2) for the first time on 22 December 1944. Later that same day Paul Bader, a test pilot from the Luftwaffe's Rechlin Test Center, flew the same plane again. He was generally satisfied, but pointed at some instability.[61] The development program moved on at full speed, and as customary in the German aviation industry, by mid–November 1944 there were already a dozen further developments of the basic airframe on the drawing board, including two with different jet engines, one with a V-tail and the two aforementioned two-seat trainer versions.[62] The V-tail version was considered to be an important development because of the enhanced stability and controllability it theoretically offered. Heinkel asked DFS in late September to test this tail on his old He 280V8 jet fighter prototype. Due to different difficulties DFS could not run the test until April 1945, and by that time the original twin tailfin configuration was considered satisfactory enough.[63]

All the flight-testing of the new fighter was carried out in Vienna. Earlier, new prototypes were usually taken after the first flights to one of the Luftwaffe's test centers — usually Rechlin — and were test-flown there. At that time of the war most flight-testing was carried out at the firms' airfields in order to save time and fuel. Test personnel from Rechlin were usually sent to observe flight-testing and to try new planes for themselves.[64] At the beginning of January 1945 the Luftwaffe ordered the establishment of a special operational trials unit for the aircraft. *Erprobungskommando 162* was expected to start operations at the Lärz airbase by February.[65] However, the program kept suffering setbacks, and as a result the trial unit was never established. During advanced flight-testing of more prototypes the plane displayed poor flying characteristics. Prominent among them were general instability — particularly on the lateral axis, ineffective rudders, excessive side slip and leaking wing fuel tanks. The aerodynamic difficulties dictated additional wind tunnel tests in the research facilities of the DVL in Berlin in January 1945. Heinkel also consulted Dr. Alexander Lippisch, one of Germany's top aerodynamicists and designers. Further tests were conducted in the water tunnels of the Hamburgische Schiffbau Versuchsanstalt in Hamburg—a shipping research center.[66] These tests provided valuable data about the airflow around the airframe, and as a result the wings and tail assembly were redesigned. One of the most important changes was the addition of new wing tips canted down at a 45° angle, as suggested by Lippisch, and accordingly named "Lippisch Ears."[67] Another problem that was difficult to solve was the leaking of the wing fuel tanks. The wooden wing was of the "wet" type. In such wing no separate fuel tanks were installed within the wing structure. Instead, the wing itself was a fuel tank. It was extremely difficult to make the wood skinning of the wing completely leak-free and the problem was solved only at the beginning of March after some improvement of the skinning lamination process.[68] The redesign and more problems that surfaced during flight-testing of the prototypes delayed the delivery of the first aircraft to the Luftwaffe by more than three months.

The third and the fourth prototypes (M3 and M4)[69] of the plane flew for the first time in early January 1945, and more planes, including the two first A-0 pre-production planes, joined the flight-testing program during this month. A strengthened wing, which was tested earlier in the water tunnels in Hamburg, was installed on them, along with other improvements. Further tests of the improved aircraft proved, however, that it was still far from perfect. By mid February 1945 several high-ranking Luftwaffe officers expressed again their reservations considering the usefulness of the new fighter. This cold attitude resulted mostly from the unsatisfactory performance of the plane once Heinkel delivered some of the development aircraft to the Luftwaffe's main flight-test center at Rechlin. General Kammhuber, the Commissioner for Combating Four-Engine Enemy Aircraft (*Kommissar für die Bekämpfung 4-motoriger Feindflugzeuge*— another pompous title invented by Hitler), estimated that the He 162 would come to nothing. Gollob, General of the Fighters, and Colonel Petersen, chief of the Luftwaffe's flight-testing, complained that the aircraft used the same long runways as the Me 262. General Eckhard Christian, chief of the Luftwaffe's operations division (*OKL/Führungsstab*), thought that the He 162 was extremely difficult to fly.[70] Gollob, the future user of the plane, flew it for the first time on 10 February. He and the chief technical officer in his staff flew prototype M3, and after landing left the impression that they liked the plane. Obviously Heinkel's personnel hosting the two received the wrong impression.[71] Therefore, the Luftwaffe was highly skeptical about the aircraft it was about to receive in huge numbers.

Despite these reservations, flight-testing progressed quickly in January and February 1945, testing different redesign features, but they repeatedly revealed more problems. As Francke summed up in his weekly report on 28 January, the series of changes required to solve these problems was no reason to stop the project, but they caused repeated delays.[72] As a result, by early February only 4 prototypes were ready; 14 were still under construction or receiving the latest updates. Earlier plans to deliver 20 planes to different flight-test centers in January therefore failed.[73] Another fatal accident happened on 4 February, when test pilot Georg Wedemeyer crashed with prototype M6 in Vienna. Consequent investigation of the wrecked plane again revealed severe deficiencies in the bonding of the wooden parts. As investigators discovered, "The ailerons of the M6, with which Wedemeyer has crashed, were beyond contempt: hardly any rib was glued, the layers of the plywood skinning were loose from each other, without any bonding by the glue film."[74] Ernst Heinkel was furious and demanded that those responsible be found. The faulty manufacture was traced to the firm Albrecht of the Kalkert production district and Heinkel ordered that all the parts received from this source be scrapped.

Another finding of Wedemeyer's crash investigation also enraged Heinkel. One additional novel feature of the He 162 was an ejection seat developed earlier by Heinkel. The ejection seat was practically the only way to get out of the aircraft in case of an in-flight emergency because the engine was mounted on the back of the aircraft behind the cockpit. Following several accidental pulls of ejection-seat handles while aircraft were still on the ground (mostly in the He 219, the first series aircraft equipped with an ejection seat), a safety wire secured the handle and was removed only before flight. The investigators found this wire still in place in the remains of Wedemeyer's aircraft — a major breach of pre-flight checks and safety routines. Heinkel discovered that this accident was only part of a general ignorance of safety rules, many of them issued personally by him after the crash of the first prototype. One of them concerned lack of communication between the test pilots and flight test engineers on the ground due to faulty equipment and poor training. As Heinkel pointed out, test pilots did not bother to report regularly in real time about the progress of their flights, thus making it difficult for investigators to find out what happened in case of an accident.[75]

Frequent Allied bombing raids on targets in and around Vienna added to the difficulties of the flight-testing. Daily work breaks in Heinkel's facilities due to air-raid alarms averaged 5 hours in mid–February. In the week of 11–18 February there were six attacks, each of which lasted 1.5 to 3 hours. These attacks caused power cuts, transportation disruption and bomb damage to the Schwechat settlement, where most German skilled workers lived. The factory at Schwechat-Heidfeld was also hit, and Francke estimated that by that time 80 percent of its facilities lay in rubble. During the last attack aircraft number M31 was destroyed and two other aircraft were slightly damaged.[76]

Despite these hardships most of the aerodynamic and technical problems were solved by the third week of February. There were still some issues of instability on the lateral axis, which had handicapped the plane from the start, but it was not as bad as it was before the redesign.[77] Nevertheless, more misfortunes plagued the flight-testing program. These were caused mainly by the crude construction of the plane and by the unreliabilty of some of the new items installed in it. On 25 February M3, the third prototype, flown by civilian test pilot and engineer Full, crashed. Full used the aircraft's ejection seat, but was killed when his parachute failed to open. The cause of this crash was an engine fire. Heinkel's technical director Francke also experienced a mishap with M20 when its landing gear collapsed during landing.[78] On 2 March, M25, with an experimental lengthened fuselage, was heavily damaged after a faulty fuel gauge failed to warn the pilot that his fuel supply was exhausted. This accident was therefore not caused by a design fault.[79] By that time, however, it looked like most of the problems were solved and the aircraft was finally ready for series production.

One constant source of problems that had nothing to do with the plane's design was the engine. The BMW 003 engine was in series production, but still suffered from poor reliability, as reported on 6 November by test pilot Peter after visiting Arado's flight test center. Peter summed up pessimistically: "An emergency landing should be always taken into account when flying an aircraft equipped with the BMW 003A-1." He recommended equipping the aircraft with different fire protection means and with a braking parachute in order to ease emergency landings.[80] After reading this and other negative reports, Heinkel demanded that the designers do all they could to reduce the risk of fire and reminded them that Göring gave Heinkel the contract on the condition that the plane would not burn.[81] Although the reliability of the engines constantly improved, as the loss of the third prototype demonstrated, early jet engines remained unreliable and prone to different in-flight malfunctions until the end of World War II. Their unsteadiness was demonstrated again by two accidents that happened in the first half of March. On 8 March a series of flights tested the stability of the aircraft. The last flight of that day, however, ended with a slight incident. The engine flamed out as a result of high-G maneuvers and the pilot was forced to perform a dead-stick landing, which lightly damaged the aircraft.[82] Four days later M8, the 8th prototype, was destroyed as a result of another engine flameout. The aircraft crashed during landing and burst into flames. Sergeant Wanke, an operational pilot with II/JG1 converting to the new plane, was rescued from the wrecked plane with burns.

Although the design work of the basic model of the He 162 was largely finished by that time, by the end of the war several further variations were in different stages of study, design and development. Some of them were equipped with different engines, including the Jumo 004, which was considered slightly more reliable than the BMW 003. Another development was an He 162 made completely out of welded steel. Blohm & Voss, which was highly experienced with welded steel construction, finished almost 50 percent of the construction of a prototype before the war's end.[83]

Although the developers encountered various problems during the He 162 development and testing, considering the extremely tight schedule and compared to other development programs, this program was quite efficient. It took much longer to develop conventional fighters, let alone jet fighters like the Me 262. One element that helped to cut short the development was, of course, the availability of a serially produced jet engine — something that was not available to the Me 262 for two years. In this respect the short development period of the He 162 highlights not only the level of German aviation technology at the end of 1944, but also its limits up to this time, specifically the development of jet technology. Surprisingly, pilots considered the production He 162 to be a flyable aircraft and potentially an effective light fighter, at least in the hands of a well-trained pilot. Emil Demuth, one of the few experienced Luftwaffe pilots to fly the plane operationally, considered it to be a first-class combat aircraft.[84] British test pilot Eric Brown, who test-flew the He 162 in England after the war, thought that "it could have run rings around the contemporary Meteor."[85] Therefore, its development can be viewed as a quite successful one, especially when considering the tight schedule and the extremely difficult circumstances.

Initial Production Planning and Organization

While the engineers designed and manufactured the prototypes, Heinkel, the RLM and the main committees working under the Rüstungsstab made preliminary preparations for the mass production of the aircraft. Before the end of September they worked out an initial network of firms which could participate in the program. The industry-wide character of the project was made clear when on 27 September representatives of Heinkel, Mimetal, Gotha, Siebel, Junkers and Focke-Wulf met at Heinkel's design bureau in Vienna to discuss different aspects of the aircraft's construction and manufacture. Junkers and Focke-Wulf representatives were invited especially because of the experience gained by their firms with the manufacture of largely wooden aircraft (Ju 352 and Ta 154, respectively).[86]

Capacity limitations dictated the sharing of production between Heinkel and Junkers — the two former bomber producers that were left with large unused capacity after their bomber production was terminated. Heinkel's share was one-third of production and Junker's, due to larger capacity, two-thirds.[87]

The most important location in the early phase of the project was the Vienna-Schwechat complex, where most of the development and prototype manufacture took place (although early fuselages were produced at Helling, near Rostock). Even before it was decided to produce the He 162, Heinkel relocated some of its Vienna facilities to several underground galleries of a collapsed and flooded gypsum mine at Hinterbrühl, southwest of Vienna, which became a subterranean lake and a tourist attraction. The place was commandeered in May 1944 and received the code name "Languste," but it took time to pump out the required amount of water in order to convert it into a 12,000-square-meter facility. Around 2,400 workers, mostly inmates from Mauthausen, worked on the construction site.[88]

In order to accommodate them a sub-camp code named "Lisa" was constructed nearby and became part of a growing web of camps serving different Heinkel enterprises in and around the former Austrian capital. These camps were administered by an SS staff under the leadership of Anton Streitwieser based in the Vienna-Florisdorf camp.[89] Converting the Hinterbrühl galleries into a factory was not an easy task. Work dragged on mainly because of air-conditioning difficulties, the danger of ceiling collapse, and shortages of building materials and pieces of equipment.[90] In order to fill the gap, Heinkel turned some of the barracks of the labor camp into workshops. In early September, Heinkel decided to place its light-fighter prototype manufacture (*Musterbau*) center in these workshops. In a meeting about the P 1073 project on 11 September, Francke, head of the technical department in Vienna, asked for an allocation of 200 Germans and 200 inmates to "Lisa," as well as 200 square meters of floor space in the camp until 1 October, in order to start manufacturing the prototypes. He also asked for 50 guards, either from the SS or from the Luftwaffe.[91] The plant was organized as a concentration camp and the first detachment of inmates arrived from Mauthausen in late September.

Gradually most of the workshops were moved from the camp into the underground galleries. By 15 November a total of 568 men worked in "Languste," 427 of whom were prisoners. While prisoners formed 69 percent of the manpower allocated to prototype manufacture, their share in the series production plans was 87 percent.[92] According to testimony by former Polish inmate Marian Siczyńsky, most of the inmates came from Poland and the Soviet Union, but there were also groups of Frenchmen and Italians. Most of the inmates came from one of the other Heinkel camps in the vicinity.[93]

At the same time, work on the prototypes and initial series production was carried out in two additional factories: Vienna-Heidfeld (the current location of Vienna's Schwechat international airport) and "Julius"— a Junkers concentration camp factory near Schönebeck.

Out of 1,315 people working in these factories at that time, 835 were inmates. In series production 72 percent of the total of 757 workers were inmates and 2 percent were POWs. In "Julius" 88 percent of the 261 workers were inmates and in "Languste" 87 percent of the 195 workers.[94] One month later, on 15 December, the total number of workers increased to 1,352, and the number of inmates increased to 841. A largely unfulfilled demand for 500 additional inmates that were supposed to arrive from mid–November was the main cause for this insignificant increase.[95] The inmates in the underground factories worked in three 8-hour daily shifts.[96] Later statistics show that "Languste's" 2,000-strong workforce included around 1,700 inmates.[97] When series production started, "Languste" produced some 198 fuselages and delivered them to a final assembly plant near a Luftwaffe base at Vienna-Heidfeld. Later some of the workshops of this factory were also moved underground, to the nearby storage cellars of a local brewery. These workshops were code named "Santa I to III" and they also hosted the workshops of two Austrian aero-engine manufacturers. According to an unverified source, most of the workforce of the Schwechat factory came from the Schwechat I and II concentration camps, where some 2,500 people were locked up.[98]

The Mittelwerk complex also played a role in the He 162 story. Already in early October someone suggested opening an additional production line in the *Nordwerk* of the huge underground factory complex in the Harz Mountains. During October, Mittelwerk representatives visited Heinkel's office in Vienna and were fully briefed about the He 162 project.[99] On 19 October, Mittelwerk's planning director Alwin Sawatzki submitted to Kessler a production plan for 1,000 planes and 2,000 BMW 003 engines in a 25,500-square-meter facility code named "Schildkröte." The completed aircraft were supposed to be fly out from the nearby Nordhausen airfield. Sawatzki planned to employ in the new factory 8,000 workers — 7,000 of them were to come from the concentration camp reservoir.[100] On 8 November, Kessler authorized the plan and gave his formal go-ahead.[101] However, due to different logistical difficulties it became clear in early December that Mittelwerk would deliver its first aircraft only in May instead of March.[102]

Mittelwerk, as one of the most brutal slave labor high-tech industrial complexes, was an excellent choice for the He 162 program, which was also based right from the start on slave labor as its main workforce.[103] Looking at the entire early production plan of the He 162, it is clear that the share of prisoners was exceptionally high right from the beginning. This planning reflects the prominence of slave labor in aviation production at this phase. Slave labor was no more an improvisation, but an integral part of the planning process. As Francke's suggestions during the 11 September meeting clearly demonstrates, slave labor became a normal and accepted measure not only at the "Nazi" leadership level, but also by technical personal and factory executives.

On 15 November 1944, Heinkel submitted a detailed project plan, including lists of required raw materials and workforce, for the manufacture of the prototypes and for subsequent serial production. Production was supposed to take place in highly dispersed complexes, which required a massive construction and transportation effort, mostly by OT.[104] The way the aircraft was constructed — as was obvious from the plans submitted by Heinkel — dictated the way it was supposed to be produced. While the fuselage was constructed largely from aluminum, the wings and the tail surfaces were made of wood. As was already known from the experience with the ill-fated Ta 154, constructing high-performance aircraft out of wood was not an easy task. These difficulties were further aggravated by the

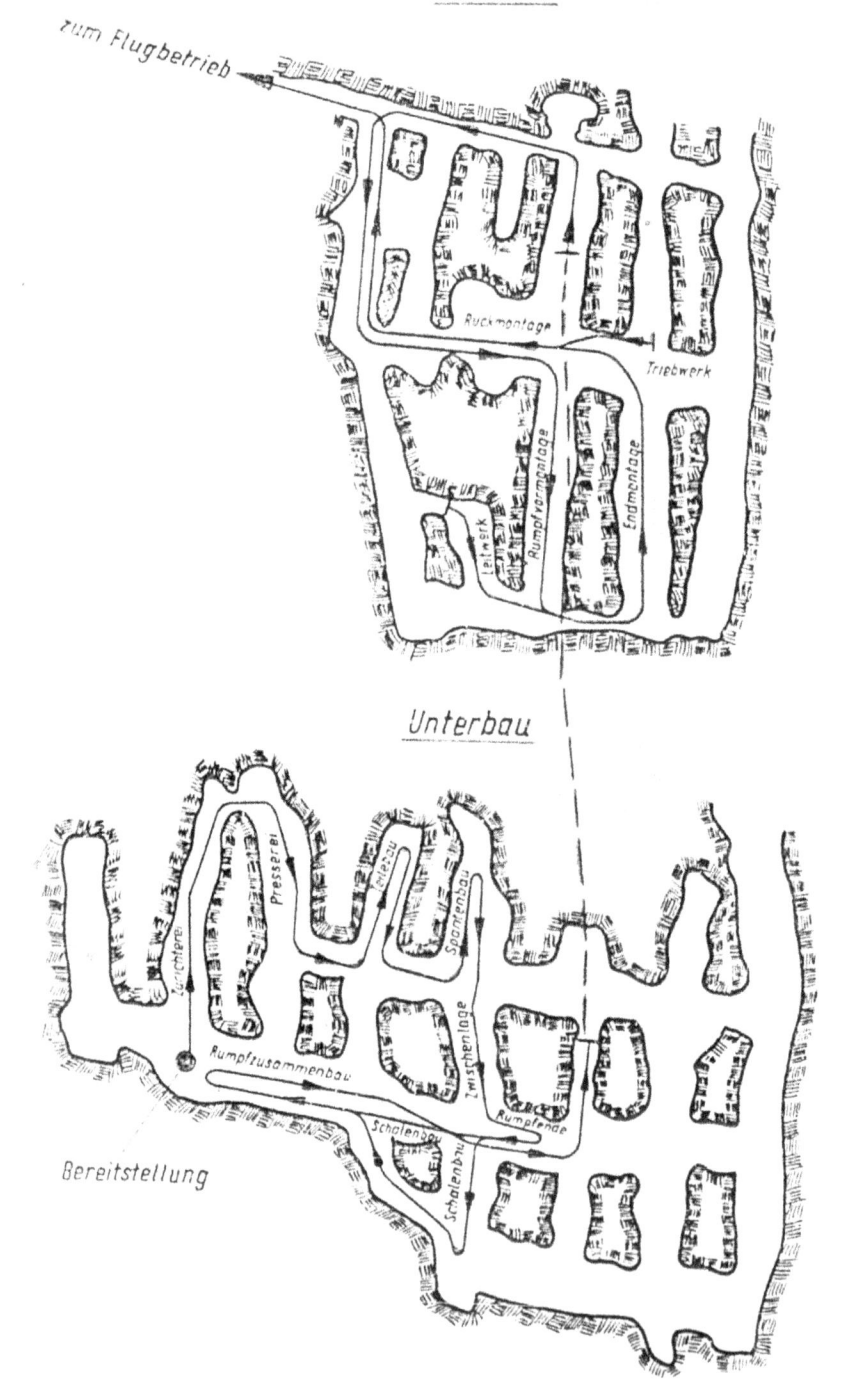

loss of much of the know-how of wood high-performance aircraft manufacture after metal aircraft became dominant before World War II.[105] Therefore, the project management decided that while Heinkel and Junkers would produce the fuselage, production of the wooden parts must be allocated to the traditional cabinetmaking industry. The elaborate wood components production network will be discussed separately.

Towards the end of 1944 the production plan of the new fighter expanded and included more factories. The two main contractors started constructing more production lines for the He 162. Heinkel's Marienehe and Junkers' Bernburg factories established final assembly lines. The Junkers production was directed by Bernburg's director Richard Thiedemann, who after the war became Junkers' general director. Heinkel's concentration-camp factory in Barth and a Junkers underground factory near Stassfurt (code named "Ludwig II") established assembly lines for fuselages, tail surfaces and engine cowlings. These factories were supposed to deliver components to the four final assembly factories. Each final assembly plant was allocated its own range of series numbers: aircraft produced at Marienehe were allocated a serial number beginning with 120000; those produced at "Languste" were numbered 200000; the Bernburg planes were numbered 300000 and those from Mittelwerk 310000.[106]

During January 1945 the He 162 production became more widely dispersed as the main contractors as well as new subcontractors opened more production facilities in different places. This dispersal was termed *Absicherung* (safeguard) and included part of the fuselage manufacture in a new workshop near Marienehe and fuselage assembly in the "Berta" labor camp near Düsseldorf. Construction of production lines in a large forest factory code named "Robert" started in November 1944 near Rostock.[107] Around 1,500 slave workers were employed on this construction site, but it seems that no plane was ever produced there.[108] Construction of a second forest factory code named "Karl" started in early February. Some production workshops were also established at the Perchim, Neubrandenburg and Neustadt-Glewe airbases.[109] In March 1945 Junkers' Aschersleben factory and the "Tarthun" underground factory in a salt mine in Egeln, near Magdeburg (also known as "Salzwerke" and "Maulwurf"), were also converted to produce He 162 fuselages. Around 2,500 men and women worked in "Tarthun," 950 of them were foreigners and around 500 were inmates. The workers lived in camps, in villages around the mine, or in the nearby town of Schönebeck.[110] Both factories produced fuselages from parts manufactured in other Junkers factories in Leopoldshall, Halberstadt and Bernburg.[111] The Stassfurt underground factory was located less than 2 Km north of the Leopoldshall factory and formed part of its dispersal. Although most of the He 162-related production was carried out in the underground factory, the nearby above-ground factory provided it with parts and logistical support.[112]

The Germans also planned from mid–January 1945 to start producing the fighter in a WNF factory in Klagenfurt by terminating Me 109 production there. Production was to start in August 1945. One of the cabinetmaking districts in Austria was placed under this old Austrian aircraft manufacturer. The required drawings and blueprints were handed over to the firm in early April. It was too late and WNF never produced a single He 162.[113]

Opposite: "Languste." A Heinkel blueprint depicting the two levels of the "Languste" as well as its production flow, showed by the arrowed line (courtesy National Air and Space Museum, Smithsonian Institution, USN220/408).

BMW provided the BMW 003 engine for the He 162 from its widely dispersed network of factories. A special version of the engine designated 003E-1 was specifically developed for mounting on the fighter's fuselage.[114] Early in 1945 it was planned to supplement BMW's production by opening a 003 production line in the Mittelwerk complex, therefore turning this factory into one of the few centralized aircraft factories, where both airframe and power plant were produced under the same roof. The RLM/Rüstungsstab planned to initiate engine production in a section of Mittelwerk called *M-Werk* in April 1945 with a monthly output of 30 engines, and then to gradually increase the monthly output to 1,500 engines by December 1945. This output, combined with BMW's monthly output of 2,000 engines, was expected to fully cover the monthly requirement of engines for new aircraft, as well as providing reserves for overhaul depots, maintenance centers and operational units.[115] At the time the 003E-1 entered series production, BMW relied heavily on slave labor. For example, 55 percent of the 6,000 workers of the Springen underground factory, producing parts for the 003 engine, were slave workers.[116]

Even without taking into account the cabinetmaking industry, the He 162 production was extremely dispersed. While it was logical to use such a scheme in order to keep production running even if some factories were damaged by bombing raids, this layout certainly ignored the problems posed by the badly disrupted German transportation system at that time of the war. Once series production started, this shortcoming became a major handicap.

The Cabinetmaking Industry and the He 162

The Volksjäger, as a low-cost and lightweight fighter, was largely constructed of wood. Among the wood components were the wings, the vertical stabilizers, the nose cap, landing gear doors, ribs, spars and several smaller fuselage parts. Since the project management had decided earlier to let the traditional cabinetmaking industry produce the wooden parts,[117] its organization as a major supplier for the program became an important and urgent task.

The traditional wood industry was not under the direct jurisdiction of the RLM nor the armaments Industry, but at that stage of the war it was easy to mobilize it. The organization of the wood parts production was entrusted to the hands of *SS-Hauptsturmführer* Dr. Kurt May. May and the SS became involved in the program mainly because of their experience with cabinetmaking businesses. May already managed for the SS the German Noble Furniture firm, and therefore knew the cabinetmaking industry from within. Furthermore, his SS owned firm had worked once as a sub-contractor of Messerschmitt, so he also knew the aviation industry.[118] His organization was designated *Organisation May* and it controlled a wide array of woodcraft and furniture-making industry located mostly in south and central Germany and lower Austria. These manufacturers were organized into 3 "construction districts" (*Baukreise*): Neustadt/Orla, Erfurt (both in Thuringia), and Stuttgart/Esslingen (Swabia). Each district was managed by the director of the most important regional firm.[119] Next to his overall supervision, May also managed the last district.

It should be noted that this organization was loosely modeled on the production plan intended for the abortive Ta 154 multi-role fighter. The *Baukreise* functioned in a similar way to the *Fertigungsringe* planned for the Focke-Wulf wood-plane project. They were expected to supply the final assembly factories in their region with the necessary wooden parts and components, therefore creating several autonomic and self-sustaining production complexes, relying mostly on local suppliers and local transportation. However, before starting production, several technical problems involving the use of wood in a fast jet fighter had to be solved. For this purpose the specialists of the Engineering School in Esslingen were contacted and incorporated into Dr. May's organization. They were ordered to develop the special wing construction and the required wood processing procedures. After completion of the study they handed over the know-how to Heinkel's design bureau in Vienna, which designed the wings, and to May's organization.[120] Several persons involved in the design work reported, however, that the designers never used the methods developed in Esslingen.[121]

Besides the cabinetmaking industry, two other extraordinary organizations became involved in the Volksjäger program because of their experience with wooden aircraft. In October, Saur and the NSFK's leader Keller agreed to produce some wood parts at the organization's workshops.[122] May was happy to integrate the NSFK capacity into his organization because of the experience this organization gained in manufacturing and repairing wooden training aircraft and gliders.[123] The main task of the NSFK was to manufacture the two training glider versions of the He 162. The NSFK was also expected to train a large proportion of the new pilots for this aircraft using its own training aircraft fleet and facilities.[124] The Oranienburg workshops of the now-idle national airline Lufthansa were also contracted to produce the two-seat glider trainer and to retrofit completed aircraft.[125]

The conversion of the wood industry to aviation production encountered several difficulties. By late November 1944 May was behind schedule with the supply of wing sets and tail surfaces for the prototypes. Under time pressure with the He 162 and considering mass production of the "Julia" fighter in the near future, Heinkel's technical director Francke suggested allocating the Heinkel-Vienna branch its own cabinetmaking facilities and personnel at least for the initial production runs. Among other possibilities, he suggested locating and moving a wood-processing workshop from a concentration camp into "Languste," or alternatively to confiscate a couple of Viennese cabinetmaking firms and move their workshops to "Languste."[126] Francke's way of thinking and planning demonstrate again the way concentration-camp production had become embedded in the minds of aviation industry executives by this time. It became a fully acceptable way to solve logistical-industrial problems.

Towards the end of November it became clear that things were not going well with *Organisation May*. On 25 November, Francke wrote to Frydag and bluntly expressed his misgivings considering May and his ability to manage production of the wood parts. He asked Frydag to send an inspector to Esslingen and see what was wrong there. He also asked Frydag to summon the "woodmen" (*Holzmänner*) May, Kalkert and Wächter (leaders of the other two districts) to Berlin and ask them to report in person.[127]

This was only the beginning. Poor workmanship and sloppy quality control in this sector became obvious even before the first aircraft took to the air. As consequences of the crash of the first prototype, the spotlight was turned to the "woodmen" and underlined Francke's misgivings. An inspection tour of five Austrian cabinetmaking firms in early January 1945 uncovered a series of defects, mainly in the lamination process. All the firms failed to separate the different sections of their workshops, and as a result, sawdust from the sawmills sunk into the lamination material and weakened the lamination.[128] Upon receiving the report Heinkel repeated an earlier order to tighten quality control on the wooden components and to integrate the smaller firms into the larger in order to simplify quality control.[129] During the next couple of months poor quality of the wooden parts persisted despite the measures ordered by Heinkel.

Series Production

Production of the He 162 started before the prototypes took to the air. Some problems appeared early, while Heinkel and Junkers still manufactured the pre-production series. The high degree of dependence on a functioning transportation system was the major drawback of a highly dispersed industrial network designed to produce a single product. The production of the plane started at a time Allied air power systematically targeted the German transportation system and supply of parts and components was affected immediately. As mentioned before, finishing the development of the aircraft proved lengthier than expected and caused further delays. Thus the target output of 30 series production aircraft projected for January in Heinkel's main factory at Rostock-Marienehe proved to be totally unrealistic. Components like wings and engines failed to arrive due to the "general situation" ("*allgemeinen Lage*").[130] As a result, the He 162 production line could operate only one shift a day. Only on 27 January, when enough parts and components arrived and 16 aircraft were assembled, was a second shift initiated. Other factories also suffered from lack of parts and components. On 23 January 1945, Francke sent an urgent telegram to Kessler complaining about severe delays caused by slow delivery of major components, like engines, wooden fins and wooden nose caps. At that time there were components for only 14 aircraft in the Vienna complex.[131] Final assembly was carried out at Heidfeld while fuselage manufacture was performed in "Languste." Finished fuselages were taken out of the underground factory through a narrow exit — which caused several accidents — and were loaded on flatbed trucks and transported to Heidfeld. Some 133 Germans and 52 inmates working in the final assembly department mated the metal fuselages with the wooden parts and with the engines.[132] Bottlenecks in the supply of the required jigs to the different factories involved in the He 162 caused further delays. It is unclear, though, if this difficulty was caused by poor transportation or due to difficulties in the production and supply of jigs for the production lines, as happened a year before with the Me 262 production tooling.[133]

The initiation of series production long before the development of the aircraft was finished caused problems specific to this program. Experience gathered during flight-testing dictated constant changes and upgrades to the basic design. A weekly "Changes Committee" discussed these changes and coordinated their incorporation into the aircraft. The "Changes Committee" regularly sent lists of changes to Heinkel and Junkers, which subsequently integrated them into the production process.[134] The biggest problem was how to integrate changes into completed aircraft or to aircraft in advanced manufacture phase. Since aircraft without the latest updates were considered unfit for operational service, it was necessary to deliver them to special workshops, where they received the different updates. The need to retrofit the constant changes was another reason for delays in delivery of the fighter to the Luftwaffe.

Initially Frydag decided to carry out retrofit of early series aircraft at Heidfeld. At the end of January 1945 he ordered to allocate to Heidfeld 50 skilled workers from Lufthansa and 200 to 300 skilled workers from WNF solely for the retrofit work.[135] However, in February he decided to use an outside contractor instead. Heinkel thus contracted the Amme Luther Sack (ALS) engineering firm, which started retrofitting the first 30 aircraft in mid to late February. These included 10 prototypes and 20 pre-production planes. Heinkel evacuated Production Hall 47 in Heidfeld and made it available to ALS, which employed, among others, 20 foreign workers from the East.[136] While this arrangement was satisfactory for the

Austrian production, Heidfeld was located too far from the production centers in Germany. Therefore, the project managers in Berlin decided in early February to perform the retrofit of aircraft produced in Marienehe in an airbase at Ludwigslust and at Lufthansa's workshops in Oranienburg, while Junkers' Bernburg factory retrofitted the Junkers aircraft.[137] As a result of the need to retrofit early production aircraft, the entire production network became even more complicated and therefore more dependent on functioning transportation. This enlargement also further complicated manpower allocations.

During the second week of February it became clear to the project management that as a result of the need to further test the prototypes and to retrofit all the changes to aircraft currently on the production line, the delivery of the He 162 was going to start later than planned. On 10 February, Francke declared that he expected to deliver the first operational plane to the Luftwaffe only in the second half of April.[138] Later a less pessimistic timetable was announced. Heinkel promised Gollob and Lieutenant Hachtel, the commander of the first Volksjäger squadron, to deliver the first operational aircraft by 20 March.[139]

At least some of the problems affecting the He 162 production were caused by mismanagement. The initial problems with Kurt May's organization led Kessler to reorganize *Organisation May* in January. May was demoted and became leader of the Stuttgart/Esslingen production district. Kessler basically accepted the scheme suggested by Francke back in late November 1944 and reorganized the wood parts production by placing the wood firms directly under the airframe manufacturers in their geographic regions. In correlation with the expansion of the airframe producer network, the cabinetmaking network also expanded. By early February the number of wooden parts production districts had increased to 7; three were placed under Junkers, three under Heinkel and one under WNF — the new Austrian partner in the He 162 program.

Even after the reorganization, some bottlenecks in the supply of wood parts remained unsolved. It seems that the main reason for delays in the supply of wood parts was what directors of two wood firms described as "catastrophic" supply of the P600 glue used to bond the wooden parts. More serious bottlenecks existed in the supply of nose landing gears and the BMW engines. As a result of engine shortage, the development of a version powered by a Jumo 004 engine was rushed forward.[140] In early February the production plan of the Volksjäger was revised according to the latest developments. Output share was redistributed among the involved firms: Heinkel was to produce 800 aircraft per month, Junkers 500 and WNF 200 aircraft.

Besides general problems affecting the program, each factory involved in the He 162 production encountered its own unique problems, which provide a glimpse at the state of aircraft production in Germany in the last couple of months of World War II. At Heidfeld, Heinkel failed to deliver the first operational aircraft on 20 March as promised. The main problem here was with the retrofit of the finished aircraft. ALS, the firm contracted to perform the retrofit, was unable to keep the schedule. Surprisingly, at this time of the war and while working on such a high-priority project, its workers failed to show up for work on Sundays.[141] As a result, Heinkel workers were brought in and eventually performed 50 percent of the work ALS's workers were supposed to do.[142] The situation was aggravated by the loss of 500 POWs working at the Heinkel complex. The local military authorities confiscated them at the beginning of March for rubble-clearing tasks in bombed Vienna, completely ignoring their work for an important military project.[143] As a result, by the end of March

Heinkel delivered only 15 of the 20 aircraft it was supposed to deliver during this month. Even then, bad weather impaired their delivery flights.[144]

The first Junkers-produced He 162 flew in Bernburg in the week of 19–25 February and was flown to Vienna on the 26th for a thorough inspection by Heinkel's engineers.[145] Nevertheless, production in the Junkers complex also suffered delays, and the first Junkers series plane flew only on 23 March. The Junkers underground factory in Stassfurt was not ready for production until the end of February 1945, especially due to difficulties in converting the mine in which it was located. A total of 210 men worked in this factory; these included 116 POWs of different nationalities. The workers worked in two daily shifts of 12 hours each, with shift changes at 6:00 and 18:00 o'clock. Oddly, even at this time the German workers of the plant were allowed every other Sunday off. Even after work started in Stassfurt, production proceeded slowly, and by the end of the war this factory had delivered only 50–60 tail surfaces and 7 engine cowlings. This output was far short of its projected monthly output of 600 tail surfaces.[146] The "Tarthun" underground factory fared somewhat better. By the time it was captured by the U.S. Army, it completed 70 fuselages and was still manufacturing 56 additional units.[147]

General deterioration of infrastructure contributed significantly to the low output. Electric power outages became frequent in early 1945 and affected work in most factories. Directors of the Heidfeld factory reported during a meeting with Ernst Heinkel on 12 February 1945 that between 17 January and 10 February their factory was left without power for a total of 104 hours. As a result, development and upgrade of series aircraft done in Heidfeld suffered significantly.[148] In early February, Kessler's staff tried to solve some of the transportation problems affecting the He 162 production by establishing two regional transportation centers—*West* and *Südwest*. Each center possessed a fleet of trucks, which stood ready to transport urgently needed materials to cut-off factories.[149] This solution was not always effective. It generally took longer to deliver the wooden wings by rail or truck to the final assembly plants than to produce the complete aircraft under normal conditions. Many smaller parts had to be hauled by trucks or delivered by special couriers, and even then travel time was up to five times the normal duration of the rides.[150] The transportation crisis also caused a general drop in the production of BMW engines, including the 003 engine. In late 1944–early 1945 it took cargo trains carrying engine parts up to 14 days to travel from middle Germany to the area of Munich. At the end of January 1945 the Armaments Ministry was compelled to allocate BMW trucks as a replacement, despite the acute fuel shortage.[151]

Several serious bottlenecks were created at the final assembly factories due to lack of components provided by outside contractors, like flaps, hydraulic systems, instruments and other parts. The loss of Posen and its industrial complex in early February caused a severe shortage in the supply of the Mk 108, 30mm cannon and forced Heinkel to equip the He 162 with two less powerful 20mm cannons, which the General of the Fighters viewed as unsatisfactory.[152] The deadly bombing of Dresden on 13–15 February 1945 destroyed, among others, a factory producing the new EZ 42 gun-sight intended for use on the Me 262 and the He 162. As a result all the available sights were allocated to the Me 262 and the He 162 received older sights. Heinkel's technical director had to beg Kessler for a monthly allocation of 20 to 50 sights in order to avert what he called a possible "moral crisis" (*Stimmungskriese*) due to equipping of the He 162 with inferior gun-sights.[153] In numerous cases half-completed

U.S. soldiers found these He 162 fuselages outside the Bernburg factory. These aluminum components could not be finished because of failure to deliver the required engines and wood components from outside suppliers (courtesy U.S. National Archives and Records Administration).

aircraft were left unfinished because missing components got stuck on the way. Tooling was also in short supply in March 1945, especially spiral drills.[154]

Adding to the general confusion of these days was the Main Aircraft Development Committee's decision in mid–February to approve part-time work on "Julia," because design staff became available as the He 162 development approached its end. On 5 March, Frydag called Heinkel in Vienna and approved full-time work on the tiny rocket-fighter. He ordered Heinkel to complete the development work in the shortest time possible. Heinkel was ordered to move the development work to its new facility in Neustadt, southwest of Vienna. Part of the work involved in the new project was allocated to some of Heinkel's subsidiaries, particularly from the cabinetmaking sector. The renewed development required further wind tunnel tests, especially in a high-speed tunnel, further design work, completion of two almost finished unpowered M2 prototypes, and construction of two powered M4 prototypes by the Austrian wood firm Schaffer-Linz.[155] In this way the Development Main Committee revived an already abandoned project, which was of little use in this phase of the war, and ordered Heinkel to work on it while it was still trying to sort out problems with the He 162 — definitely a more useful type. It is a prime example of the way research and development kept running as usual even at times of extreme streamlining and type reduction. The development work of "Julia" dragged on and was never finished. The tiny last-resort rocket fighter never took to the air.

Towards the end the advance of Allied armies, especially the Soviet army, heavily disrupted production of the He 162. On 31 March, Speer ordered the evacuation of the Vienna complex to Bad Gendersheim. Chaos was the order of the day during the evacuation. Among others, the Vienna complex director Dr. Schüngel was arrested by the police in Salzburg after a coworker complained that he stole a large sum of money from firm's funds he was supposed to transfer to the new location. He was released after it was found out that the charge was baseless. Eventually Dr. Schüngel and most of the Vienna workers ended the war with Ernst Heinkel and others at Landsberg in Bavaria.[156] Some German and Austrian workers were not evacuated and were drafted instead by the *Volkssturm* in order to put up a last stand against the Soviets.

The evacuation order also set in motion a general evacuation of the Heinkel camps within the greater Vienna area. The "Lisa" camp served as a collection point for some 1,884 inmates from most other camps. On 31 March, camp commander Streitwieser ordered the execution of all ill inmates that could not walk. Fifty-two inmates were consequently executed and buried in the camp's grounds. On 1 April the camp's surviving inmates started the long, exhausting march towards the Mauthausen main camp, during which 152 inmates died.[157] According to one source, 84 of them were shot by their guards.[158]

The "Languste" factory, where most early production run took place, was largely destroyed by German engineers, who exploded 7 bombs in its top level.[159] The Soviets soon captured the complex, dismounted all the functioning machinery, and transported it to the Soviet Union. Around 12 aircraft that were finished in Vienna by the time of its evacuation were flown westwards, but not all of them made it. Components for several aircraft were also evacuated to a small factory near Salzburg, but these aircraft were probably never completed. Later Heinkel's Rostock factory was also evacuated. The Junkers factories, because of their central location, were less affected by the general evacuation until the very last days of the war. Around 20 aircraft of the total of 34 He 162 fighters produced in March were completed there and were delivered to the Luftwaffe.[160] At least three fuselages produced by Junkers were converted into Link simulators, but it is unknown if they were ever used to train pilots.[161] During April the staffs supervising the production of the He 162 were also evacuated: Kessler left Berlin with his staff for Munich-Allach, and Saur went with the Rüstungsstab to Blankenburg, Mittelwerk and finally to Salzburg.

At the end of March or very early in April, Kammler issued the order to curb the production of the He 162 in favor of the Me 262 — the aircraft it was supposed to replace. This decision was in contrast to the contemporary opinion of the Luftwaffe, which thought that production-wise it made more sense to increase the production of the little fighter. Ironically, Kammler did now exactly what the Luftwaffe, Messerschmitt and the RLM had sought so desperately to achieve in 1943 and 1944 — concentrate every effort on the production of Messerschmitt's jet. Kammler's order, however, came too late to influence aircraft production during the twilight of the Third Reich, and production of the He 162 dragged on until the very end. By 10 April 1945, a total of 124 He 162 fighters had been produced, but only 56 of them were delivered to the Luftwaffe.[162] Junkers executives reported after the war that Junkers was able to complete only 29 aircraft by mid–April. Only 15 of these were delivered to the Luftwaffe and 4 were damaged during their delivery flight; the rest were disabled by Luftwaffe personnel on the firm's airfields.[163] It is impossible to tell how many units of Heinkel's last fighters were produced until the Allies overran their factories, but the final figure could not be much higher than the aforementioned 124.

The He 162 in Service

The He 162's entry into operational service represents a microcosm of the interface between the industry and the Luftwaffe when introducing new combat aircraft into service. In spite of the notions of Keller, Saur and others, it was not seriously considered to let inexperienced pilots fly the new fighter—at least not initially, and definitely not with HJ boys. Due to the delays with the delivery of series planes, the Luftwaffe decided to skip operational testing by a dedicated outfit and to commence operational testing within a normal fighter wing. Initially the Luftwaffe planned to establish a completely new fighter wing numbered JG80 to fly the new fighter. However, after the first delays were encountered, the plan was changed, and the Luftwaffe decided instead to convert JG1—an existing propeller fighter wing that had recently lost a large number of its FW 190 fighters.[164] The first contingent of 10 pilots from II/JG1 (the second group of the wing) arrived at Vienna on 2 March. Under command of Lieutenant August Hachtel, the pilots quickly familiarized themselves with their new plane and started flying it. Heinkel made aircraft M19 available to them, and they were limited by the flight restrictions imposed on all pilots flying the plane following the loss of the first prototype: maximum speed of 500 km/h and maximum altitude of 3,000 meters. On 7 March, Hachtel telexed Gollob that eight pilots completed conversion training but that the ground personnel had not yet arrived.[165] Another detachment of 5 pilots from III/JG1, led by Lieutenant Emil Demuth, arrived at the same time from the Perchim airbase but was sent away because no aircraft was available for conversion training due to the demands of the flight-test program.[166]

At least one more prototype, M8, was made available to the Luftwaffe pilots, but it crashed on 12 March with the pilot severely injured. The conversion program suffered another setback on 14 March, when M19 crashed during landing and burst into flames. The pilot, *Unteroffizier* Tautz from the staff of II/JG1, was killed instantly. The cause of the crash was clearly human error. The pilot failed to line up the runway, touched down next to it, and hit a group of barrels.[167]

During March another group of pilots from JG1, commanded by Captain Paul-Heinrich Dähne, arrived at Rostock to begin conversion, but delivery of the first planes was expected in Vienna. Second Lieutenant Hachtel became the Luftwaffe's permanent representative with Heinkel and was expected to accept the first operational plane by 20 March. After Heinkel failed to keep up the schedule, the frustrated Hachtel notified his superiors in Berlin and threatened to leave Vienna. Francke persuaded him to stay but stressed to all others involved in the production of the plane the urgency of delivering the first planes as soon as possible despite the prevailing bottlenecks.[168] Then came another problem. Hachtel received an order from his superiors to accept only aircraft with engines converted to use the standard J2 fuel instead of B4 fuel.[169] B4 was an aviation-grade gasoline in general use by the Luftwaffe and specified originally to simplify logistics by using a single fuel type for all aircraft. J2 was a diesel-type fuel, which was more adequate for jet engines. For one thing, it decreased fuel consumption and therefore increased the endurance of jet-powered aircraft—an important consideration with early fuel-hungry jets. Its main disadvantage was a slow throttle response—a problem considered by Francke as a meaningful deficiency.[170] The conversion required only the replacement of the fuel pump, but it took more precious time.[171] The aircraft finally received by Hachtel proved to be disappointing. Their speed and endurance were below expectations—probably due to poor workmanship and sloppy quality assurance. Hachtel left JG1 soon after accepting the planes and took no additional part in the He 162 program.[172]

The story of He 162 flying with JG1 and its operational record is extremely patchy, especially because no operational record of this unit survived; and since Heinkel evacuated Vienna in early April, there are no records of its technical department afterwards. Obviously a trickle of He 162 fighters was delivered to Perchim by Heinkel and Junkers during the first week of April. Clearly, the plane was not yet fully ready for operational deployment. A speed limit of 600 km/h was set for the early operational training flights because of control difficulties at higher speeds. Pilots who went above this speed experienced instability and loud noises that sounded like "someone was hammering the canopy with a blunt object."[173] Generally, operational pilots of JG1 were not happy with the plane. They judged the workmanship of the planes they received as poor. The poor polish of the wings limited their speed, and speed difference of up to 100 km/h between individual planes was observed. Rudder control was problematic and it seems that the wet wings kept leaking despite all efforts to cure this problem. Pilots who flew both the He 162 and the Me 262 were united in judging the latter to be much superior. On the other hand, pilots converting from piston fighters were more satisfied with Heinkel's fighter.[174] Following a heavy bombing of Perchim on 7 April, I/JG1 relocated to Ludwigslust and continued its training with around 15 aircraft. At around the same time II/JG1 also completed its conversion in Rostock and received new aircraft from the Marienehe production line. During training several German pilots encountered low-flying enemy planes. However, deterioration of the fighter control network, together with the plane's short endurance and the ever-critical fuel situation, hampered the operations. According to some sources, on 19 April an He 162 pilot shot down a British fighter, but was killed when his own fighter crashed on the way back to base. On the following day another pilot successfully used the fighter's ejector seat for the first time. *Luftflotte Reich,* the air command responsible for the aerial defense of the Reich, took over I/JG1 on 23 April and cleared it to commence combat operations. On the following day, the unit's commander, Captain Dähne, was killed in a training accident over Leck, when his plane crashed into a marsh after he went over the speed limit. He tried to eject, but forgot to release the canopy first and was probably killed when his ejection seat broke through the Plexiglas. His death demonstrated again the fact that this was not a plane for beginners.[175] Accidents were quite frequent, and the plane's short endurance contributed to several of them. Between 13 April and the end of the war I/JG1 lost 13 aircraft and 10 pilots — probably only one of them to enemy action. On 22 April, for example, pilot Erwin Steeb became the second person to eject successfully from an He 162 after his horizontal stabilizer disintegrated — probably because of faulty welding — during a test flight. Another pilot was killed after his plane entered a bomb crater during a take-off run.[176]

Because of the advance of the Allied armies, JG1 was forced to constantly relocate its bases in northern Germany. It ended up in the Leck airbase in Schleswig-Holstein. On 26 April an He 162 pilot claimed a second kill of an unspecified aircraft. The two fighter groups flying the plane were merged into one unit under command of Colonel Ihlefeld. On 4 May a pilot claimed a third kill of a British fighter — the only verified victory of the He 162. Second Lieutenant Rudolf Schmidt, a quite inexperienced pilot, shot down a British Tempest fighter near Rostock. Its pilot was captured and was bought to Leck. The following day a cease-fire was declared and JG1 prepared to surrender to the British Army.[177]

The Allies captured scores of completed or partially completed He 162 fighters in underground facilities, in depots and in factory airfields all over Germany. The Americans found several aircraft, including some pre-production aircraft, at the Munich-Riem airbase, and a number of incomplete aircraft near the Bernburg factory. These were mostly completed fuselages found in sheds waiting for their wings and engines.[178] The Americans also captured Heinkel and most of the staff involved in the He 162 project at Landsberg. The Technical Intelligence Section of the USAAF was aware of the intelligence value of these people. In June it enabled them to work for a couple of months at an improvised design office under chief designer Siegfried Günter, and reconstruct the blueprints of some of their advanced projects. In October the Americans collected all the available material, shipped it to the USA ,and dispersed the group.[179] The British Army captured most of the operational aircraft on 6 May in JG1's last base in Leck. The British shipped eleven aircraft to England for detailed evaluation, and the Americans took three aircraft to the USA. The British also handed over seven of their aircraft to the French. Several captured He 162s were flown by western test pilots, and one crashed during an air display of captured German aircraft in England on 9 November 1945.[180] A structural failure in the tail during a low-altitude rolling maneuver plunged the plane into the ground with the pilot inside. The Soviets also captured and tested at least one He 162. In contrast to the more mature Me 262, the He 162 left little impression on the Allies. As the Royal Aircraft Establishment in Farnborough determined after flight-testing the He 162 and other German jets: "The He 162 falls short of the 490 mph ground level and 522 mph at 20,000 feet quoted by the German sources of information, but this is mostly due to the poor thrust of the engine."[181] It had no influence whatsoever on the design of postwar jet fighters.

The Volksjäger program epitomized to the extreme all the main trends of the late-war German aviation industry. It was a desperate creation enabled by Germany's well-established research and development infrastructure. Its production scheme was dispersed from the outset to the limits in order to increase its survivability. Several of the factories producing the aircraft or its components were located underground. The production process of the aircraft was made less complex by simplifying the airframe, and this in turn enabled production on highly streamlined production lines. It was powered by a single engine and was constructed largely of non-strategic wood. This simplicity, together with the high segregation of the production in different locations, enabled the Germans to produce it easily with a mostly unskilled workforce. Since at that phase of the war a large portion of the workforce of the aviation industry was made of slave workers, the He 162 was a "slave workers friendly" product. In any case the He 162 was indeed a cheap and cost-effective weapon system when considering the amount of work involved in its production. Considering all the difficulties and complications described so far, the Volksjäger was economically a reasonable production project at the desperate last months of World War II. It was clearly easier and cheaper to manufacture than the Me 262. Frydag estimated that the production of each aircraft required only 1,500 man hours.[182] This figure compared very favorably with the 9,000-plus man hours required to produce a single Me 262 in early 1945.[183] A single He 162 cost 75,000RM — exactly half the unit price of the Messerschmitt fighter (150,000RM) and almost half the unit price of the conventional Ta 152 fighter (144,000RM).[184]

At the same time the He 162 project represented the total loss of realism prevailing in the German leadership towards the end of the war. It should be remembered that its production plans were prepared after the transportation situation became intolerable. Nevertheless, total dispersal was chosen and in contrast to the trend of clustering production of certain types in a specific region, the He 162 production was dispersed practically all over the Third Reich. Consequently it can be said that the He 162 production scheme was doomed even before it had started.

The He 162 and the concepts behind it were an oddity, but an oddity with its own rationality. This rationality represented everything the German aviation industry became during World War II.

Conclusion

The German aviation industry was a unique and central element of the Third Reich. Its story represented not only part of its economic and industrial history, but also other aspects of its short-lived, but intense history. These included military, political, scientific and social aspects. Understanding the inner works of this industry can therefore shed light on some important aspects of the history of the Third Reich. Aviation industry historian Lutz Budrass wrote: "Arming the Luftwaffe was the largest industrial project of the Third Reich. No element of the national-socialist economy, from synthetic petrol through the 'Four Years Plan' with its 'substitute materials' strategy, to the use of slave labor, can be understood without this project. Out of it came an armament industry, which is unique in the German history due to its size and significance."[1] This research sought to look at the way the German aviation industry evolved and functioned in the latter part of World War II, sketching as a byproduct a broader view of the Third Reich at the time of its downfall.

The German aviation industry and its scientific infrastructure fascinated Allied intelligence officers surveying it after the war. They noticed some of its main characteristics in numerous reports they composed about it:

> The scale on which science and engineering have been harnessed to the chariot of destruction in Germany is indeed amazing. There is no shortage of technical personnel or material facilities, no stinting of financial assistance even for apparently long term and complex developments. There is a unity of purpose in the Germans' application of science to the needs of war; they are not interested in civil aviation.[2]

This sort of observation hit the mark and missed the mark at the same time. It is true that under Nazi rule Germany mobilized and expanded its aviation industry to an unparallel magnitude. This mobilization was aimed foremost at building a mighty air power, which was to support the march of conquest dictated by Nazi ideology. Although most firms also considered civilian aviation at the time of expansion, and even well into World War II, the Reich's leadership was focused on military applications.

The aviation industry became the locomotive of the German wartime economy. Its total revenue, the sheer number of its workers, the proliferation of its factories and the political power of its leaders made this industrial branch a crucial player in Germany's economy even before the war. Like other branches, the aviation industry also transformed dramatically during the war.

Perhaps the most striking change was the gradual and massive shift from an "elite" German workforce to slave labor and largely forced foreign labor. Although National-Socialist ideology and manpower shortage were the primary forces behind the general shift from a German workforce to slave labor, in the aviation industry several other factors played a significant role in the implementation of slave labor as a primary workforce. Foremost was the acute manpower shortage experienced by this sector due to its rapid expansion and due to the highly skilled nature of its workforce. In no other industrial sector this shortage constitute the main obstacle to expansion and dramatic output increase. The RLM recognized this bottleneck even before Udet sent a bullet through his brain, and tried to solve it in different ways. Under Milch the RLM intensified its effort to find alternative sources of manpower. From here the road was short to a close and successful cooperation with the SS and its concentration camp empire.

Within this general framework, particular corporate interests and business strategies contributed to the early energetic employment of foreigners and slave labor. This was also the place where technology and specific products made their own contribution. Heinkel is the main example in this regard. It was the firm owner's ambitions to turn his firm into the leading German aircraft producer that pushed the firm so early to cooperate with the RLM and the SS and to employ concentration camp inmates. Heinkel tried to expand his firm in order to carry out its biggest contract ever at a time of an escalating manpower crisis. Therefore, foreigners and inmates formed a reasonable solution to his business difficulties. Early experiments with SS inmates failed mainly because they could not perform their work efficiently. This experience should have caused Heinkel to give it up. The fact that eventually Heinkel gave it a second try only proves the point that Heinkel badly needed those extra workers to put into effect his business strategies.

The other model employed by Heinkel, in occupied Poland, served the same purpose and also proves how Heinkel accepted an unorthodox solution in order to expand the production capacity of his firm in difficult times. It is important to note that slave labor was not necessarily the first and preferred solution firms chose in order to solve their manpower problems.

Messerschmitt chose to use slave labor after failing to solve its manpower problem by other means. Messerschmitt recruited an above-average number of women, but this alternative workforce was not productive enough due to the structure of wartime employment in Germany. Most women preferred to stay at home or were tied to different domestic obligations, which frequently kept them away from work. Other firms also experienced this problem with female workers, but Messerschmitt's difficulties were bigger because of the larger scale of women employment in its factories and due to its size. Firms like Focke-Wulf and Junkers came relatively late to slave labor because they sought other ways to solve the manpower shortage, mostly through outsourcing and business cooperation with foreign firms.

Technology and specific production programs also played a central role in decision-making regarding the use of slave labor. Most striking is the He 177 program, whose fate practically dictated the timetable of acceptance of slave workers by Heinkel. The Me 262 was also a significant program in utilizing slave labor, although it came later, when Heinkel had already proved the usability of slave labor in manufacturing complicated weapon systems. The decision to mass-produce the Me 262 as the Luftwaffe's next fighter came at a time when Messerschmitt had reached the end of the line with its women employment experiment. It high priority and the extent of its production plan made it extremely urgent to allocate it all available manpower.

The complexity of these two aircraft, together with the plans to produce large numbers of them, made them labor-intensive products, which demanded more manpower. Besides struggling with the manpower shortage, management of this huge industry was a major challenge. Its central role in arming the German war effort demanded strict and centralized management, which largely failed up to the end of 1941. In contrast to the widespread image of the Germans as extremely efficient, the aviation industry and its controlling organs proved to be extremely inefficient in many instances. This trend continued even after the rationalization measures initiated by Milch and after the Armaments Ministry took over aviation production. One of the most serious structural problems of the German aviation industry was its inability to move smoothly from the development stage to series production. In some cases the reason for this problem was the advanced nature of the product and its technical immaturity. In other instances inefficiency contributed to the difficulties. Even development programs controlled closely by the RLM sometimes failed to mature and reach series production. As demonstrated by the Messerschmitt case, power struggles, politics and the general administrative chaos prevailing in the German polycracy were the main reasons for this inefficiency.

The failures and problems plaguing German development and production programs are quite striking, but these problems should be set in the right context. Development and production of modern military aircraft and technologies became a complicated undertaking, which only relatively few industrial nations were able to accomplish successfully. Maintaining a large and functioning aviation industry demanded large research, development and industrial infrastructures and the cooperation of numerous organizations and firms. Furthermore, the complicated engineering and systems integration involved in development and production of modern warplanes prolonged the time it took to introduce them. Through this complexity a lot could have gone wrong, and not only in Germany.

During World War II even the USA, whose huge aviation industry enjoyed complete immunity from enemy action and huge economic backing, suffered several development and production fiascos. One of the biggest American fiascos was the P-75 fighter program. General Motors proposed to develop this high-performance heavy fighter using components from existing aircraft, therefore cutting down significantly the time it would take to make it ready for mass production. The USAAF was highly interested in this design, especially after intelligence reports received from Europe in 1943 suggested that new versions of German fighters were superior compared to contemporary American fighters. It was thus decided to use the proposed P-75 as a long-range escort fighter. The USAAF was so eager to get the new fighter that it placed an order for 2,500 aircraft even before the aircraft took to the air for the first time. Unfortunately, development of the aircraft encountered problems from early on. First it was found out that the scheme of using ready-made assemblies from aircraft already in production was not working because of integration difficulties. Changes were made in the original design and each change demanded redesign of the production jigs and fixtures. When prototypes finally began flying in November 1943 (far behind schedule) they performed disappointingly. More redesign was required, which led in turn to new problems. The new engine intended for the plane also proved to be troublesome and too weak for the large airframe. Despite the odds, the USAAF stuck to the project and ordered GM to make any effort to fix the problems. When it became clear in autumn 1944 that the aircraft would be ready for series production only from mid–1945, the project was terminated. The failure cost the USAAF 49.75 million dollars.[3] Nevertheless, operationally this failure was completely insignificant because the intelligence upon which it was largely based proved to be wrong. Besides, better alternative designs were introduced while it was developed.

Another problematic aircraft, which eventually entered service and played a crucial rule in World War II, was the Boeing B-29 Superfortress. In many respects it was the American equivalent of the He 177, but more complicated and more advanced. Proposed in 1940 as a long-range heavy bomber, the B-29 was one of the most ambitious aeronautical projects of World War II and certainly the biggest and the most complicated wartime bomber. Just as did the He 177, it suffered from a protracted and troubled development. A large number of technical problems repeatedly set back its development. As with the German bomber, the main culprit was the engine, which also tended to catch fire in mid-flight. At a certain point Boeing was ready to give up the project, but under pressure of the USAAF's commander-in-chief General Arnold, it continued working out the problems. Unlike the Germans, the Americans enjoyed the advantage of being capable to work out the problems of the B-29 in relative leisure. The huge capacity of the American aviation industry enabled it to solve the problems of the bomber one by one without compromising other activities. When the aircraft finally began slowly to enter service in early 1944 it was far from being in its final form. Its accident rate was well above average throughout much of its operational career.[4]

Manufacturing this big and complicated aircraft also proved to be difficult. Several firms were contracted to manufacture the aircraft and its sub-systems. Different firms, besides Boeing, performed final assembly. Each producer used different standards and manufactured at its own pace. This multiplication of producers and standards resulted in delays caused by bottlenecks in the supply of different parts and components. It was difficult to find a qualified workforce for the large B-29 factories, and in spring 1942 Boeing recruiters toured colleges and universities to obtain graduating engineers. Early aircraft were unsatisfactory and had to be upgraded before going overseas to operational units. One of the main contractors, Bell of Marietta, Georgia, became infamous for the poor quality of its aircraft, caused by slack quality control and by a poorly trained workforce. Repeated complaints brought a series of reviews, inspections and reshuffles.[5] Eventually, however, the U.S aviation industry manufactured 3,432 of these aircraft and they turned out to be a war-winning weapon system by bombing Japan into surrender.

Faulty policy-making and mismanagement at top levels were not restricted to the RLM and the Nazi system. Administrative chaos in the British government and faulty priorities caused a grand failure in the development and production of aircraft for the Royal Navy. As a result, from 1942 most aircraft based on British aircraft carriers were American.

While the Germans were probably unaware of the troubles experienced by the Allies, they were well aware of the fact that they were losing the air armament war by being outproduced. At the end of June 1944 the Technical Office of the RLM acknowledged that Allied production dwarfed Germany's production and that new ways must be found at least to use available capacity more efficiently.[6] A report submitted in September 1944 by the Planning Office of the Armaments Ministry compared the aircraft production rate of Germany with that of the Allies. The figures were staggering: while the Planning Office estimated that Germany would produce 36,500 aircraft in 1944, the projection for the USA alone stood at 200,500.[7]

Nevertheless, the loss of touch with reality prevailing in the German leadership and in the industry is evident almost to the end. In early March 1945 representatives of the Armaments Ministry, Messerschmitt and Special Committee F2 (overseeing Messerschmitt) still expected an output of 750 Me 262 fighters in that month. Me 109 output for March was set more realistically at 500. Afterwards it was planned to terminate Messerschmitt's production of the Me 109 and let it concentrate its entire capacity on the Me 262. Through this reshuffle the Germans expected to increase its monthly output to 1,000 in July and August 1945. German planners, however, were well aware that there were not enough engines for these aircraft. Since the maximum monthly output of the Jumo 004 engine was 1,200, it was possible to complete only 500 aircraft each month, because a portion of the engine production was intended for maintenance depots, spare stores and research. Final assembly of the Me 262 at this stage took only 14 days as long as all the components arrived at the final assembly factories at Regensburg, Schwäbisch Hall, Leipheim, Kuno I and Kuno II.[8] In most cases the parts and components trickled in through the crumbling transportation system, or never arrived.

Another sign of the problematic "situational awareness" prevailing in the German state and industrial leadership is the staggering number of advanced technology projects. German aviation research and development during World War II was well ahead of the Allies in most respects. The main problem was to turn these developments and inventions into working operational weapon systems. It was one thing to perfectly run an advanced jet engine on the test stand, and a completely different matter to use it operationally on a military aircraft. Not only laymen in the German leadership, but also technical experts failed to realize this gap and to fully appreciate the amount of time and effort it takes to close it. This type of unrealistic approach to industrial planning was evident also in other sectors of the German armaments industry, but it was extremely evident in the aviation industry.[9] Allied intelligence officers visiting Messerschmitt's design office in Oberammergau after the war summarized their observations of the research and development activities of this team: "There is evidence of their difficulties in trying to adapt relatively long term and revolutionary but imperfectly developed achievements to the practical shifting emergencies of war. In other words, their planning was poor but their application of science to the arts of war was on the whole superior to that of the Allies and much can be learnt from a study of it."[10]

Although German research and development was superior, most of it found little practical military use. Even the celebrated jet engines and jet planes had little impact on Germany's strategic and tactical situation. The German leadership largely overlooked these deficiencies and continued to rely largely on advanced technology as a miracle solution to Germany's strategic problems. In contrast, the Allies realized the immature nature of these technologies and relied instead on a gradual and continued improvement of existing technologies. The USA and Great Britain invested in the development of advanced aviation technologies, but without relying on them as the backbone of their military power in the current conflict.

Another gap the Germans and other nations found difficult to close was between visions of aviation mass production and its practice. It may seem surprising that the Germans accomplished modern mass production of aircraft only from late 1942, but it is less surprising considering the difficulties the Allies experienced, particularly of the Americans, in implementing mass production methods of the car industry to aviation industry. Nevertheless, although far from being a smooth process, mass production of aircraft was achieved. Furthermore, new production techniques brought with them completely new workflow and work processes. The workflow of mass production enabled the use of less-skilled or unskilled

The end of the line. Workers moving a DB 605 engine in the Gauting forest factory. This is what large portions of the German aviation industry looked like in the last six months of World War II (courtesy U.S. National Archives and Records Administration).

workforce and enabled the mass employment of slave labor in the aviation industry. These factors enabled the Germans to open the main bottlenecks obstructing increased output in the post–Udet era and reach the highest output ever in 1944.

The introduction of modern production methods and the solution of the manpower shortage were long-term developments. Although they picked up pace from 1942, their roots can be traced already in the prewar period.

In line with recent research trends,[11] the long-term nature of these and other factors make the contribution of Speer, Saur and the Jägerstab to the "Production Miracle" less significant than was thought in the past. There is no doubt that their management style and the reorganization they initiated contributed to the increased production in 1944, but they were riding a wave that was already there. Conversion to mass production lines, the solution of the manpower crisis, and more rationalized construction and production of aircraft started long before "Big Week" and its aftermath.

Göring was therefore right in pointing to long-term processes as crucial to the understanding of late-war events and developments.

Chapter Notes

Introduction

1. For critical views of the Heinkel exhibition, see Karl Heinz Jahnke, *Ernst Heinkel und die Stadt Rostock: Eine Dokumentation* (Rostock: Ingo Koch, 2002); Lutz Budrass, "Zur Heinkel-Ausstellung," *Zeitgeschichte Regional* 2, no. 6 (2002): pp. 91–96. Budrass was one of the members of the advisory committee hired by the city of Rostock in order to review the exhibition.

2. See an open protest letter to the Rostock city authorities in Lutz Budrass, "Hans-Joachim Pabst von Ohain: Neue Erkenntnisse zu seiner Rolle in der nationalsozialistischen Rüstung," *Technikgeschichte kontrovers: Zur Geschichte des Fliegens und des Flugzeugbaus in Mecklenburg-Vorpommern*, Beiträge zur Geschichte Mecklenburg-Vorpommern, Bd. 13 (Schwerin, 2007), pp. 52–69.

3. National Archives and Records Administration (NARA), RG243/6/Box 158, USSBS, Over-all Report (European War), 30 September 1945, p. 11.

4. Ralf Schabel, *Die Illusion der Wunderwaffen: Die Rolle der Düsenflugzeuge und Flugabwehrraketen in der Rüstungspolitik des Dritten Reiches* (München: Oldenbourg, 1994), p. 111.

5. Bundesarchiv-Berlin (BA-B), R3/3009, Rümi/Zentralamt/Abt. Arb. E Wi Stat: Ergebnisse der Beschäftigtenmeldung. Stichtag 31 Dezember 1944, 23 Februar 1945.

6. United States Strategic Bombing Survey (USSBS), *Airframes Plant Report No. 1. Junkersflugzeug und Motorenwerk, Dessau, Germany* (n.p., 1945), p. 1.

7. Ibid., pp. 3–4.

8. NARA, RG243/6/Box 223, Speer an Hitler, 20 Juli 1944.

9. Michael Thad Allen, *The Business of Genocide: The SS, Slave Labor, and the Concentration Camps* (London: Chapel Hill, 2002), p. 63.

10. Hans Mommsen, *Der Mythos von der Modernität: Zur Entwicklung der Rüstungsindustrie im Dritten Reich* (Essen: Klartext, 1999), p. 11.

11. Allen, *The Business*, pp. 63–64.

12. Frank Bajohr, "Dynamik und Disparität. Die nationalsozialistische Rüstungsmobilisierung und die Volksgemeinschaft," in *Volksgemeinschaft: Neue Forschungen zur Gesellschaft des Nationalsozialismus*, ed. Frank Bajohr and Michael Wildt (Frankfurt/M: Fischer, 2009), pp. 78–79.

13. NARA, RG243/31/Box 1, USSBS: Interview No.56, Reichsmarshal Hermann Goering, 6/7/1945, p. 1.

14. *The Simpsons*, episode CABF01, "Lisa the Tree Hugger."

15. Wolfgang Wagner, *Kurt Tank: Focke-Wulf's Designer and Test Pilot* (Atglen: Schiffer, 1998), p. 174.

16. See, for example, Christoph Vernaleken and Martin Handig, *Junkers Ju 388: Entwicklung, Erprobung und Fertigung des letzten Junkers-Höhenflugzeugs* (Ochsenfurt-Hohestadt: Aviatic, 2003), especially the remarks on pp. 194–195.

17. Edward M. Homze, *Arming the Luftwaffe: The Reich Air Ministry and the German Aircraft Industry 1919–1939* (Lincoln: University of Nebraska, 1976); Richard Overy, "German Aircraft Production 1939–1942" (diss., Cambridge, 1978).

18. Lutz Budrass, *Flugzeugindustrie und Luftrüstung in Deutschland 1918–1945* (Düsseldorf: Droste, 1998).

19. Schabel, *Die Illusion*.

20. See, for example, Klaus W. Müller and Willy Schilling, *Deckname Lachs: Die Geschichte der unterirdischen Fertigung der Me 262 im Walpersberg bei Kahla 1944/45* (Zella-Mehlis: H. Jung, 1995); Peter Müller, *Das Bunkergelände im Mühldorfer Hart* (Mühldorf a. Inn: Kreismuseum, 2000); Edith Raim, *Die Dachauer KZ-Außenkommandos Kaufering und Mühldorf: Rüstungsbauten und Zwangsarbeit im letzten Kriegsjahr 1944/1945* (Landsberg a. Lech: Landsberger Verl. Anst., 1992); Jan-Christian Wagner, *Produktion des Todes: Das KZ Mittelbau-Dora* (Göttingen: Wallstein, 2004).

21. See especially: Neil Gregor, *Daimler-Benz in the Third Reich* (New Haven: Yale University Press, 1998); Hans Mommsen and Manfred Grieger, *Das Volkswagenwerk und seine Arbeiter im Dritten Reich* (Düsseldorf: Econ, 1996); Constanze Werner, *Kriegswirtschaft und Zwangsarbeit bei BMW* (München: Oldenbourg, 2006).

22. On the USSBS see: David Madsaac, *Strategic Bombing in World War Two: The Story of the United States Strategic Bombing Survey* (New York: Garland, 1976).

Chapter 1

1. John H. Morrow, "Defeat of the German and Austro-Hungarian Air Forces," in *Why Air Forces Fail: The Anatomy of Defeat*, ed. Robin Higham and Stephan J. Harris (Lexington: University Press, 2006), p. 129.

2. Homze, *Arming the Luftwaffe*, p. 93.

3. Schabel, *Die Illusion*, pp. 70–71.

4. Christoph Buchheim, "Introduction: German Industry in the Nazi Period," in *German Industry in the Nazi Period*, ed. Christoph Buchheim (Stuttgart: Franz Steiner, 2008), pp. 17–18.

5. In order to understand the immense organization placed under Udet, see Schabel, *Die Illusion*, p. 72.

6. Ibid., pp. 65–69.

7. Williamson Murray, *Strategy for Defeat: The Luftwaffe 1933–1945* (Maxwell AFB: Air University Press), 1983, p. 12.

8. Ibid., pp. 2–3.

9. United States Strategic Bombing Survey (USSBS), *Aircraft Division Industry Report*, 2nd edition (n.p., 1947), p. 16.

10. Homze, *Arming the Luftwaffe*, pp. 65–68.

11. Budrass, *Flugzeugindustrie*, pp. 344, 376.
12. NARA, RG243/6/Box158, USSBS, Over-all Report (European War), 30 September 1945, p. 11.
13. Jörg Armin Kranzhoff, *Arado: History of an Aircraft Company* (Atglen: Schiffer, 1997), p. 101; NARA, T177/39/3729838, Arado Flugzeuge GmbH: Geschäftsjahr 1938.
14. NARA, RG243/6/Box241, 13a6 Production Data, p. 25.
15. Air Ministry (Great Britain), *The Rise and Fall of the German Air Force, 1933–1945* (New York: St. Martin's Press), 1983, p. 28.
16. Ernst Stilla, *Die Luftwaffe im Kampf um die Luftherrschaft: Entscheidende Einflussgrössen bei der Niederlage der Luftwaffe im Abwehrkampf im Westen und über Deutschland im Zweiten Weltkrieg unter besonderer Berücksichtigung der Faktoren "Luftrüstung," "Forschung und Entwicklung" und "Human Ressourcen"* (diss., Bonn, 2005), p. 86.
17. Richard Overy, *War and Economy in the Third Reich* (Oxford: Clarendon, 1994), pp. 305–306.
18. Schabel, *Die Illusion*, p. 106.
19. USSBS, *The Effect of Strategic Bombing on the German War Economy* (Washington, D.C., 1945), appendix, table 102.
20. Murray, *Strategy for Defeat*, p. 101.
21. Deutsches Museum (DM), FA001/904, Dr. Heinkel: Besprechung zwischen den Herren Dr. Heinkel, Benz, Beu, Günter, Hilber, Hohbach, 7 Juli 1942.
22. National Air and Space Museum Archive (NASM), 3237/560, Kommando der Erprobungsstellen an R.d.L u. Ob.d.L./GL üb., 17 August 1942.
23. Peter Schmoll, *Nest of Eagles: Messerschmitt Production and Flight-Testing at Regensburg 1936–1945* (Hersham: Ian Allen, 2010), pp. 40–41.
24. Schabel, *Die Illusion*, p. 119; Budrass, *Flugzeugindustrie*, p. 737. Both Heinkel and Messerschmitt continued to waste efforts trying to cure their troubled creations. When they finally succeeded in 1944, bombers and multi-role fighters were not needed anymore.
25. David Irving, *The Rise and Fall of the Luftwaffe: The Life of Field Marshal Erhard Milch* (Boston: Little, 1974), p. 126.
26. Murray, *Strategy for Defeat*, pp. 102–103.
27. USSBS, *Aircraft Division*, p. 43; NARA, RG243/31/Box1, USSBS: Interview with Dr. Karl Frydag and Mr. Heinkel, 19 May 1945, p. 4.
28. Murray, *Strategy for Defeat*, p. 88.
29. Bajohr, "Dynamik und Disparität," pp. 84–86.
30. USSBS, *Aircraft Division Industry Report*, Figures II-1 & VII-1.
31. Homze, *Arming the Luftwaffe*, pp. 84–85. In 1933 some 20,000 people worked in the British aviation industry.
32. USSBS, *Messerschmitt AG Augsburg*, Exhibit R-I.
33. Klaus Hesse, *1933–1945 Rüstungsindustrie in Leipzig*, 1. Teil 1 (Leipzig: Eigenverlag, 2000), p. 93.
34. Homze, *Arming the Luftwaffe*, p. 89.
35. Peter Kohl and Peter Bessel, *Auto Union und Junkers: Die Geschichte der Mitteldeutschen Motorenwerke GmbH Taucha 1935–1948* (Wiesbaden: Franz Steiner, 2003), pp. 86–89.
36. *Ibid.*, pp. 92, 97.
37. NARA, RG243/6/Box247, MIMO: Geschäftsbericht über das Geschäftsjahr vom 1. Januar 1943 bis 31 Dezember 1943, 4 Dezember 1944, p.2.
38. Budrass, *Flugzeugindustrie*, pp. 449–450.
39. *Ibid.*, pp. 445–446. For a statistical overview see especially the table on page 447. For a graphical view of the growth in this region between 1933 and 1936 see the maps in: USSBS, *Junkersflugzeug*, Exhibit A.
40. *Ibid.*, pp. 450–451.
41. See the plans of this expanded complex in: DM, FA001/275.
42. Dieter H. Koehler, *Ernst Heinkel-Pionier der Schnellflugzeuge: Eine Biographie* (Bonn: Bernard & Graefe, 1999), pp. 131–132.
43. Lutz Budrass, "Der Schritt über die Schwelle: Ernst Heinkel, das Werk Oranienburg und der Einstieg in die Beschäftigung von KZ-Häftlingen," in *Zwangsarbeit während der NS-Zeit in Berlin und Brandenburg*, ed. Winfried Meyer (Potsdam: Berlin-Brandenburg, 2001), p. 157.
44. Koehler, *Ernst Heinkel*, pp. 131–132; DM, FA001/325.
45. BA-B, R3/3693, P.A. Unbehagen, "Unsere soziale Betreuung," *20 Jahre Focke-Wulf: Werkzeitschrift der Betriebsgemeinschaft Focke-Wulf* 6 Jahrgang, Heft 1 (Januar–März 1944): p. 15.
46. BA-B, R3/3693, Dr. Med. H. Warnig, "Unser Gesundheitswesen," *20 Jahre Focke-Wulf: Werkzeitschrift der Betriebsgemeinschaft Focke-Wulf*, 6 Jahrgang, Heft 1 (Januar–März 1944): pp. 12–13.
47. Junkers Flugzeug- und Motorenwerke AG, *Vier Jahre sozialer Aufbau* (Dessau, 1937), especially pp. 52–60.
48. *Ibid.*, p. 68.
49. BA-B, R3/3693, Hauptbetriebssportwart A. Plümer, "Unser Betriebssport," *20 Jahre Focke-Wulf: Werkzeitschrift der Betriebsgemeinschaft Focke-Wulf*, 6 Jahrgang, Heft 1 (Januar–März 1944), p. 20.
50. NARA T83/11/3353570, Betriebssportgemeinschaft Focke-Wulf. Werk Sorau, 28 Juni 1943.
51. Budrass, *Flugzeugindustrie*, p. 463.
52. "Mensch und Werk," *Junkers Nachrichten*, Jahrgang 10, Heft 5/6 (Mai/Juni 1939): p. 146.
53. Junkers Flugzeug- und Motorenwerke AG, *Vier Jahre sozialer Aufbau*, p. 14. See also p.16 for the way the *Führerprinzip* was viewed and implemented in Junkers.
54. *Schönheit durch Arbeit* was a Nazi Party organization dedicated to the promotion of Germanic aesthetics.
55. "Wiener Neustadt—Flugzeugwerk," www.geheimprojekte.at.
56. Allen, *The Business of Genocide*, p. 63.
57. Jahnke, *Ernst Heinkel*, p. 37.
58. Junkers Flugzeug- und Motorenwerke AG, *Vier Jahre sozialer Aufbau*, p. 50.
59. Budrass, *Flugzeugindustrie*, pp. 801–803.
60. "Mensch und Werk," p. 147.
61. *Werkszetischrift der Betriebsgemeinschaft Focke-Wulf Flugzeugbau GmbH*, Nr. 6 (Juni 1939): p. 12.
62. BA-B, R3/3693, Unbehagen, "Unsere soziale Betreuung," p. 15.
63. NARA, T177/39/3729921, Arado Flugzeugwerke: 1940, 31 Juli 1941; Kranzhoff, *Arado*, p. 104.
64. Rainer Fröbe, "'Wie bei den alten Ägyptern.' Die Verlegung des Daimler-Benz Flugmotorenwerk Genshagen nach Obrigheim am Necker 1944/45," in *Das Daimler-Benz-Buch. Ein Rüstungskonzern im "Tausendjährigen Reich,"* ed. Hamburger Stiftung für Sozialgeschichte des 20. Jahrhunderts (Nördlingen: Delphi, 1987), p. 397.
65. See an example and explanation of these visions in Allen, *The Business of Genocide*, pp. 64–65.
66. "Mensch und Werk," p. 152.
67. Schmoll, *Nest of Eagles*, p. 24.
68. Bajohr, "Dynamik und Disparität," p. 89.
69. Mommsen, *Der Mythos*, pp. 7–8, 12, 14.
70. Overy, "German Aircraft Production," pp. 151–161.
71. Robert G. Ferguson, "One Thousand Planes a Day: Ford, Grumman, General Motors and the Arsenal of Democracy," *History and Technology* 21, no. 2 (June 2005): p. 149.
72. NARA, RG243/6/Box 263, The Aircraft Industry, p. 2175.
73. NARA, M888/6/1289, Stenographische Niederschrift der 33. Sitzung der zentralen Planung betreffend Arbeitseinsatz, 16 Februar 1943, p. 22.
74. NARA, RG243/6/Box233, Erla files: 7c, Assembly

Diagrams on Wings and Fuselages; Box 235, SA F4: Besichtigungsauswertung in Bezug auf die geplante Neukonstruktion der Takt- und Fliessstrassen für die FW 190, 8 Oktober 1943.

75. Sebastian Ritchie, *Industry and Air Power: The Expansion of British Aircraft Production 1935–1941* (London: Frank Cass, 1997), pp. 100–102.

76. Ferguson, "One Thousand Planes," p. 149.

77. Jonathan Zeitlin, "Flexibility and Mass Production at War: Aircraft Manufacture in Britain, the United States and Germany," *Technology and Culture* 36 (1995): pp. 46–79.

78. Irving Brinton Holley Jr., "A Detroit Dream of Mass-produced Fighter Aircraft: The XP-75 Fiasco," *Technology and Culture* 28 (1987): pp. 579–584.

79. Ferguson, "One Thousand Planes," p. 153.

80. *Ibid.*, pp. 154–155; David L. Lewis, "'They may save our honor, our hopes — and our necks': Michigan's Damnedest Colossus," *Michigan History* (September–October 1993). http://www.michiganhistorymagazine.com/extra/willow_run/willow_run.html

81. Zeitlin, "Flexibility and Mass Production," p. 58.

82. Mommsen, *Der Mythos*, pp. 22–25.

83. DM, FA001/260, Aktennotiz über die Besprechung bei Generalfeldmarschall Milch am 14 November 1942, pp. 11–12.

84. Budrass, *Flugzeugindustrie*, pp. 789–792.

85. NARA, RG243/6/Box 263, The Aircraft Industry, p. 2176.

86. NASM, 2497/237, Betriebsbüro Montage an Aussenstelle Bergkristall: Änderungen der Taktpläne für Rumpf-Vormontage 8-609, 7 April 1945.

87. NASM, 2497/237, Betriebsbüro Montage: Vorläufige Aufstellung der der bis zum 16 Januar 1945 erstellten Takte für Rumpf-Vormontage 8-609, 17 Januar 1945.

88. NARA, M888/5/167, Rü IV: Vermerk über Ausführungen des Reichsmarschalls in der Sitzung am 7 November 1941 im RLM.

89. NARA, T83/11/3353743-6, Direktor Werk Adelheide an Tank, 23 Dezember 1944.

90. NARA, M888/8/733, Stenographischer Bericht über die St/GL Besprechung am Freitag, dem 31 März 1944, p. 21.

91. NASM, 8177/1760, Air Technical Intelligence (W.W. Moore and D.D. Prye): Assembly Lines, July 1945, p. 1.

92. USSBS, *Aircraft Division Industry Report*, p. 19.

93. NARA, RG331/13d/Box 98, CIOS: Survey of Production Techniques used by the German Aircraft Industry, p. 4.

94. NARA, RG331/13d/Box 86, CIOS: Administration, Plastics, Production Tooling, Spare Parts and Servicing in German Aircraft Industry, p. 14.

95. NARA, RG331/13d/Box 98, CIOS: Survey of Production Techniques used by the German Aircraft Industry, p. 20.

96. Loc. cit.

97. USSBS, *Aircraft Division Industry Report*, Figure II-5.

98. *Ibid.*, pp. 20–21.

99. United States Strategic Bombing Survey, *Focke-Wulf Aircraft Plant Bremen*, 2nd edition (n.p., 1947), p. 3.

100. USSBS, *Aircraft Division Industry Report*, exhibit III-D.

101. Overy, *War and Economy*, p. 363.

102. Budrass, *Flugzeugindustrie*, p. 780.

103. NARA, M888/6/1332, Stenographischer Bericht über die Jägerstab Besprechung am Freitag, 26 Mai 1944, p. 93.

104. NASM, 8177/1759, Air Technical Intelligence (W.W. Moore and D.D. Prye): Jigs and assembly procedure, July 1945, pp. 9–10.

105. Hesse, *1933–1945 Rüstungsindustrie in Leipzig*, Teil I, pp. 84–85.

106. DM, FA001/277, Heinkel: Leistungsbericht, 1942, pp. 3–4.

107. NARA, M888/6/719, Stenographischer Bericht über die Jägerstab Besprechung unter Vorsitz von Generalfeldmarschall Milch am Montag 20 März 1944, p. 49.

108. Schmoll, *Nest of Eagles*, p. 75.

109. NARA, M888/6/326, Stenographischer Bericht über die Besprechung mit den Flotteningenieuren und Oberquartiermeistern unter dem Vorsitz von Generalfeldmarschall Milch am Sonnabend, 25 März 1944, p. 14.

110. Schmoll, *Nest of Eagles*, p. 30.

111. Kohl and Bessel, *Auto Union und Junkers*, p. 62.

112. Hans-Joachim Braun, "Aero-engine Production in the Third Reich," *History of Technology* 14 (1992): p. 1.

113. Some producers, like BMW, viewed the rationalization also as essential for withstanding competition in the postwar world. Werner, *Kriegswirtschaft*, p. 160.

114. NARA, M888/6/213-215, Niederschrift zur 21 Besprechung der "Zentralen Planung" am 31 Oktober 1942, p. 15.

115. Gregor, *Daimler-Benz*, p. 123.

116. Werner, *Kriegswirtschaft*, pp. 158–160.

117. Kohl and Bessel, *Auto Union und Junkers*, p. 62.

118. NARA, RG243/31/Box1, USSBS: Minutes of meeting with Reichsminister Albert Speer, Flensburg, 18 May 1945, pp. 5–6; see also: Braun, "Aero-engine Production," p. 10.

119. NARA, T177/34/3723010-3723042, RLM/GL1: Übersicht über die Fluggerät herstellenden Werke von Frankreich (Stand: Juni 1940), 20 November 1940.

120. Murray, *Strategy for Defeat*, pp. 96–98. See also an account by a French factory director about this policy in: NARA, RG331/13d/Box90, CIOS, German Activities in the French Aircraft Industry, p. 34.

121. NARA, M888/13/000-005, Ansprache des Generalluftzeugmeisters vor den Vertretern der französischen Luftfahrtindustrie in Paris am 30 Januar 1941.

122. Walter Hertel, *The German Air Force: Aircraft Procurement* (Karlsruhe: Studiengruppe Geschichte des Luftkrieges, 1955), pp. 431–432.

123. NARA, RG331/13d/Box 93, CIOS, Junkers 004 (203), Jet Propulsion Engines, p. 3.

124. USSBS, *Aircraft Division*, Figure VII-2.

125. Hertel, *The German Air Force*, p. 438.

126. NARA, T177/2/3683644, GL1: Kurzbericht Nr.27 über die Einschaltung der besetzte Gebiete und des Auslandes, 5 Mai 1941.

127. NARA, T83/11/3353731, Büro Paris an Tank: Französische Konstruktionsbüros, die im Auftrag deutscher Flugzeugwerke arbeiten, 29 November 1941.

128. Schmoll, *Nest of Eagles*, pp. 34–35.

129. Buchheim "Introduction," pp. 18–19.

130. NASM, 8031/16, Mitteilung: Bauprogramm der FW 206, 12. März 1941.

131. NARA T83/15/3358414, Tank: Aktenvermark. Arbeiten in Chatillon, 8 August 1941. Typical of the way things run under Udet, Focke-Wulf continued to work on this and other civilian projects despite Udet's order.

132. NARA, RG177/37/3727728-30, Tank: Ausnutzung der Kapazität der Konstruktions- und Baustunden im Werk Chatillon, Paris, 15 Dezember 1942; see also RG177/37/3727881-4, Tank an die Gruppe Technique de Chatillon: Bericht über die Betriebsführung der GTC, 12 Juli 1941.

133. NARA, RG331/13d/Box90, CIOS, German Activities in the French Aircraft Industry, pp. 45–46.

134. NARA, T83/6/3747530, RLM GL/LC2d: Programm FW-190 in Frankreich, 23 August 1943.

135. NARA, T83/6/3747464, Vermark: SNCASO Chatillon — Streik am 13 Dezember 1943.

136. Schmoll, *Nest of Eagles*, p. 35.
137. NARA, RG331/13d/Box 90, CIOS, German Activities in the French Aircraft Industry, pp. 29–30.
138. NARA, M888/6/1371, Stenographischer Bericht über die VIII Reise des Unternehmens Hubertus vom 1 bis 3 Juni 1944, p. 9.
139. See its blueprints in: NARA, T83/6/3747683-4.
140. NARA, T83/6/3746712-21, Planungsunterlagen für Konstruktionsbüro Bad Eilsen und Chatillon, 8 April 1942.
141. NARA, T83/6/3747527, RLM GL/C-E an FW: Einschaltung der französischen Entwicklungskapazität für die deutsche Luftrüstung, 14 August 1943.
142. NARA, T83/6/3747500, Besprechungsprotokol: Besprechung zwichen der Leitung SNCASO und der Leitung Focke-Wulf über Projekte Lage und Personalfragen, 22–23 August 1943.
143. NASM, KB Mitteilung: Richtlinien für den Anlauf der Arbeiten im Konstruktionsbüro für den Rumpf FW 300a, 31 January 1944.
144. NASM, 8013/1185, GL/C-E2: Aktenvermark, 10 July 1943.
145. NARA, T177/16/3700469, Besprechung über die Behälter-Anlage der FW 300A, 10 Juni 1944.
146. NARA, T83/6/374456, Aussenstelle Chatillon an Tank: Planung für Konstruktion Stahlflügel Ta 152, 20 Juni 1944.
147. Bundesarchiv-Militärarchiv (BA-MA), RL3/9, Stenographischer Bericht über die Jägerstabbesprechung vom 3 Juli 1944, p. 192.
148. NARA, RG331/13d/Box 90, CIOS, German Activities in the French Aircraft Industry, p. 34.
149. Werner, *Kriegswirtschaft*, pp. 283–289.
150. NARA, RG331/13d/Box 90, CIOS, German Activities in the French Aircraft Industry, especially the table on pp. 79–80; NARA, RG331/13d/Box86, CIOS Report: Aircraft–Paris Zone.
151. Schmoll, *Nest of Eagles*, p. 35.
152. USSBS, *Aircraft Division Industry Report*, p. 22a.
153. NARA, RG331/13d/Box 90, CIOS, Gas Turbine and Jet Propulsion Work in Paris.
154. Phil Butler, *War Prizes: An Illustrated Survey of German, Italian and Japanese Aircraft Brought to Allied Countries During and After the Second World War* (Leicester: Midland Counties, 1994), pp. 279–284.
155. Murray, *Strategy for Defeat*, p. 99.
156. NARA, RG243/6/Box 216, Exhibit IIIF, German Aircraft Production.
157. Hertel, *The German Air Force*, pp. 433–434.
158. NARA, T83/16/3359559, Fertigungsring F4, 11 Juni 1942.
159. USSBS, *Aircraft Division Industry Report*, p. 22a.
160. Butler, *War Prizes*, pp. 288–290.
161. NARA, RG243/6/Box 227, CIOS, Aircraft Works of Koninkijke Maatschappij "De Schelde," pp. 5 ff.
162. Vernaleken and Handig, *Junkers Ju 388*, p. 193.
163. NARA T83/16/3360233-40, Schuchardt, Abteilung für Sonderaufgaben der Fertigung: Denkschrift vom 24 August 1944 zur Fw-Holland-Angelegenheit; T83/16/3360242, Schuchardt an Tank: Holland Verlagerungs Angelegenheit. 24 August 1944.
164. Schuchardt estimated at the end of August 1944 that Holland was quite safe from land invasion and was not threatened by air attacks more than the Reich. NARA T83/16/3360246, Schuchardt and die Techn. Direktion: Fw Holland Angelegenheit, 29 August 1944.
165. NARA, T83/6/3747396, Besprechung mit Herrn Prof. Tank und Herrn Kaether am 22 Dezember 1943.
166. NARA, T83/6/3747366-69, Aktenvermark. Besprechung mit Herrn Prof. Tank am 5 Febr. 1944.
167. NARA, T83/6/3747301, Aktenvermark. Konstruktionsaufgaben Italien, 20 April 1944; T83/6/3747201, Aktenvermark. Abwicklung der Konstruktionsaufträge bei den Firmen Breda, Fiat und Piaggio, 8 September 1944.
168. NARA, M888/6/1371-1377, Stenographischer Bericht über die VIII Reise des Unternehmens Hubertus vom 1 bis 3 Juni 1944, pp. 9–15.
169. NARA, T83/6/3747263-68, Büro Italien: Vorplanung für den Bau von 250 Flugzeugen des Baumusters Ta 152B im oberital. Raum, 23 Juni 1944. See a plan of the Po Tunnel in *ibid.*, T83/6/3747279, Tunnel bei Cimena (Po Tunnel).
170. NARA, T83/6/3747134, Monatsbericht Januar 1945 Büro Italien.
171. See, for example, NARA, T83/6/3747253-54, Der Beauftragte für Flugzeugzellenbau: Allgemeine Situation in Raum Turin, 6 Juli 1944.
172. NARA, T83/6/3747179-87, Bericht über den Einsatz in Italien, 12 September 1944.
173. NARA, RG243/6/Box 224, Besprechung beim Führer vom 6–8 Juli 1944, p. 9.
174. Schmoll, *Nest of Eagles*, p. 45.
175. Quoted in Vernaleken and Handig, *Junkers Ju 388*, p. 193.
176. Including tips about several important aircraft, like the FW 190, Ta 154, Me 262, Ar 234. Christopher Staerck and Paul Sinnot, *Luftwaffe: The Allied Intelligence Files* (Washington D.C.: Brassey's, 2002), pp. 33–34, 36, 77, 84–85.

Chapter 2

1. For a good overview of the air war in World War II see Richard J. Overy, *The Air War 1939–1945* (New York: Stein & Day, 1980).
2. Murray, *Strategy for Defeat*, p. 7.
3. *Ibid.*, pp. 88–89, 94.
4. Overy, *The Air War*, p. 178 ff.
5. See, for example, the negligible statistics of early RAF attacks in: USSBS, *Junkersflugzeug*, pp. 6b–6c, 6e.
6. NARA, RG243/6/Box 215, Tons of Bombs Dropped on German Aircraft Industry, Table V.
7. USSBS, *Aircraft Division Industry Report*, p. 51.
8. Murray, *Strategy for Defeat*, pp. 170–171.
9. See examples of such reports in NARA, RG243/6/Boxes 181–182, Target Potentiality Reports on the Aircraft Industry.
10. Unless otherwise stated, all information about attacks used in this section is from lists published in USSBS, *Aircraft Division Industry Report*, pp. 58–69.
11. USSBS, *Focke-Wulf*, pp. 6–7.
12. Among other problems, this bombing caused a delay to the Me 262 jet fighter program by destroying crucial fuselage jigs and acceptance gauges. See Richard J. Smith and Eddie J. Creek, *Me 262*, vol. 2 (Burgess Hill: Classic Publication, 1998), p. 233.
13. USSBS, *Focke-Wulf*, p. 8.
14. Irving, *The Rise and Fall*, p. 228.
15. United States Strategic Bombing Survey, *Messerschmitt AG Augsburg, Germany. Overall Report* (n.p., 1945), p. 10.
16. Wesley Frank Craven and James Lea Cate, eds., *The Army Air Forces in World War II*, vol. 3 (Washington, D.C.: Office of Air Force History, 19830, p. 8.
17. NARA, M888/6/326, Stenographischer Bericht über die Besprechung mit den Flotteningenieuren und Oberquartiermeistern unter dem Vorsitz von Generalfeldmarschall Milch am Sonnabend, 25 März 1944, p. 14.
18. Murray, *Strategy for Defeat*, p. 237.
19. Craven, *The Army*, vol. 3, pp. 30–31.
20. *Ibid.*, pp. 43–44.
21. See the statistics in: NARA, RG243/6/Box 181, Statistical Summary of Eighth Air Force Operations, European Theatre, 17 August 1942–8 May 1945, p. 52.
22. *Ibid.*, pp. 30 ff.

23. USSBS, *Aircraft Division*, pp. 61–62, 69.
24. USSBS, *Messerschmitt AG. Overall Report*, p. 10.
25. Kommando der Erprobungsstelle der *Luftwaffe*: Kurzbericht über besondere Schwierigkeiten Anlauf Me 262, 1 März 1944. Published in Smith and Creek, *Me 262*, vol. 2, p. 234.
26. Craven, *The Army*, vol. 3, p. 45.
27. See, for example, NARA, RG243/31/Box1, USSBS: Minutes of meeting with Reichsminister Albert Speer, Flensburg, 18 May 1945, p. 6.
28. USSBS, *Messerschmitt AG. Overall Report*, pp. 11–12.
29. For a comprehensive discussion of Heinkel's early involvement in these projects see Schabel, *Die Illusion*, pp. 37–45; Koehler, *Ernst Heinkel*, pp. 148–178.
30. On the prewar development of these technologies, see especially Michael J. Neufeld, "Rocket Aircraft and the 'Turbojet Revolution': The *Luftwaffe's* Quest for High-Speed Flight, 1935–1939," in *Innovation and the Development of Flight*, ed. Roger D. Launius (College Station: Texas A&M, 1999), pp. 207–232. On jet propulsion development, see Anthony L. Kay, *German Jet Engine and Gas Turbine Development 1930–1945* (Shrewsbury: Airlife, 2002).
31. Schabel, *Die Illusion*, pp. 182–183.
32. Alfred Price, *The Last Year of the Luftwaffe: May 1944 to May 1945* (London: Arms and Armour, 1991), pp. 175–176.
33. Richard J. Smith and Eddie J. Creek, *Arado 234 Blitz* (Sturbridge: Monogram, 1992), p. 166.
34. DM, FA001/335, Flugabteilung — Peter: Erfahrungsaustausch mit Flugerprobung Arado und Nachfliegen Ar 234 im Hinblick auf 162, 6 November 1944.
35. NARA, RG243/6/Box 218, A.D.I(k): Jet Propulsion. German Progress in the Field of Gas Turbines, Athodydes and Turbojet Units, 22/7/1945, p. 7.
36. NARA, RG72/116/Box 7, Technical Report N0.236-45. General Survey of Rocket Motor Development in Germany, August 1945, p. 80.
37. Bernd Krag, "Erfahrung bei der Entwicklung und Erprobung der ersten Strahlflugzeuge mit Pfeilflügeln," in *Die Pfeilflügelentwicklung in Deutschland bis 1945*, ed. Hans-Ulrich Meier (Bonn: Bernard & Graefe, 2006), pp. 303–364.
38. See especially: NARA, RG72/116/Box 7, Technical Report N0.237-45 Survey of German Activities in the Field of Guided Missiles, Vol. I–II, August 1945.
39. See especially Schabel, *Die Illusion*, pp. 292–293.
40. On the development of TV guidance systems in Germany see Joseph Hoppe, "Fernsehen als Waffe: Militär und Fernsehen in Deutschland 1935–1950," in *Ich diente nur der Technik. Sieben Karrieren zwischen 1940 und 1950*, ed. Museum für Verkehr und Technik (Berlin: Nicolai, 1995), pp. 67–76. On anti-radiation guidance see NARA, RG72/116/Box 7, Technical Report No. 237-45 Survey of German Activities in the Field of Guided Missiles, vol. 1, August 1945, p. 251.
41. Schabel, *Die Illusion*, p. 188. Besides Schabel's research, the four-volume work of Smith and Creek is the best Me 262 reference. Smith and Creek, *Me 262*.
42. NASM, 2831/637, Arado: Ar 234B Programm 225, 16 November 1945.
43. Schabel, *Die Illusion*, p. 231. A comprehensive book about the Me 163: Stephan Ransom and Hans-Hermann Cammann, *Me 163 Rocket Interceptor*, 2 vols. (Crowborough: Classic Publications, 2001).
44. NASM, 3391/16, Technische Direktion: Bericht über die wichtigsten Entwicklungsfragen in Wien (Stand Jahresende 1944), 29 Dezember 1944, p. 5.
45. On the development of the pulse-jet engine see Kay, *German Jet Engines*, pp. 238–257.
46. The most detailed history of the V-1 can be found in Heinz Dieter Hölsken, *Die V-Waffen: Entstehung, Propaganda, Kriegseinsatz* (Stuttgart: Deutsche Verlags Anstalt, 1984), pp. 33 ff.

47. This section is based mostly on William Green, *The Warplanes of the Third Reich* (London: Arms & Armor, 1970).
48. The aircraft is also widely known as the Bf 109. This designation originated from the name of the Bayerische Flugzeugwerke firm, which produced the aircraft before the war and was subsequently merged with Messerschmitt. The Germans used both designations until the end of the war. In order to keep things simple I will use here only the Me 109 designation.
49. Just to put things in context, the Soviet MiG 21 fighter, which is considered one of the most successfully marketed fighter designs of the postwar era, was produced in 11,496 units between 1959 and 1985.
50. NARA, RG243/6/Box228, ATG Maschinenbau GmbH, Leipzig, Exhibit A, Sheet 1.
51. USSBS, *Aircraft Division*, p. 77. There is a list of all the 20 types on p. 81.
52. NARA, RG243/31/Box 1, USSBS: Interview No. 56, Reichsmarshal Hermann Goering, 6 July 1945, pp. 2–3.
53. NARA, RG243/32/Box 2, USSBS: Interview no. 35, General Galland (Part II), 4 June 1945, p. 3.
54. USSBS, *Aircraft Division*, p. 92.
55. Obb. Forschungsanstalt Oberammergau: Besprechungsniederschrift, 17 Juni 1944. Published in Smith and Creek, *Me 262*, vol. 1, p. 194. Hand-made prototypes were almost always of higher quality than series production aircraft.
56. Peter Schmoll, *Messerschmitt-Giganten und der Fliegerhorst Obertraubling 1936–1945* (Regensburg: MZ Buchverlag, 2002), p. 150.
57. Dietmar Hermann, *Focke-Wulf Ta 152: The Story of the Luftwaffe's Last Variant High-Altitude Fighter* (Atglen: Schiffer, 1999), pp. 117–119.
58. Smith and Creek, *Arado 234 Blitz*, p. 157.
59. *Ibid.*, p. 174.
60. USSBS, *Aircraft Division*, p. 90.
61. DM, FA001/827, E. Heinkel an Francke: He 162, 30 September 1944.
62. NARA, RG331/13d/Box86, CIOS: Administration, Plastics, Production Tooling, Spare Parts and Servicing in German Aircraft Industry, p. 15.
63. See, for example, NARA T83/11/3353650, Kontrol-Leitung Kreising an Tank: Bericht Nr. 16, 17 August 1944.

Chapter 3

1. USSBS: Interview No. 22 with Dr. Saur, General Buhle, 18 May 1945, p. 2.
2. NARA, RG243/31/Box 1, USSBS: Interview No. 44, Dr. Karl Frydag, Chief of Air Frame Industry, 9 July 1945, p. 20.
3. Zeitlin, "Flexibility and Mass Production," p. 47.
4. NARA, RG243/6/Box 224, Nachtrag zu den Besprechungen beim Führer am 18–20 August 1944.
5. Schabel, *Die Illusion*, pp. 138–139.
6. After the war he remarked humorously to his American interrogators: "A man like Messerschmitt is an artist. One should never put an artist at the head of such a large firm (unless a minister over the armament) (laugh)." NARA, RG243/31/Box1, USSBS: Minutes of meeting with Reichsminister Albert Speer, Flensburg, 18 May 1945, p. 9.
7. USSBS, *Aircraft Division*, p. 39.
8. BA-B, R3/3349, Sonderausschuss F3 (Heinkel-Zellen) im Hauptschuss "Zellen" beim Reichsminister für Rüstung und Kriegsproduktion.
9. Schabel, *Die Illusion*, pp. 139–141.
10. NARA, RG177/3727650-53, Tank an Milch, 17 März 1942.
11. Irving, *Rise and Fall*, pp. 197, 223. The Ju 288 was supposed to replace the older medium bombers of the Luftwaffe from January 1942, but its development was never completed.

12. Ransom and Cammann, *Me 163*, p. 99.
13. NARA, RG243/31/Box1, USSBS: Minutes of meeting with Reichsminister Albert Speer, Flensburg, 18 May 1945, p. 9.
14. Irving, *Rise and Fall*, p. 267.
15. Hermann Kaienburg, "KZ-Haft und Wirtschaftsinteresse: Das Wirtschaftsverwaltungshauptamt der SS als Leitungszentrale der Konzentrationslager und der SS-Wirtschaft," in *Konzentrationslager und deutsche Wirtschaft 1939–1945*, ed. Hermann Kaienburg (Opladen: Leske & Budrich, 1996), pp. 56–58; NARA, M888/5/267-9, Pohl an Himmler: Eingliederung der Inspektion der Konzentrationslager in das SS-Wirtschafts-Verwaltungshauptamt, 30 April 1942.
16. Allen, *The Business*, pp. 34–35.
17. *Ibid.*, pp. 202–203.
18. Peter Longerich, *Heinrich Himmler: Biographie* (München: Siedler, 2008), pp. 652–655.
19. Arguably the best monograph about this topic is Michael J. Neufeld, *The Rocket and the Reich: Peenemünde and the Coming of the Ballistic Missile Era* (New York: Harvard University Press, 1995), especially chapters 6 and 7.
20. *Ibid.*, p. 265.
21. Budrass, "Hans Joachim," pp. 1, 3–6.
22. NARA, RG72/116/Box 14, Technical Report No. 550-45 History of German Jet Engine Developments, October 1945, pp. 20–21.
23. Yad Vashem Archives (YVA), TR.2/PS-1584, Himmler an Göring, Einsatz von Häftlingen in der Luftfahrtindustrie, 9 März 1944 (also in NARA, M888/6/544-547).
24. Irving, *The Rise and Fall*, p. 236.
25. NARA, M888/6/334-335, Stenographischer Bericht über die Besprechung mit den Flotteningenieuren und Oberquartiermeistern unter dem Vorsitz von Generalfeldmarschall Milch am Sonnabend, 25 März 1944, pp. 22–23.
26. Irving, *The Rise and Fall*, p. 255.
27. NARA, M888/6/565, Vernehmung des Dip. Ing. Karl Otto Saur durch Mr. Koch, 13. November 1945, p. 4.
28. NARA, RG243/31/Box 1, USSBS: Minutes of meeting with Reichsminister Albert Speer, Flensburg, 18 May 1945, p. 8.
29. BA-B, R50II/46a, Anordnung des Reichsministers für Rüstung und Kriegsproduktion vom 1 März 1944 über die Errichtung des Jägerstabes.
30. NARA, RG243/6/Box 224, Punkte aus der Besprechung beim Führer am 5 März 1944, p. 1.
31. NARA, RG243/32/Box 5, Minutes of the interview with Herr Saur, 7 June 1945, p. 24.
32. NARA, RG243/6/Box 221, Decision taken at the 56th meeting of Central Planning, 5 April 1944.
33. NARA RG73/30/3159057, Rüstungskommando Augsburg: Liste über die neue Rangfolge für die Zuweisung von Arbeitskräften, 18 April 1944.
34. NARA, RG243/31/Box 1, USSBS: Interview no. 48, Dr. Saur, 23 May 1945, pp. 1–2.
35. Speer attended a *Jägerstab* meeting for the first time only on 26 May 1944.
36. NARA, M888/6/591, Stenographischer Bericht über die Jägerstab Besprechung am 6 März 1944, p. 5.
37. NARA, RG243/31/Box1, USSBS: Interview No. 44, Dr. Karl Frydag, Chief of Air Frame Industry, 9 July 1945, p. 21.
38. NARA T73/30/3159045, "Einige Angestellte des Hauses, die diesem Treiben nicht mehr zusehen können" an den Beauftragten des Reichsministers für Rüstung u. Kriegsproduktion: Benzinvergeudung im gemeinsten Masse, 20 August 1944.
39. NARA, BDC SSO A3343/182A/1181, BDC SSO, Kloth, Albert, geb. 3 Januar 1905. In Kloth's SS personal file there is no mention of this appointment.
40. NARA T73/30/3159008, Der Reichsminister für Rüstung und Kriegsproduktion/Rüstungsstab an Rüstungskommando München: Auszüge aus Kurz- und Schnellberichten des Rüstungsstabes, 25 November 1944.
41. NARA T73/30/3159025, Der Kommandant/Rüstungskommando Augsburg: Führung der Fa. Messerschmitt AG, 22 Februar 1945.
42. NARA, M888/6/318-9, Stenographischer Bericht über die Besprechung mit den Flotteningenieuren und Oberquartiermeistern unter dem Vorsitz von Generalfeldmarschall Milch am Sonnabend, 25 März 1944, pp. 6–7.
43. NARA, RG243/31/Box 1, USSBS: Interview No. 22 with Dr. Saur, General Buhle, 18 May 1945, p. 2.
44. NARA, M888/6/795, Stenographischer Bericht über die Jägerstab Besprechung unter Vorsitz von Hauptdienstleiter Saur am 12 April 1944, p. 1.
45. *Trials of War Criminals Before the Nuremberg Military Tribunals Under Control Council, October 1946–April 1949*, Washington D.C.: U.S. Government Printing Office, 1949–1953 (NMT), vol. 2, p. 589, doc. R-124, Minutes of Discussion with the Fuehrer on 6 and 7 April 1944.
46. Irving, *The Rise and Fall*, p. 243.
47. *Ibid.*, pp. 258–259.
48. Hölsken, *Die V-Waffen*, p. 91.
49. Raim, *Die Dachauer*, p. 45.
50. No protocol of this meeting was found, but after the war some of the surviving participants reconstructed it. A good reconstruction of this event can be found in Schabel, *Die Illusion*, pp. 227–230.
51. *Ibid.*, p. 228; Irving, *The Rise and Fall*, p. 283.
52. NMT, vol. 2, pp. 570–571.
53. YVA, TR.2/PS-1584, Himmler an Göring, Einsatz von Häftlingen in der Luftfahrtindustrie, 9 März 1944 (also in NARA, M888/6/544-547).
54. NARA, BDC SSO Kammler, Hans, A3343/151A, Chef OKL/TLR (Diesing): Vorschlag für die Verleihung des Ritterkreuzes zum Kriegsdienstkreuz mit Schwertern, 20 August 1944.
55. *Ibid.*, Speer an Kammler, 17 Dezember 1943.
56. Raim, *Die Dachauer*, pp. 53–57.
57. Kohl, *Auto Union und Junkers*, p. 228.
58. Allen, *The Business*, p. 203.
59. NARA, RG243/6/Box 235, SA F4: Planungsunterlagen Nachtjäger Ta 154, 15 Mai 1943. More about this project in a later chapter.
60. NARA, M888/6/1082-1083, Stenographischer Bericht über die Jägerstab Besprechung am Freitag, 5 Mai 1944, pp. 4–5.
61. NARA, RG243/6/Box 224, Punkte aus Führer Besprechungen am 3–5 Juni 1944, pp. 5–6.
62. BA-B, R3/1827, Speer an die Amtchefs, 23 Juni 1944.
63. NARA, RG243/31/Box 1, USSBS: Interview no. 1, Dr. Kurt Tank, 17 & 24 April 1945, p. 10. As was pointed out correctly by the interviewing officer, Tank was no production expert.
64. The following revealing lines are found in Saur's postwar interrogation: "Question: But where did the thousands of planes get to? Answer: That, I do not know." NARA, RG243/32/Box 5, Minutes of the interview with Herr Saur, 7 June 1945, p. 28.
65. See especially his remarks in: BA-B, R3/1749, Extract from minutes of a Rüstungsstab meeting on 2 Oktober 1944 & 1 November 1944.
66. NARA, RG243/31/Box 1, USSBS: Interview No. 56, Reichsmarshall Hermann Goering, 6 Juli 1945, p. 3. Nevertheless Göring praised Saur's ability to increase production.
67. BA-B, R3/1749, Extract from minutes of a Rüstungsstab meeting on 6 November 1944.
68. NARA, RG243/180/3391832, Rede Reichminister Speer am 3 Dezember 1944 in Rechlin, p. 3.
69. NARA, M888/6/1263, Stenographischer Bericht über die Jägerstab Besprechung am Freitag, 26 Mai 1944, p. 15.
70. BA-B, R3/1597, Speer an Saur, 17 Juli 1944.

71. NARA T73/30/3159064, Der Reichsminister für Rüstung und Kriegsproduktion: Erlass über die Bildung des Rüstungsstabes, 1 August 1944.
72. NARA, RG243/6/Box 235, Report of Field Team no. 81: Focke-Wulf Flugzeugbau GmbH, Bremen, IIIb.
73. Schabel, *Die Illusion*, pp. 202–203. See also the organizational table of the armament committees on p. 206.
74. NARA, T73/132/3295143, Rüstungsinspektion VII/Gruppe Fertigung: Sonderkommando Nebel, Metallbau Offingen, 2 März 1945.
75. NARA, T73/132/3295139-41, Rüstungsinspektion VII/Gruppe Fertigung an das Luftgau Kommando VII: Sonderkommando Nebel, Offingen/Donau, 2 März 1945.
76. NARA, T73/132/3295145, Rüstungsinspektion VII/Abt. Fertigung: Aktenvermark über den Besuch beim Sonderkommando Nebel am 23 Februar 1945.
77. NARA, RG243/6/Box 224, Speer an Hitler, 30 Juni 1944 & 29 Juli 1944.
78. BA-B, R3/1944, Der Rüstungsminister: Dringlichkeitsfolge der Fertigungen, 20 Juni 1944. Other highest-priority production programs were assault guns and new types of submarines.
79. BA-B, R3/1554, Rede Speers auf der Sitzung des Rüstungsstabs am 21 August 1944.
80. *Ibid.*, Punkte für die Rede des Reichsministers Speer vor den Betriebsführern, Produktionschefs, Hauptwerk- und Werkbeauftragten sowie Firmenvertretern und Hauptausschuss- und Sonderausschussleitern, 24 August 1944.
81. NARA T73/30/3159063, Der Führer: Aktion Hochleistungsflugzeug, 12 Oktober 1944.
82. NARA, RG243/6/Box 222, Saur: Sicherstellung der Vorrichtungen für Hochleistungsflugzeuge, 15 Dezember 1944.
83. NARA, RG243/6/Box 226, Besprechung beim Führer am 14 Februar 1945, p. 2.
84. NARA, BDC SSO Kammler, Hans, A3343/151A, Chef OKL/TLR (Diesing): Vorschlag für die Verleihung des Ritterkreuzes zum Kriegsdienstkreuz mit Schwertern, 10 Oktober 1944. Interestingly Himmler was angered by the fact that the recommendation came from the Luftwaffe, and he objected.
85. BA-MA, RL3/2567, Bl.2567; Martin Moll, *Führer-Erlasse 1939–1945* (Stuttgart: Franz Steiner, 1997), p. 488.
86. Schabel, *Die Illusion*, p. 283.
87. Overy, *The Air War*, pp. 155–156.
88. *The Strategic Air War Against Germany, 1939–1945: Report of the British Bombing Survey Unit* (BBSU) (London: Frank Cass, 1998), p. xxxv.
89. See especially: Lutz Budrass, Jonas Scherner, and Jochen Streb, "Demystifying the German 'Armament Miracle' During World War II: New Insights from the Annual Audits of German aircraft Producers," *Economic Growth Center Discussion Papers No. 905* (Yale University, January 2005).
90. Over, *The Air War*, p. 229.
91. *Ibid.*, p. 230.
92. Irving, *The Rise and Fall*, pp. 232, 237.
93. Schabel, *Die Illusion*, pp. 160–170.
94. Budrass, *Flugzeugindustrie*, pp. 858–859.
95. NARA, RG243/6/Box 224, Führer-Besprechung vom 11/12 September 1943.
96. USSBS, *Aircraft Division*, pp. 44–45.
97. Loc. cit.
98. Irving, *The Rise and Fall*, p. 244.
99. NARA, M888/6/614, Stenographischer Bericht über die Jägerstab Besprechung am 6 März 1944, p. 31.
100. NARA, M888/6/1426-1427, Stenographischer Bericht über die VIII Reise des Unternehmens Hubertus vom 1 bis 3 Juni 1944, pp. 63–64.
101. NARA, RG243/32/Box 2, USSBS: Interview n0.35, General Galland (Part II), 4 June 1945, p. 5.

102. NARA, RG243/6/Box 224, Führer Besprechungen vom 19–22 Juni 1944, p. 10.
103. NARA, T321/59/807004, Punkte aus Besprechung beim Herrn Reichsmarschall am 1 Juli 1944. It was decided, in fact, to hasten the termination of this aircraft.
104. USSBS, *Aircraft Division*, pp. 45–46.
105. *Ibid.*, Exhibit IIIa; USSBS, *The Effect*, p. 150.
106. *Ibid.*, Figure VI-I.
107. NARA, RG243/6/Box 211, Statistische Schnellbericht zur Kriegsproduktion. Rüstungsfertigung, Stand Februar 1945, 15 März 1945, p. 13.
108. USSBS, *Aircraft Division*, Exhibit IIIb.
109. Steve Coates, *Helicopters of the Third Reich* (Hersham: Classic Publications, 2002), pp. 109, 112–113; NARA, M888/6/614, Stenographischer Bericht über die Jägerstab Besprechung am 6 März 1944, p. 31.
110. NARA, M888/8/947, Stenographischer Niederschrift der Besprechung beim Reichsmarschall über die Flugzeug Programm Entwurf am Montag, 22 Februar 1943, p. 52.
111. NARA, RG243/6/Box 235, SA F4: Planungsunterlagen Nachtjäger Ta 154, 15 Mai 1943.
112. NASM, 3252/975, Absprache über Montage Ta 154 A-Serie in Posen-Kreising, 17 November 1943.
113. Budrass, *Flugzeugindustrie*, pp. 792–794; NARA, RG243/6/Box 235, SA F4: Fertigungsmappe Ta 154, 30 Juni 1943.
114. NARA T83/11/3353641-6, Betr. Ltg. Detmold an Kommissar Diehle, Gestapo Bielfeld: Fertigungsablauf der Ta 154 in Detmold und bei dem angeschlossenen Nachbaufirmen, 28 August 1944.
115. Green, *The Warplanes*, p. 241.
116. Wagner, *Kurt Tank*, p. 193.
117. NARA, M888/6/1272, Stenographischer Bericht über die Jägerstab Besprechung am Freitag, 26 Mai 1944, p. 24.
118. Ferenc A. Vajda and Peter Dancey, *German Aircraft Industry and Production 1933–1945* (Bath: SAE, 1998), p. 198.
119. NARA, T83/11/3353897, Werk Kreising an RLM/GL/Planungsbauamt P13: Bauvorhaben und Fabrikationsmittelbeschaffung, 15 Mai 1943.
120. Mommsen and Grieger, *Das Volkswagenwerk*, pp. 682–686, 817.
121. NARA T83/7/3748860, Vorplanung KB Bauüberwachung und Personal Stand im KB-Bü am 1 September 1943.
122. NARA, RG331/13d/Box 86, CIOS: Aircraft Production Activity of the Peugeot Organization, p. 4; Mommsen and Grieger, *Das Volkswagenwerk*, pp. 670–673.
123. Vajda and Dancey, *German Aircraft*, pp. 196–197.
124. BA-B, NS19/3652, RFSS/PS an SSHA: Schriftwechsel des RFSS und Minister Speer: Führungsmängel in Luftwaffe und Luftindustrie, 24 Oktober 1944.
125. BA-B, R3/3034b, Rüstungsstab: Kurzbericht Flugzeugenbau und Reparatur, 19 Oktober 1944; dito, 8 November 1944.
126. NARA T83/11/3353641-6, Betr.Ltg. Detmold an Kommissar Diehle, Gestapo Bielfeld: Fertigungsablauf der Ta 154 in Detmold und bei dem angeschlossenen Nachbaufirmen, 28 August 1944.
127. Heinrich Beauvais, Heinrich Kössler, Max Mazer, and Christoph Regel, *Flugerprobungsstellen bis 1945* (Bonn: Bernard & Graefe, 1998), p. 260.
128. Key, *German Jet Engines*, pp. 246–248.
129. Hölsken, *Die V-Waffen*, p. 47.
130. *Ibid.*, p. 59; Mommsen and Grieger, *Das Volkswagenwerk*, p. 691.
131. NARA, M888/6/1263-1266, Stenographischer Bericht über die Jägerstab Besprechung am Freitag, 26 Mai 1944, pp. 15–18.

132. *Ibid.*, p. 22.
133. Irving, *The Rise and Fall*, p. 280.
134. BA-B, R3/1959, RLM/GL/C-Fertg. XI: Bericht zur Fertigungsführung: Leistungssteigerung durch Betriebskonzentrazion und Typenverminderung im Zellenbau, 30 Juni 1944.
135. NARA, T321/59/807004, Jägerstab: Punkte aus Besprechung beim Herrn Reichsmarschall am 1 Juli 1944.
136. USSBS, *Aircraft Division*, p. 46; OKL/Gen. Qu. 6 Abt.: Anlage. Flugzeuglage Monatsmeldung Januar 1945. A copy kindly provided by Manfred Boehme via Richard Eger.
137. BA-B, R3/1749, Extracts from minutes, letters and memoranda on aircraft production planning connected with the chief of the Rüstungsstab Sauer. Translation by the Foreign Office, German Economy Dept., 15 Januar 1945.
138. NARA T73/30/3158999, Bezirksbeauftragter der Hauptausschüsse für Luftwaffenfertigung im Bereich der Rü In. VII, 30 Dezember 1944.
139. USSBS, *Aircraft Division*, Exhibit IIIA.
140. NARA, T73/180/3392326, Speer/Burchard an Göring, 10 Juli 1944.
141. NARA, RG243/180/3391838, Rede Reichminister Speer am 3 Dezember 1944 in Rechlin, p. 8.
142. Schabel, *Die Illusion*, p. 283.
143. NARA, M888/6/323-4, Stenographischer Bericht über die Besprechung mit den Flotteningenieuren und Oberquartiermeistern unter dem Vorsitz von Generalfeldmarschall Milch am Sonnabend, 25 März 1944, pp. 11–12.
144. *Ibid.*, p. 13.
145. Schmoll, *Messerschmitt-Giganten*, p. 150.
146. BBSU, p. xxxiv: USSBS, *Aircraft Division*, p. 92.
147. Zeitlin, "Flexibility and Mass Production," p. 71.
148. Schmoll, *Nest of Eagles*, p. 38.
149. Stubner, Helmut, *Der Kampfflugzeug Heinkel 177 Greif und seine Weiterentwicklung*, Zürich: Eurodoc, 2005, p. 90.
150. NARA, M888/6/346, Stenographischer Bericht über die Besprechung mit den Flotteningenieuren und Oberquartiermeistern unter dem Vorsitz von Generalfeldmarschall Milch am Sonnabend, 25 März 1944, p. 43.
151. NARA, M888/6/679, Stenographischer Bericht über die Jägerstab Besprechung am Freitag 20 März 1944, p. 3.
152. NARA, M888/8/739, Stenographischer Bericht über die St/GL Besprechung am Freitag, dem 31 März 1944, p. 27.
153. NARA, RG243/6/Box 227, Punkte aus der Besprechungen beim Führer am 8 März 1945, p.2; 22 März 1945, p. 4.
154. See orders for new variants of the FW 190 in late 1944–early 1945 in: BA-B, R3/3723.
155. NARA, RG72/116/Box 2, Technical Report N0.123-45 German Aircraft Maintenance, July 1945, pp. 13–14. The Ta 183 was never completed.
156. BA-B, NS19/3652, RFSS/PS an SSHA: Schriftwechsel des RFSS und Minister Speer: Führungsmängel in Luftwaffe und Luftindustrie, 24 Oktober 1944. See also Wagner, *Kurt Tank*, p. 198.
157. NARA, T83/6/3749066, Techn. Leitung/Bad Eilsen: Terminierung von Änderungen, 22 März 1944.
158. BA-B, R3/3723, Klemm and Keather: Entwicklungsmitteilung. Fw 190 Sturmflugzeuge, 20 Oktober 1944. Torpedo bombs were streamlined bombs designed originally for maritime attack. Later it was planned to use them against hardened land targets. They were much easier and cheaper to produce than normal torpedoes. As far as is known, they never entered operational service.
159. BA-B, R3/1749, Extract from minutes of a Rüstungsstab meeting on 7 September 1944.
160. It is interesting to note that Klemm never produced the FW 190 under license.
161. NARA, T83/6/3749009, KB-Verbindungen, 28 August 1944.

162. NARA, T83/6/3749005-6, Neuorganisation der Serienbetreuung, 17 November 1944.
163. William Green, *The Augsburg Eagle: Messerschmitt Bf 109* (London: Jane's, 1980), pp. 124–127.
164. Zeitlin, "Flexibility and Mass Production," p. 60.
165. *Ibid.*, p. 47.
166. For example, Overy, *The Air War*, pp. 227–228.
167. See especially: Schabel, *Die Illusion*.
168. *Ibid.*, p. 211.
169. NARA, RG243/6/Box 224, Führer Besprechungen vom 19–22 Juni 1944, p. 5.
170. Irving, *The Rise and Fall*, p. 267.
171. The somewhat bizarre obsession with this topic is well exemplified by a website called "Luft 46," which is dedicated to the description of what could have been German airpower in 1946: www.luft46.com. See also several books written by author David Myhra about different German secret aircraft projects as well as his History Channel film, *The Nazi Plan to Bomb New York City*.
172. Budrass, *Zur Heinkel*, p. 2. See also Budrass's valuable comments regarding the a-historical character of the "what if" school and the postwar German "fantasy" aircraft literature industry.
173. NARA, RG331/13d/Box 89, CIOS, Focke-Wulf Designing Offices and General Management, Bad Eilsen, p. 21.
174. NARA, RG243/31/Box 1, Auszug eines Briefes an Generalfeldmarschall Milch von Prof. Dr. Tank am 7 Juli 1942, p. 1.
175. NARA, RG72/116/Box 12, Technical Report No. 501-45 Comments on the German Aeronautical Research Program, September 1945, Figures 2–3.
176. Lutz Budrass, "Zwischen Unternehmen und Luftwaffe. Die Luftfahrtforschung im Dritten Reich," in *Rüstungsforschung im Nationalsozialismus*, ed. Helmut Maier (Göttingen: Wallstein, 2002), pp. 168–170, 176.
177. NARA, RG331/13d/Box 95, CIOS, Messerschmitt Company's Design and Development Department Oberammergau, p. 4.
178. Budrass, "Zwischen Unternehmen und Luftwaffe," pp. 180–181.
179. Smith and Creek, *Arado 234 Blitz*, pp. 139–140.
180. DM, FA001/924, Francke: Strahlbomber, 17 Januar 1944.
181. On the evolution of the P 1068 see DM, FA001/924; LRD/2442.
182. DM, FA001/924, v. Pfistermeister: Aktennotiz über telefon. Besprechung mit Herrn Scheibe RLM/GL/C-E am 23 Januar 1944.
183. DM, LRD/2442, DFS/Abteilung S: 1068 Zeichnungen.
184. DM, FA001/924, Hauptverwaltung: P 1068 Beginn der Ausbringung Januar 1945, 27 Januar 1944.
185. NASM, 3391/16, Technische Direktion (Francke): He 343, 22 November 1944.
186. Smith and Creek, *Arado 234 Blitz*, pp. 140–145. The talents of Kosin and Lehmann were fully appreciated by the Allies. After the war, Kosin was taken to the USA, while Lehmann was forced to work 10 years for the Soviets. Moreover, British designers used their crescent wing design in the Victor nuclear bomber, which was in service with the RAF from the mid-1950s until 1993.
187. Ransom and Cammann, *Me-163*, vol. 1, pp. 357–361.
188. BA-B, R3/3015, Burg, Ausfertigungen: Dringlichkeitseinstufung der Flugzeuge, 29 August 1944; Schabel, *Die Illusion*, pp. 228–231.
189. NASM, 3391/16, Technische Direktion: Bericht über die wichtigsten Entwicklungsfragen in Wien (Stand Jahresende 1944), 29 Dezember 1944, p. 1.
190. NARA, RG331/13d/Box 91, CIOS, German High Speed Airplanes and Design Development, pp. 57–65.

191. NARA, RG243/6/Box 224, Besprechungen beim Führer am 7 Juni 1944, p. 2.
192. NARA, RG243/6/Box 224, Niederschrift über die Besprechungen beim Führer am 22 März 1945, p. 4.
193. NARA, RG243/6/Box 243, JFM/FZA: Besprechung Rüstungsstab am 28 März 1945.
194. Helmut Trischler, "Historische Wurzeln der Grossforschung: Die Luftfahrtforschung vor 1945," in *Grossforschung in Deutschland*, eds. Margit Szöllösi-Janye and Helmut Triscler (Frankfurt: Campus, 1990), p. 30.
195. Burghard Ciesla, "German High Velocity Aerodynamics and their Significance for the U.S. Air Force 1945–52," in *Technology Transfer Out of Germany After 1945*, eds. Matthias Judt and Burghard Ciesla (Amsterdam: Harwood, 1996), pp. 95–96. For a detailed technical description of the German tunnels see: NARA, RG331/13d/Box 92, CIOS, High Speed Tunnels and Other Research in Germany.
196. Trischler, "Historische Wurzeln", pp. 32–33.
197. Ernst-Heinrich Hirschel, Horst Prem, and Gero Madelung, *Luftfahrtforschung in Deutschland* (Bonn: Bernard & Graefe, 2001), p. 87.
198. See a detailed description of DFS and its projects in NARA, RG331/13d/Box 88, CIOS, Deutsche Forschungsanstalt für Segelflug, Ainring.
199. NARA, RG72/116/Box 13, Technical Report No. 543-45 Facilities, Instrumentation and Methods of the Hamburgische Schiffbau Versuchsanstalt, September 1945.
200. NARA, RG331/13d/Box 86, File XXV-22, CIOS, Aerodynamische Versuchanstalt and Kaiser Wilhelm Institut Gottingen, pp. 5–6.
201. NARA, RG72/116/Box 7, Technical Report No. 237-45 Survey of German Activities in the Field of Guided Missiles, vols. 1–2, August 1945.
202. Krag, "Erfahrung bei der Entwicklung," p. 326.
203. NARA, RG331/13d/Box 90, CIOS, German Aircraft Industry Bremen-Hamburg Area, p. 27.
204. For a detailed survey of the test centers see Beauvais, *Flugerprobungsstellen*.
205. Prandtl is being generally viewed today as father of aerodynamics science.
206. NARA, RG72/116/Box 12, Technical Report No. 272-45 The Organization of Research in Germany, September 1945, pp. 24–26; Technical Report No. 501-45 Comments on the German Aeronautical Research Program, September 1945, p. 13.
207. DM, FA001/261, RLM/TA/GL an Heinkel, 14 Oktober 1943. On the same occasion he also supported the work of A. Lippisch, who also worked on flying wings.
208. Robert McLarren, "Captured Tunnel Advances U.S. Research," *Aviation Week* 51, no 1 (August 1949): pp. 19–20; Reimar Horten and Peter F. Selinger, *Nurflügel: Die Geschichte der Horten-Flugzeuge 1933–1960* (Graz: Weishaupt, 1983), pp. 135–148.
209. NASM, 3984/85, Seiler: Notiz über eines Besprechung mit Prof. Messerschmitt, General Direktor Frydag, und Rechtsanwalt Merkel am 11 April 1945 in Murnau.
210. Combined Intelligence Objectives Sub-Committee Report: Horten Tailless Aircraft, by K.G. Wilkinson, October 1945, pp. 21–22. Copy kindly provided by Russell Lee of the National Air and Space Museum.
211. NARA, RG72/116/Box 11, Technical Report No. 436-45 Junkers Developments in High Speed Tunnels, October 1945.
212. The problems associated with practical military flying wings were solved only during the 1980s with the development of the B-2 stealth bomber.
213. Schabel, *Die Illusion*, p. 203.
214. BA-B, R3/1583, Der Reichsminister für Rüstung und Kriegsproduktion: Bildung der Entwicklungs-Hauptkommission Flugzeuge, 15 September 1944; Moll, *Führer-Elasse*, pp. 421–422.

215. BBSU, p. xxxv.
216. Overy, *The Air War*, p. 192.
217. NARA, M888/6/679, Stenographischer Bericht über die Jägerstab Besprechung am Freitag 20 März 1944, p. 3.
218. NARA, RG331/13d/Box 98, CIOS, Survey of Production Techniques used by the German Aircraft Industry, p.11.
219. Overy, *The Air War*, p. 208.
220. NARA, RG243/6/Box 182, Embassy of the USA, Enemy Objectives Unit/Economic Warfare Division: Dispersal of Aircraft Production, 28 December 1944. Noteworthy, this report was prepared in order to gain some insight about possible patterns of German dispersal.
221. NARA, RG243/31/Box 1, USSBS: Interview no. 1, Dr. Kurt Tank, 17 & 24 April 1945, p. 13.
222. NARA, RG243/6/Box 229, Ministry Report: Focke-Wulf Design Offices in Bad-Eilsen, Germany, IV Appendix A.
223. See an example of such a plan in: NARA T83/17/3361078, Focke-Wulf an das RLM/GL Planungsamt: Verlagerung einzigartiger Fertigungen, 28 November 1942.
224. NARA T83/17/3361080-83, Der Leiter des Sonderausschusses F4: Ausweichplanung der dem Sonderausschuss F4 angehörenden Firmen bei Fliegerschäden, 13 November 1942.
225. USSBS, *Aircraft Division*, p. 23.
226. NARA, RG331/13d/Box 91, CIOS, German Manufacture of Airscrews, p. 4.
227. NARA, RG243/6/Box 242, USSBS, *Wiener Neustaedter Flugzeugwerke, Wiener Neustadt, Austria*, 28 September 1945, p. 11.
228. Schmoll, *Nest of Eagles*, p. 66.
229. NARA, M888/8/808-809, Stenographischer Niederschrift über die Besprechung beim Reichsmarschall am Donnerstag, dem 4 November 1943 in den Junkers-Werken zu Dessau, pp. 3–4.
230. Irving, *The Rise and Fall*, p. 252.
231. YVA, TR.2/NOKW-180, Extracts from stenographic notes on the conference of the Reichsmarshall on Thursday, 4 November 1943, at the Junkers plant in Dessau.
232. NARA, T83/6/3747360, Der Reichsminister der Luftfahrt: Bombensichere Fertigungsräume, 26 Januar 1944.
233. The Allies knew that the Germans knew through "Ultra" code encryption of German radio traffic. See an example in Murray, *Strategy for Defeat*, pp. 236–237.
234. NARA, RG243/31/Box 1, USSBS: Minutes of meeting with Reichsminister Albert Speer, Flensburg, 18 May 1945, p. 10.
235. NARA, RG243/10/Box 1, Headquarters USSBS: Intelligence Notes No. 6, 1 June 1945, p. 2.
236. NARA, RG 243/6/Box 244, Junkers Bernburg: Beantwortung des am 9 Mai 1945 besprochenen Fragebogen, 11 Mai 1945, p. 6.
237. NARA, RG243/32/Box 5, Minutes of the interview with Herr Saur, 7 June 1945, p. 27.
238. United States Strategic Bombing Survey, *Gerhard Fieseler Werke GmbH, Kassel, Germany* (n.p., 1945), p. 7.
239. NARA, RG243/31/Box 1, USSBS: Interview No. 22 with Dr. Saur, General Buhle, 18 May 1945, p. 3.
240. NARA T83/16/3360447-52, Der Reichsminister d. Rüstung u. Kriegsproduktion: Grundsätze für den Neu-, Um-, und Erweiterungsbau und die Verlegung von Fertigungsstätten der Rüstungs- und Kriegsproduktion auf Grund der Luftkriegserfahrungen, Fassung August 1944.
241. See especially Milch's remarks in NARA, M888/8/514, Stenographischer Bericht über die GL Besprechung am Dienstag, dem 5 Mai 1942, p. 61.
242. NARA, M888/8/507-508, Stenographischer Bericht über die Jägerstab Besprechung am Montag, dem 19 Juni 1944, pp. 53–54.

243. USSBS, *Aircraft Division*, pp. 24–25.
244. See a list of locations used for Messerschmitt's Regensburg production in USSBS, *Messerschmitt AG. Appendix*, Exhibit R-D.
245. NARA, RG331/13d/Box 87, CIOS, Bayrische Motorenwerke AG Munich-Oberwiesenfld, p. 15.
246. NARA, T83/11/3353796, Tank an Dir. Schubert Bremen Flughafen: Verlagerungsbefehl, 14 März 1944.
247. United States Strategic Bombing Survey, *Gothaer Waggonfabrik AG, Gotha, Germany*, 2nd edition (n.p., 1947), Exhibit B.
248. USSBS, *Aircraft Division*, p. 25.
249. NARA, RG243/31/Box 1, USSBS: Interview N0.22 with Dr. Saur, General Buhle, 18 May 1945, p. 3.
250. USSBS, *Junkersflugzeug*, p. 12.
251. NARA, RG243/6/Box 242, Report of USSBS Field Team No. 43 on Junkers Flugzeug- und Motorenwerk AG, 12–14 May 1945, p. 4.
252. NARA, RG243/6/Box 241, 13a6 Production Data, unnumbered summery titled "Production Loss."
253. USSBS, *Aircraft Division*, p. 86.
254. Loc. cit.
255. See dispersion maps *ibid.*, Figures II-2, VII-2.
256. USSBS, *Gothaer*, pp. 3–4 and Exhibit B. See also the dispersion map in Exhibit C.
257. NARA, RG331/13d/Box 93, CIOS, Junkers 004 (203), Jet Propulsion Engines, pp. 3, 7–8. See also the map in USSBS, *Aircraft Division*, Figure VII-2.
258. USSBS, *Aircraft Division*, Figures II-3, II-4.
259. *Ibid.*, Figure II-6.
260. NARA, M888/6/1147, Stenographischer Niederschrift der 6 Reise des "Unternehmen Hubertus" vom 8 bis 10 Mai 1944, p.19.
261. NARA, RG243/6/Box 224, Punkte aus der Besprechung beim Führer am 30 April 1944, p. 3.
262. NARA, RG243/6/Box 242, USSBS, *Wiener Neustaedter Flugzeugwerke, Wiener Neustadt, Austria*, 28 September 1945, pp. 12–13, Exhibit A.
263. USSBS, *Messerschmitt AG. Part B*, p. 8.
264. BA-B, R3/1959, RLM/GL/C-Fertg. XI: Bericht zur Fertigungsführung: Leistungssteigerung durch Betriebskonzentrazion und Typenverminderung im Zellenbau, 30 Juni 1944.
265. NARA, RG243/32/Box 6, USSBS: Interrogation of Albert Speer, 2 August 1945, p. 2.
266. USSBS, *Gerhard Fieseler*, p. 2.
267. Schmoll, *Nest of Eagles*, pp. 49, 134.
268. USSBS, *Messerschmitt AG. Part B*, Exhibits R-E, R-F.
269. Schmoll, *Messerschmitt-Giganten*, p. 143.
270. These main factories kept producing conventional fighters until the end of the war.
271. Smith and Creek, *Me 262*, vol. 2, p. 284.
272. USSBS, *Messerschmitt AG. Appendix*, p. 2.
273. USSBS, *Messerschmitt AG Augsburg, Germany*, Appendix I–III (n.p., 1945), p. 2. See a map of the factory in: NARA, RG243/6/Box 239, Lageplan Horgau.
274. *Ibid.*, Exhibit H-A. The rudimentary character of this factory is well illustrated by the 18 photos in Exhibit H-B.
275. Geoffrey P. Megargee, ed., *The United States Holocaust Memorial Museum Encyclopedia of Camps and Ghettos, 1933–1945*, vol. 1, Par A (Bloomington: Indiana University Press, 2009), pp. 452–453.
276. USSBS, *Messerschmitt AG Augsburg*, pp. 2–4; Smith and Creek, *Me 262*, vol. 2, pp. 384–385.
277. NARA, RG243/6/Box 238, Messerschmitt AG Augsburg, Gebäude Beschreibung, 30 April 1945.
278. NASM, 3984/65, Denkschrift betreffend die Produktion, Entwicklung und Verwendung von Jagdeinsitzern der Firma Messerschmitt, pp. 9–10.

279. USSBS, *Aircraft Division*, pp. 31–32; USSBS, *Messerschmitt AG Augsburg, Germany. Overall Report* (n.p., 1945), p. 3.
280. See, for example, aerial imagery interpretation report of the Leipheim airfield, which completely missed the nearby "Kuno II" forest factory in NARA, RG341/217/Box 278, Military Intelligence Photographic Interpretation (MIPI) Report: Report No. L64: Status and Activity at Leipheim Airfield Following Bombing Attacks, 25 July 1944.
281. Schmoll, *Messerschmitt-Giganten*, p. 143.
282. NARA, RG243/31/Box 1, USSBS: Minutes of meeting with Reichsminister Albert Speer, Flensburg, 20 May 1945, p. 1.
283. NARA, RG243/6/Box 224, Besprechung beim Führer am 28 und 29 November 1944, p. 4.
284. NARA, RG243/31/Box 1, USSBS: Interview with Prof. William [sic] Messerschmitt on 11–12 May 1945, p. 10.
285. NARA, RG331/13d/Box 87, CIOS: Bavarian Motor Works (BMW). A production Survey, p. 45; Werner, *Kriegswirtschaft*, pp. 170–171.
286. DM, LRD 275.
287. Werner, *Kriegswirtschaft*, p. 173.
288. About these bunkers, see Gordon Williamson, *U-Boat Bases and Bunkers 1941–1945* (Oxford: Osprey, 2003).
289. NARA T83/6/3747440-42, Aktenvermark: Besprechung mit Herrn Baurat Elten vom Jägerstab Abteilung Leistungsteigerung in St. Astier am 19–20 Juli 1944.
290. NARA, RG331/13d/Box 90, CIOS, German Activities in the French Aircraft Industry, pp. 48–49.
291. Willi Boelcke, *Deutschlands Rüstung im Zweiten Weltkrieg: Hitlers Konferenzen mit Albert Speer 1942–1945* (Frankfurt am Main: Athenaion, 1969), p. 247.
292. Hölsken, *Die V-Waffen*, pp. 50–51.
293. Allen, *The Business*, p. 215.
294. Neufeld, *The Rocket*, pp. 197–201; NARA, RG331/13d/Box 99, CIOS, Underground Factories in Central Germany, pp. 7–8.
295. Hölsken, *Die V-Waffen*, pp. 50–51.
296. Budrass, *Flugzeugindustrie*, p. 794.
297. Raim, *Die Dachauer*, pp. 53–55.
298. BA-B, R3/1580, Göring an Speer, 10 Oktober 1943.
299. NARA, M888/8/893, Stenographischer Niederschrift über die Besprechung beim Reichsmarschall am Dienstag, dem 2 November 1943 in den Messerschmitt-Werken in Regensburg, p. 50.
300. USSBS, *Junkersflugzeug*, p. 13.
301. NARA, M888/8/817-818, Stenographischer Niederschrift über die Besprechung beim Reichsmarschall am Donnerstag, dem 4 November 1943 in den Junkers-Werken zu Dessau, pp. 12–13.
302. Werner, *Kriegswirtschaft*, pp. 174–177.
303. Hermann Kaienburg, *Die Wirtschaft der SS* (Berlin: Metropol, 2003), p. 643.
304. Wagner, *Produktion des Todes*, pp. 205–206.
305. *Ibid.*, pp. 221–222.
306. *Ibid.*, p. 223.
307. NARA, T83/6/3747360, Der Reichsminister der Luftfahrt: Bombensichere Fertigungsräume, 26 Januar 1944.
308. NARA, T83/6/3747359, Der Reichsminister der Luftfahrt: Bombensichere Fertigungsräume, 1 Februar 1944.
309. Fröbe, "'Wie bei den alten Ägyptern,'" pp. 401–402.
310. YVA, TR.2/PS-1584, Himmler an Göring, Einsatz von Häftlingen in der Luftfahrindustrie, 9 März 1944 (also in NARA, M888/6/544-547).
311. BA-B, R3/1509, Protokoll der Führerbesprechung am 5 März 1944.
312. NARA, RG243/6/Box 224, Besprechung beim Führer am 13 Mai 1944, p. 2.
313. NARA, T73/123/3285000, Rüstungsinspektion VII, Abt. Luftwaffe: Bestätigung. Teilverlagerung Firma Messerschmitt, 2 März 1944.

314. BA-B, R3/1827, Rumi/WF-Fin 1: Merkblatt über Finanzierungshilfen für die Rüstungswirtschaft, 20 August 1944.
315. BA-B, R7/1192, Sonderstab Höhlenbau: Liste der A- und B-Vorhaben der SS, 12 März 1944.
316. Wagner, *Produktion des Todes*, pp. 101–104; BA-B, R7/1192, RLM: Liste der U-Verlagerung der 1 und 2 Welle, 3 März und 4 April 1944.
317. NARA, M888/6/887-890, Stenographischer Bericht der Jägerstab Besprechung am 25 April 1944, pp. 17–20.
318. Fröbe, "'Wie bei den alten Ägyptern,'" pp. 402–405.
319. NARA, RG331/13d/Box 99, CIOS, Underground Factories in Central Germany, pp. 16–17.
320. NARA T83/16/3360447-52, Der Reichsminister d. Rüstung u. Kriegsproduktion: Gründsätze für den Neu-, Um-, und Erweiterungsbau und die Verlegung von Fertigungsstätten der Rüstungs- und Kriegsproduktion auf Grund der Luftkriegserfahrungen, Fassung August 1944.
321. Raim, *Die Dachauer*, p. 116.
322. Wagner, *Produktion des Todes*, p. 117.
323. NARA, RG243/6/Box 224, Punkte aus dem Einzelvortrag beim Führer am 6 März 1944, p. 2.
324. NARA, M888/6/642, Stenographischer Bericht über die Jägerstab Besprechung unter Vorsitz von Generalfeldmarschall Milch am Freitag, dem 17 März 1944, p. 16.
325. NMT, vol. 2, p. 589, doc. R-124, Minutes of Discussion with the Fuehrer on 6 and 7 April 1944.
326. NARA, M888/6/946-948, Stenographischer Bericht der Jägerstab Besprechung am 2 Mai 1944, pp. 16–18.
327. NARA, RG243/6/Box 242, Report of USSBS Field Team No. 43 on Junkers Flugzeug- und Motorenwerk AG Motorenbau Zweigwerk at Magdeburg Neustadt, 8–14. Mai 1945, pp. 12–13.
328. USSBS, *Junkersflugzeug*, p. 13. See also NARA, RG331/13d/Box 86, CIOS: Aircraft Gas Turbine Engine Developments at Junkers, Dessau and Associated Factories, p. 7.
329. NARA, RG331/13d/Box 99, CIOS, Underground Factories in Central Germany, pp. 8, 59 and aerial photo with overlaid sketch of these factories, p. 32.
330. Kammler an Himmler: Zusammenstellung über den Stand der Sonderbaumassnahmen der Waffen SS, 23 September 1944. Copy in author's possession.
331. Wagner, *Produktion des Todes*, p. 228.
332. NARA, RG331/13d/Box 92, CIOS, Group 2 Targets in Nordhausen Area, p. 7.
333. *Ibid.*, p. 39.
334. BA-B, NS19/3832, Kammler an Brandt: Aufstellung der bis Ende 1944 fertiggestellten Fertigungsflächen in den A- und B-Projekten, 11 Januar 1944.
335. NARA, M888/8/892-893, Stenographischer Niederschrift über die Besprechung beim Reichsmarschall am Dienstag, dem 2 November 1943 in den Messerschmitt-Werken in Regensburg, pp. 49–50.
336. Kommando der Erprobungsstelle der Luftwaffe: Kurzbericht über besondere Schwierigkeiten Anlauf Me 262, 1 März 1944. Published in Smith and Creek, *Me 262*, vol. 2, p. 234.
337. Megargee, *Encyclopedia of Camps and Ghettos*, vol. 1, part B, p. 1042.
338. NARA, RG243/6/Box 238, Messerschmitt AG Augsburg, Produktions-Stätten, 30 April 1945. A map of the Me 262 production web: USSBS, *Messerschmitt AG. Appendix*, Exhibit M-B.
339. NARA, RG331/13d/Box 90, CIOS, German Aircraft Industry Bremen-Hamburg Area, p. 22.
340. NARA, RG331/13d/Box 99, CIOS, Underground Factories in Central Germany, pp. 109–113.
341. USSBS, *Aircraft Division*, p. 28.
342. USSBS, *Gothaer*, Exhibit B.
343. Hölsken, *Die V-Waffen*, p. 71; USSBS, *Aircraft Division*, pp. 114–114a.
344. As Milch told the Americans after the war. USSBS, *Aircraft Division*, p. 97. See also Saur's remarks in a *Jägerstab* meeting on 12 April 1944 regarding the urgent move of the DB603 engine production to underground facilities. NARA, M888/6/826, Stenographischer Bericht über die Jägerstab Besprechung unter Vorsitz von Hauptdienstleiter Saur am 12 April 1944, p. 32.
345. NARA, RG243/10/Box 1, Headquarters USSBS: Intelligence Notes No. 3, 5. May 1945, pp. 5–6.
346. NARA, RG331/13d/Box 87, CIOS, Bayrische Motorenwerke AG Munich-Oberwiesenfld, p. 13.
347. NARA, RG331/13d/Box 90, CIOS, Gas Turbine Development BMW, Junkers, Daimler Benz, p. 5.
348. NARA, RG331/13d/Box 94, CIOS, MAN, Augsburg and Harburg. The Franziskaner Keller, Munich, pp. 21–22.
349. NARA, RG243/6/Box 228, ATG Maschinenbau GmbH, Leipzig. Exhibit B: Report on Underground Manufacturing Facilities.
350. NARA, M888/6/1270, Stenographischer Bericht über die Jägerstab Besprechung am Freitag, 26 Mai 1944, p. 22.
351. NARA, RG331/13d/Box 99, CIOS, Underground Factories in Central Germany, pp. 12–13. On the construction of tunnels see p. 128ff.
352. By November 1944 Allied intelligence already knew about this underground expansion. NARA, RG243/14/Box 11, Air Ministry, London, A.I.2 (a): Schedule of Known and Suspected Aircraft and Aero Engine Factories, 11 November 1944.
353. USSBS, *Messerschmitt AG. Appendix*, Appendix II, pp. 2–4 & Exhibits K-B, K-D, K-E.
354. NARA RG73/30/3159057, Rüstungskommando Augsburg: Liste über die neue Rangfolge für die Zuweisung von Arbeitskräften, 18 April 1944.
355. NARA RG331/13d/Box 90, CIOS, German Aircraft Industry Friedrichshafen-Munich Area, pp. 7–10.
356. Alan S. Milward, *War, Economy and Society, 1939–1945* (Berkeley: University of California, 1979), p. 127.
357. Kohl, *Auto Union und Junkers*, p. 226.
358. NARA, RG331/13d/Box 87, CIOS: Bavarian Motor Works (BMW). A production Survey, p. 45. A scheme of the Stassfurt Mine is available in: RG243/6/Box 243, Junker Flugzeug Motoren Werke underground plant Mine Ludwig II.
359. NARA, RG331/13d/Box 99, CIOS, Underground Factories in Central Germany, pp. 116–117.
360. NARA, RG243/6/Box 228, ATG Maschinenbau GmbH, Leipzig. Exhibit B: Report on Underground Manufacturing Facilities.
361. NARA, M888/6/1149-1150, Stenographischer Niederschrift der 6 Reise des "Unternehmen Hubertus" vom 8 bis 10 Mai 1944, pp. 21–22.
362. YVA, Testimonies O.3/6765, O.3/5850, O.3/8069.
363. Green, *The Augsburg Eagle*, p. 121.
364. NARA, RG331/13d/Box 87, CIOS, Bartensleben Salt Mine, Morsleben (Division of Askania Werke, Berlin), p. 3.
365. NARA, RG331/13d/Box 99, CIOS, Underground Factories in Central Germany, pp. 113–114.
366. NARA, RG72/116/Box 2, Technical Report No. 124-45 A Survey of Production Techniques used in the German Aircraft Industry, July 1945, p. 9.
367. BA-B, R3/1749, Extract from minutes of a Rüstungstab meeting on 4 September 1944; Fröbe, "'Wie bei den alten Ägyptern,'" pp. 433–435.
368. USSBS, *Aircraft Division*, p. 29.
369. Fröbe, "'Wie bei den alten Ägyptern,'" p. 436.
370. USSBS, *Messerschmitt AG. Appendix*, Appendix II, p. 3.

371. Wagner, *Produktion des Todes*, p. 103.
372. NARA, M888/6/896, Stenographischer Bericht der Jägerstab Besprechung am 25 April 1944, p. 26.
373. Smith and Creek, *Me 262*, vol. 1, p. 171.
374. USSBS, *Messerschmitt AG. Overall Report*, p. 13.
375. NARA, M888/6/603, Stenographischer Bericht über die Jägerstab Besprechung am 6 März 1944, p. 20.
376. Probably Oberottmarshausen, near Augsburg. See Raim, *Die Dachauer*, p. 43.
377. NARA, M888/6/643-644, Stenographischer Bericht über die Jägerstab Besprechung unter Vorsitz von Generalfeldmarschall Milch am Freitag, dem 17 März 1944, pp. 17–18.
378. Raim, *Die Dachauer*, p. 44.
379. NARA, M888/13/000, Der Führer an Speer, 21 April 1944.
380. Raim, *Die Dachauer*, p. 56.
381. *Ibid.*, p. 48.
382. Kaienburg, *Die Wirtschaft*, p. 644.
383. NARA, T73/132/3295138, Rüstngsinspektion VII: Aktennotiz. Besprechung bei Fa. Messerschmitt und Sonderausschuss F2 am 7 März 1945.
384. See a complete list of the factory's logistical network at http://www.gusen.org/dok/b8/b8002x.htm.
385. http://www.gusen.org/gu20101x.htm.
386. Smith and Creek, *Me 262*, vol. 4, pp. 780–784.
387. USSBS, *Messerschmitt AG. Appendix*, Appendix II, p. 2.
388. This section is largely based on Raim, *Die Dachauer*; Megargee, *Encyclopedia*, pp. 488–489; NARA, RG72/116/Box 2, Technical Report No. 116-45. Bombproof Aircraft Assembly Plant, June 1945.
389. Smith and Creek, *Me 262*, vol. 2, p. 384.
390. NARA, RG243/32/Box 1, USSBS: Xaver Dorsch (chief of Amt Bau OT). General discussion of allocation of materials & control of construction, particularly during the last year of World War II, 29 July 1945, p. 13.
391. NARA, RG331/13d/Box 95, CIOS, Messerschmitt Bombproof Assembly Plant, Landsberg, pp. 5–8.
392. Smith and Creek, *Me 262*, vol. 2, pp. 385, 392.
393. NARA, T73/132/3295136-37, Rüstngsinspektion VII: Aktennotiz. Besprechung bei Fa. Messerschmitt und Sonderausschuss F2 am 7 März 1945.
394. Raim, *Die Dachauer*, p. 124.
395. NARA, RG331/13d/Box 99, CIOS, Underground Factories in Central Germany, p. 96.
396. *Ibid.*, p. 107.
397. NARA, RG341/216/MIPI Report no. 32,446, Reported Aircraft Factory at Kahla, 30 August 1944.
398. NARA, RG243/6/Box 224, Besprechung beim Führer vom 21–23 September 1944, p. 7.
399. NARA, RG331/13d/Box 99, CIOS, Underground Factories in Central Germany, p. 99.
400. Smith and Creek, *Me 262*, pp. 388, 392; http://www.reimahg.de.
401. NARA, RG331/13d/Box 99, CIOS, Underground Factories in Central Germany, p. 106.
402. NARA RG243/10/Box 1, Headquarters USSBS: Intelligence Notes No. 4, 12 May 1945, p. 5.
403. Wagner, *Produktion des Todes*, pp. 113–114.
404. NARA RG243/6/Box 826, USSBS, Oil Division: Report on Underground and Dispersal Plants in Greater Germany, pp. 11–12.
405. Mommsen, *Der Mythos*, pp. 27–28.
406. NARA T83/16/3360447-52, Der Reichsminister d. Rüstung u. Kriegsproduktion: Grundsätze für den Neu-, Um-, und Erweiterungsbau und die Verlegung von Fertigungsstätten der Rüstungs- und Kriegsproduktion auf Grund der Luftkriegserfahrungen, Fassung August 1944, p. 5.
407. USSBS, *Messerschmitt AG. Appendix*, Appendix II, Exhibit K-G.
408. NARA, RG243/6/Box 224, Punkte aus der Besprechung beim Führer am 30 April 1944, p. 3.
409. NARA, RG243/6/Box 219, AI2(a): Attack on the GAF. Recommended Target Priorities at 24 October 1944.
410. NARA, RG243/6/Box 221, Foreign Office/German Economic Dept., Report on the German Aero-Engine Industry, n.d., Part II: Factories in Detail.
411. See, for example, NARA, RG331/13d/Box 90, CIOS, German Aircraft Industry Bremen-Hamburg Area, pp. 17, 21.
412. NARA, RG331/13d/Box 99, CIOS, Underground Factories in Central Germany, pp. 13, 120 ff.
413. See, for example, NARA, RG243/14/Box 6, Aircraft Components Factories — Fighters 1 & 2, Target Information Sheets for Kahla, Langenstein and Leonberg. The Langenstein ("Malachit") file is especially interesting, because intelligence was able to provide quite accurate information about the place, based on prewar geological and geographical survey data, and to prepare a detailed model of the area. Interpreters were also able to determine with the aid of POW interrogations that Leonberg was involved in Me 262 production.
414. *Ibid.*, Target Information Sheets for Rathdamnitz nr. Stolp and Telfs near Innsbruck. See also numerous entries in: NARA, RG243/14/Box 11, Air Ministry, London, A.I.2 (a): Schedule of Known and Suspected Aircraft and Aero Engine Factories, 11 November 1944.
415. NARA, RG243/6/Box 228, A.D.I (K) Report no. 249/1945, Dispersal from Junkers, Dessau, 25 March 1945.
416. See for example information regarding the Leipheim forest factory in: NARA, RG243/6/Box 218, MAAF: Appendix, Jet Aircraft Assembly, Messerschmitt Leipheim, 13 January 1945.
417. Air Force Historical Research Archives (AFHRA), Call No. 512.619B-23, A.D.I.(K) Report no. 693/1944: Me 262 production Leipheim and associates (October 1944). A copy kindly provided by Richard Eger.
418. USSBS, *Messerschmitt AG. Overall Report*, pp. 12, 15.
419. NARA, RG331/13d/Box 93, CIOS, Junkers Aircraft and Engines Facilities, p. 10.
420. NARA, RG243/14/Box 6, Aircraft Components Factories — Fighters 2, Target Information Sheet: Leonberg nr. Stuttgart, provisional.
421. NARA, M888/6/1324, Stenographischer Bericht über die Jägerstab Besprechung am Freitag, 26 Mai 1944, p. 85.
422. NASM, 3984/65, Denkschrift betreffend die Produktion, Entwicklung und Verwendung von Jagdeinsitzern der Firma Messerschmitt, 30 Oktober 1944, p. 8.
423. NARA, RG243/14/Box 7, Armaments Works, Target Information Sheet: Nieder-Sachswerfen near Nordhausen, 23 February 1945.
424. Hölsken, *Die V-Waffen*, p. 186.
425. Albert Speer, *Der Sklavenstaat: Meine Auseinandersetzung mit der SS* (Berlin: Ullstein, 1981), p. 308.
426. Wagner, *Produktion des Todes*, p. 104. See also his remarks considering the rationality of the underground relocation scheme: pp. 116–118.

Chapter 4

1. Werner, *Kriegswirtschaft*, p. 146.
2. NARA, RG72/116/Box 2, Technical Report No. 124-45 A Survey of Production Techniques used in the German Aircraft Industry, July 1945, p. 6.
3. See a good description of these two methods and their employment by SS brick-making enterprises and Garment businesses in Allen, *The Business*, pp. 72–78.
4. Ferguson, "One Thousand Planes," p. 154.
5. Mommsen, *Der Mythos*, p. 10.
6. NARA, RG331/13d/Box 98, CIOS, Survey of Production Techniques Used by the German Aircraft Industry, pp. 5–6.

7. Lutz Budrass, "Der Schritt über die Schwelle. Ernst Heinkel, das Werk Oranienburg und der Einstieg in die Beschäftigung von KZ-Häftlingen," in *Zwangsarbeit während der NS-Zeit in Berlin und Brandenburg*, ed. Winfried Meyer (Potsdam: Berlin-Brandenburg, 2001), p. 134.
8. NARA, T177/37/3727151, Gefolgschaft am 1 April 1938.
9. Budrass, "Der Schritt," pp. 136–137.
10. Budrass et al., "Demystifying," p. 11.
11. *Ibid.*, p. 12.
12. Abelshauser, "Germany: Guns, Butter," p. 152.
13. NARA, T177/37/3727573, Tank an der Industrierat des Reichsmarschalls für die Fertigung Luftwaffengerät: Einberufung der Fachkräfte zur Wehrmacht, 10 Juni 1942.
14. USSBS, *Messerschmitt AG, Overall Report*, p. 10.
15. NARA, T177/37/3727485, Gefolgschaftsmeldung per 15 Juni 1942.
16. DM, FA001/904, Dr. Heinkel: Dringlichkeit He 219, 4 August 1942.
17. From around this time Allied intelligence began to collect valuable information about German aviation technology from POWs, who worked in the aviation industry before being drafted. Staerck & Sinnot, *Luftwaffe*, pp. 48, 69, 84.
18. Kohl, *Auto Union und Junkers*, pp. 157–158.
19. NARA, M888/6/213-215, Niederschrift zur 21 Besprechung der "Zentralen Planung" am 31 Oktober 1942, pp. 11–15.
20. NARA, RG243/6/Box 182, Embassy of the USA, Enemy Objectives Unit/Economic Warfare Division: Dispersal of Aircraft Production, 28 December 1944, p. 2.
21. Budrass, "Der Schritt," p. 139.
22. NARA, T177/37/3727151, Gefolgschaft am 1 April 1938.
23. Coates, *Helicopters*, p. 109.
24. NARA, T177/39/3729936, Mansfeld Werk GmbH: Geschäftsbericht für die Beiratssitzung am 24/25 September 1941 in Prenzlau, 20 September 1941.
25. Ulrich Herbert, *Hitler's Foreign Workers. Enforced Foreign Labor in Germany Under the Third Reich* (Cambridge: University Press, 1997), p. 257.
26. Budrass et al., "Demystifying," p. 12.
27. NARA, RG243/6/Box 248, HFM: Gefolgschaft Entwicklung, 30 Juni 1944. Interestingly, at the same time the share of women among the foreign workforce increased from 0 in 1940 to 15 percent in 1944.
28. Werner, *Kriegswirtschaft*, p. 178.
29. Budrass, *Flugzeugindustrie*, p. 797.
30. NARA, M888/8/589-590, Stenographische Niederschrift der Besprechung beim Reichsmarschall am Donnerstag, dem 28 Oktober 1943, pp. 6–7.
31. NARA, T177/39/3729793, BMW: FOB Monatsbericht, Werk I München, 7 Mai 1943.
32. NARA, T177/37/3727267, Gesamtbelegschaftsstand am 30 Juni 1943.
33. USSBS, *Messerschmitt AG. Appendix*, Exhibit K-F.
34. NARA, RG331/13d/Box86, CIOS: Administration, Plastics, Production Tooling, Spare Parts and Servicing in German Aircraft Industry, p. 6.
35. USSBS, *Junkersflugzeug*, p. 5.
36. On the *Wehrmacht's* effort to solve this problem, see Bernhard R. Kroener, "General Heldenklau: Die 'Unruh-Kommission' im Strudel polykratischer Desorganisation (1942–1944)," in *Politischer Wandel, organisierte Gewalt und nationale Sicherheit. Beiträge zur neueren Geschichte Deutschlands und Frankreichs. Festschrift für Klaus-Jürgen Müller*, eds. Ernst-Willi Hansen, Gerhard Schreiber, and Bernd Wegner (München: Oldenbourg, 1995), pp. 269–285.
37. NARA, T177/16/701101, Besprechungsniederschrift über He 177 – Besprechung bei GL/C Chef am 19 März 1943.
38. NARA T83/17/3361371-78, Mitteldeutsche Motorenwerk: Bericht über Stand und Entwicklung der Produktion zur Beiratssitzung am 4 Dezember 1944.
39. Werner, *Kriegswirtschaft*, p. 178.
40. NARA, M888/8/570-571, Stenographischer Bericht über die GL Besprechung am Mittwoch, dem 26 August 1942, pp. 23–24.
41. M888/6/574, Vernehmung des Dip. Ing. Karl Otto Saur durch Mr. Koch, 13. November 1945, p. 13.
42. Schabel, *Die Illusion*, pp. 211–212.
43. *Ibid.*, p. 216.
44. NARA, RG243/6/Box 224, Besprechung beim Führer am 12 Oktober 1944, p. 8.
45. NASM, 3984/65, Denkschrift betreffend die Produktion, Entwicklung und Verwendung von Jagdeinsitzern der Firma Messerschmitt, p. 4.
46. NARA, RG243/31/Box 1, USSBS: Interview no. 6, Prof. William Messerschmitt, 11–12 May 1945, p. 8.
47. Irving, *The Rise and Fall*, pp. 251–252.
48. NARA, M888/8/687, Stenographischer Bericht über die GL Besprechung am Mittwoch, dem 30 November 1943, p. 16.
49. NARA, M888/8/628-629, Stenographische Niederschrift der Besprechung beim Reichsmarschall am Donnerstag, dem 28 Oktober 1943, pp. 45–46.
50. *Ibid.*, p. 90.
51. NARA, RG243/31/Box 1, USSBS: Minutes of meeting with Reichsminister Albert Speer, Flensburg, 18 Mai 1945, p. 3.
52. BA-B, R3/1591, Speer an Milch, 15 November 1943.
53. YVA, TR.2/NOKW-180, Extracts from stenographic notes on the conference of the Recihsmarshall on Thursday, 4 November 1943 at the Junkers plant in Dessau.
54. BA-B, R3/1597, Speer an Sauckel, Dezember 1943.
55. *Ibid*, Speer an Milch, Auszüge aus Niederschrift über die Besprechung beim Führer am 16–17 Dezember 1943, 19 Dezember 1943.
56. NARA, M888/8/590, Stenographische Niederschrift der Besprechung beim Reichsmarschall am Donnerstag, dem 28 Oktober 1943, p. 7.
57. AFHRA, Call No. 512.619B-23, A.D.I.(K) Report no. 693/1944, Me 262 production Leipheim and associates (October 1944). A copy kindly provided by Richard Eger.
58. NARA, RG243/6/Box 190, Die Entwicklung des Volkswagenwerkes seit Kriegsbeginn 1939. I Tail: Stadt des KdF-Wagens 1944. Das Fertigungsprogramm und die Belegschaft, 1939–44; Mommsen and Grieger, *Das Volkswagenwerk*, pp. 677–682.
59. Noticeably, this consideration played exactly the opposite role in the decision to use slave labor in the A-4/V-2 production.
60. Mommsen and Grieger, *Das Volkswagenwerk*, pp. 684, 686–687.
61. Hölsken, *Die V-Waffen*, pp. 47–48. Göring, who appeared to be completely out of touch with the issue at that time, demanded an eventual monthly output of 50,000.
62. *Ibid.*, pp. 48–49.
63. Mommsen and Grieger, *Das Volkswagenwerk*, p. 687.
64. NARA, RG243/6/Box 241, Folder 13a6, Production Data, p. 28.
65. Hölsken, *Die V-Waffen*, pp. 57–58.
66. *Ibid.*, p. 59.
67. Mommsen and Grieger, *Das Volkswagenwerk*, p. 695.
68. BA-B, R3/1944, Der Rüstungsminister: Btr. Neue Rangfolge für die Zuweisung von Arbeitskräften, 17 Januar 1944.
69. Hölsken, *Die V-Waffen*, p. 64.
70. NARA, T73/123/3285157, Rüstungskommando Augsburg: Aufstellungslist, 20 Januar 1944.
71. NARA, T73/123/3285108, RLM/GL an Rü Kdo. Augsburg, 28 Januar 1944.
72. BA-B, R3/1597, Speer an Sauckel, 7 Februar 1944.

73. BA-B, R3/1591, Milch an Speer, 11 Februar 1944.
74. NARA, RG243/31/Box 1, USSBS: Interview no. 7, Mr. Seiler on 16 May 1945, p. 1.
75. Kommando der Erprobungsstelle der Luftwaffe: Kurzbericht über besondere Schwierigkeiten Anlauf Me-262, 1 März 1944. Published in Smith and Creek, *Me-262*, vol. 2, p. 234.
76. NMT, vol. 2, pp. 572–573.
77. NARA, M888/6/854, Stenographischer Bericht über die Jägerstab Besprechung unter Vorsitz von Hauptdienstleiter Saur am 12 April 1944, p. 52.
78. NARA, M888/6/667, Stenographischer Bericht über die Jägerstab Besprechung am Freitag 17. März 1944, p. 40.
79. NARA, M888/6/1092-1094, Stenographischer Bericht über die Jägerstab Besprechung am Freitag, 5 Mai 1944, pp. 14–16.
80. NARA, T73/123/3284829, Messerschmitt AG Augsburg an das Rüstungskommando Augsburg: Kräftebedarfsmeldungen, 5 Mai 1944.
81. NASM, 3984/65, Denkschrift betreffend die Produktion, Entwicklung und Verwendung von Jagdeinsitzern der Firma Messerschmitt, p. 4.
82. DM, FA001/261, Lucht an Heinkel-Wien: Einziehung 06 und jünger, 26/9/1944.
83. NARA, T73/123/3284579, Messerschmitt AG Augsburg an das Rüstungskommando Augsburg: SE V Aktion—Ihre Anordnung vom 5 Oktober 1944, 7 Oktober 1944.
84. NARA, T73/123/3284811, Rüstungskommando Augsburg an die Rüstungsinspektion VII: Antrag der Fa. Messerschmitt AG, 24 Mai 1944.
85. YVA, 0.3/4745, Testimony by Miriam Giveon.
86. NARA, M888/6/1315-1317, Stenographischer Bericht über die Jägerstab Besprechung am Freitag, 26 Mai 1944, pp. 76–78.
87. NARA, M888/6/536-6, Vernehmung von Erhard Milch durch Mr. G. Koch am 14 Oktober 1946, pp. 9–10.
88. Herbert, *Hitler's Foreign Workers*, chapter four.
89. Budrass, "Der Schritt," p. 139.
90. Herbert, *Hitler's Foreign*, p. 96.
91. NARA, T177/34/3723633, Vierteljahresbericht für den Aufsichtsrat der Junkers Flugzeug- und Motorenwerke AG, Juli–September 1940.
92. NARA, T177/34/3723604-5, Monats-Bericht für den Verwaltungsrat der JFM, August 1941.
93. Budrass, "Der Schritt," p. 139.
94. NARA, T177/39/3729937-8, Mansfeld Werk GmbH: Geschäftsbericht für die Beiratssitzung am 24/25 September 1941 in Prenzlau, 20 September 1941.
95. NARA, M888/5/153, Der Reichsarbeitsminister an die Präsidenten der Landesarbeitsämter: Einsatz von sowjet. Kriegsgefangenen, 26 August 1941.
96. See for example the distribution list in: NARA, T177/2/3683524, RLM/LC3, Dringlichkeitsliste für französische Kriegsgefangene, 30 Oktober 1941.
97. Budrass, "Der Schritt," p. 140.
98. DM, FA001/277, Heinkel: Leistungbericht 1942, p. 33.
99. NARA, T177/2/3683640, RLM/LC3, Aufstellung von Baracken, 9 September 1941.
100. See annotated Target Information Sheets in NARA, RG243/14/Box 5, Aero Engines Factories, Aircraft Factories (Bombers), Aircraft Components Factories (Bombers); Box 6, Aircraft Factories—Fighters, Aircraft Components—Fighters 1–2.
101. See for example: NARA, T177/34/3723696, Vierteljahresbericht für den Aufsichtsrat der Junkers Flugzeug- und Motorenwerke AG, April–Juni 1941.
102. *Wir arbeiten bei Junkers: Ein Bildbericht vom praktischen Sozialismus eines Industriewerkes im Kampf um das neue Europa* (München, 1943).
103. NARA, T177/34/3723605, Monats-Bericht für den Verwaltungsrat der JFM, August 1941.
104. Herbert, *Hitler's Foreign Workers*, pp. 143–150.
105. *Ibid.*, p. 145; NARA, M888/5/158, Der Chef des OKW: Kriegsgefangenen Einsatz in der Kriegswirtschaft, 31 Oktober 1941.
106. NARA, M888/5/167, Rü IV: Vermerk über Ausführungen des Reichsmarschalls in der Sitzung am 7 November 1941 im RLM.
107. NARA, T177/2/3683569, OKW/WFSt, 24 Dezember 1941.
108. NARA, T177/2/3683566, OKW/Wi Rü Amt/Rü (IIb/IVc), Reihenfolge des Einsatzes von sow. Kr. Gef. In der Rüstungsindustrie, 18 Februar 1942.
109. Kaienburg, "KZ-Haft," p. 51.
110. Bertrand Perz, "Politisches Management im Wirtschaftskonzern. Georg Meindl und die Rolle des Staatskonzerns Steyr-Daimler-Puch bei der Verwirklichung der NS-Witschaftsziele in Österreich," in *Konzentrationslager und deutsche Wirtschaft 1939–1945*, ed. Hermann Kaienburg (Opladen: Leske & Budrich, 1996), pp. 101–105; Bertrand Perz, "Der Arbeitseinsatz im KZ Mauthausen, " in *Die nationalsozialistischen Konzentrationslager—Entwicklung und Struktur*, Band II, eds. Ulrich Herbert, Karin Orth, and Christoph Dieckmann (Göttingen: Wallstein, 1998), pp. 535–539.
111. Werner, *Kriegswirtschaft*, pp. 171–172, 180; Rainer Fröbe, "KZ-Häftlinge als Reserve qualifizierter Arbeitskraft: Eine späte Entdeckung der deutschen Industrie und ihre Folgen," in *Die nationalsozialistischen Konzentrationslager: Entwicklung und Struktur*, Band II, eds. Ulrich Herbert, Karin Orth, and Christoph Dieckmann (Göttingen: Wallstein, 1998), p. 644.
112. Forschungsstelle für Zeitgeschichte in Hamburg, ed., *Der Dienstklender Heinrich Himmlers 1941/1942* (Hamburg: Christians, 1999), pp. 133, 175.
113. Irving, *The Rise and Fall*, p. 256.
114. *Der Dienstklender*, p. 325.
115. Budrass, *Flugzeugindustrie*, p. 775.
116. DM, FA001/260, Heinkel/Meschkat: Aktennotiz, 18 Februar 1942, p. 4.
117. Fröbe, "KZ-Häftlinge," p. 640; Wagner, *Produktion des Todes*, pp. 64–66.
118. NARA, M888/5/267-9, Pohl an Himmler: Eingliederung der Inspektion der Konzentrationslager in das SS-Wirtschafts-Verwaltungshauptamt, 30 April 1942.
119. Fröbe, "KZ-Häftlinge," p. 639.
120. Budrass, *Flugzeugindustrie*, p. 775.
121. DM, FA001/260, Heinkel/Meschkat: Aktennotiz, 18 Februar 1942, p. 4.
122. See aerial photos and maps in: NARA, RG243/14/Box 5, Heinkel, Oranienburg, in folders Aircraft Factories—Bombers and Aircraft Component Factories—Bombers.
123. Budrass, *Flugzeugindustrie*, pp. 775–776.
124. NARA, M888/8/565, Stenographischer Bericht über die GL Besprechung am Mittwoch, dem 26 August 1942, p. 30.
125. Budrass, "Der Schritt," p. 133.
126. NASM, 3237/560, Kommando der Erprobungsstellen an R.d.L u. Ob.d.L./GL üb., 17 August 1942.
127. NARA, T177/16/3701141, Programm für die Triebwerks-Sonderaktion He 177. Besichtigung am Sonnabend, 27 Februar 1943. Other aircraft were damaged in accidents, but were not written off.
128. DM, FA001/865, Heinkel an Vorwald, 12 September 1942, p. 8.
129. NASM, 3237/576, Erprobungsstaffel He 177: Entwicklungsarbeiten an He 177, 15 Oktober 1942.
130. Green, *The Warplanes*, pp. 340–343.
131. Budrass, "Der Schritt," pp. 142–143, 148–149.

132. NARA, M888/8/512-513, Stenographischer Bericht über die GL Besprechung am Dienstag, dem 5 Mai 1942, pp. 36–37.
133. NARA,M888/8/516, GL/A: Amtschefbesprechung bei ST/GL am 5 Mai 1942, 18 Mai 1942.
134. Herbert, *Hitler's Foreign Workers*, pp. 158–163.
135. Ulrich Herbert, *Fremdarbeiter: Politik und Praxis des "Ausländer-Einsatzes" in der Kriegswirtschaft des Dritten Reiches* (Berlin: Dietz, 1986), pp. 149–151; YVA, TR.2/NOKW-409, GL Besprechung vom 4 August 1942. Most of the workers recruited by the GBA were brought to Germany unwillingly. In March 1944 Sauckel confirmed, "Out of the 5 million foreign workers who arrived in Germany, not even 200,000 came voluntarily." See NARA, M888/6/90, Stenographische Niederschrift der 54 Sitzung der zentralen Planung betreffend Arbeitseinsatz, 1 März 1944, p. 12d.
136. NARA, RG243/6/Box221, Stenographic notes on the 21st meeting of the Central Planning Office, 30 October 1942.
137. NARA, T177/39/3729735, Generalluftzeugmeister, Verbindungsstelle Paris: Anordnung zur Arbeitsfrage. Sauckel-Aktion innerhalb der Luftfahrtindustrie, 23 September 1942.
138. NARA, T177/39/3729704, Wirtschaftgruppe Luftfahrtindustrie, Außenstelle Frankreich: Rundschreiben Nr. 82, Abzug von französischen Arbeitern nach Deutschland, 8 Februar 1943.
139. See the extensive correspondence in NARA, T177/39/3729530-3729746, especially Messerschmitt AG: Aktenvermark. Entscheidung Aussprache Gauleiter Sauckel/General von der Heyde über weitere Abzüge aus den Werken der französischen Luftfahrtindustrie, 1 März 1943.
140. NARA, T83/6/3747622, Vermark. 2 Sauckel-Aktion, SNCASO Chatillon, 16 Februar 1943.
141. Herbert, *Hitler's Foreign Workers*, pp. 275–278.
142. NARA, T83/6/3747261, Büro Italien: Produktion in Italien, 22 Juni 1944.
143. Budrass, *Flugzeugindustrie*, p. 777.
144. DM, FA1/259, Heinkel an Oberst Vorwald: Serienbau He 177 in Marienehe, 9 July 1942, S. 1, 3.
145. NARA, M888/8/565, Stenographischer Bericht über die GL Besprechung am Mittwoch, dem 26 August 1942, p. 30.
146. BA-MA, RH8/1210, Heeresanstalt Peenemünde: Besichtigung des Häftling-Einsatzes bei den Heinkel-Werken, Oranienburg, am 12 April 1943.
147. DM, F001/865, Protokoll. He 177 Besprechung bei Herrn Oberst Vorwald am 4 September 1942, p. 4.
148. Stubner, *Der Kampfflugzeug Heinkel 177*, p. 89.
149. DM, FA001/261, Aktennotiz über Besprechung bei Fl. Oberstabing. Alpers, RLM, am 17 September 1942, p. 1.
150. DM, FA001/261, Heinkel an RLM/Vorwald: Arbeiterlage, 16 Oktober 1942, p. 1.
151. *Ibid.*, Heinkel an RLM/Vorwald: 12 September 1942, p. 9.
152. DM, FA001/866, Heinkel an von Pfistermeister, 19 Oktober 1942.
153. *Ibid.*, Protokoll. Lieferplan 222-Bedarf an Menschen, Raum, Material. Sitzung des Sonderausschusses F3 am 2 November 1942.
154. See the table in Budrass, *Flugzeugindustrie*, p. 778.
155. YVA, TR.2/PS-1584, Himmler an Göring, Einsatz von Häftlingen in der Luftfahrtindustrie, 9 März 1944 (also in NARA, M888/6/544-547).
156. Budrass, *Flugzeugindustrie*, p. 779.
157. BA-MA, RH8/1210, Heeresanstalt Peenemünde: Besichtigung des Häftlings-Einsatzes bei den Heinkel-Werken, Oranienburg, am 12 April 1943.
158. DM, FA001/260, Aktennotiz über die Besprechung bei Generalfeldmarschall Milch am 14 November 1942, pp. 3–4.
159. NARA, M888/8/912-915, Stenographischer Niederschrift der Besprechung beim Reichsmarschall über die Flugzeug Programm Entwurf am Montag, 22 Februar 1943, pp. 1–4.
160. DM, FA001/261, Heinkel an RLM/GL/C-E Chef: Abgabe von Kostrukteuren, 2 April 1943.
161. NARA, T177/16/3701142, He-177 — Besprechung beim GL/C Amtchef, 22 Februar 1943.
162. NARA, RG243/6/Box 232, Verluste durch Programmänderungen im Flugzeugbau, Anlage 25–28.
163. Budrass, "Der Schritt," p. 153.
164. Green, *The Warplanes*, p. 343.
165. Stubner, *Der Kampfflugzeug Heinkel 177*, Anlage 39.
166. Allied intelligence immediately noticed this development. See NARA, RG243/6/Box216, AI2(a): Report No. 24/44. German Aircraft Production for September 1944, 27 October 1944.
167. BAMA, RL3/9, Stenographischer Bericht über die Jägerstab Besprechung am Freitag, 13 Juli 1944, p.437.
168. Budrass, *Flugzeugindustrie*, pp. 779–780.
169. Gregor, *Daimler-Benz*, p. 122.
170. DM, FA001/260, Heinkel/Meschkat: Aktennotiz, 18/2/1942, p. 3.
171. *Ibid.*, Heinkel an Milch: Verlagerung, 18 Juni 1942, p. 3.
172. NARA, RG243/14/Box 5, Aircraft Factories — Bombers folder. Target Information Sheet: Mielec (Chorzelow), 11 May 1944.
173. DM, FA001/260, Heinkel an Milch: Verlagerung, 18 Juni 1942, p. 4.
174. Magergee, *Encyclopedia of Camps and Ghettos*, vol. 1, Part B, p. 869.
175. DM, FA001/260, Heinkel an Milch: Verlagerung, 18 Juni 1942, p. 6.
176. YVA, O.3/7158, Testimony by Ruth Tetarko.
177. NARA, M888/8/534, Stenographischer Bericht über die GL Besprechung am Dienstag, dem 7 Juli 1942, p. 28.
178. DM, FA001/261, Aktenntiz über Besprechung bei Fl. Oberstabsing. Alpers, RLM, am 17 September 1942, p. 4.
179. Birgit Weitz, "Der Einsatz von KZ-Häftlinge und jüdischen Zwangsarbeitern bei der Daimler-Benz AG (1941-1945). Ein Überblick," in *Konzentrationslager und deutsche Wirtschaft 1939–1945*, ed. Hermann Kaienburg (Opladen: Leske & Budrich, 1996), p. 171.
180. DM, FA001/259, Heinkel an Dr. Passwaldt: He 219, 30/7/1942.
181. DM, FA001/261, Aktenntiz über Besprechung bei Fl. Oberstabsing. Alpers, RLM, am 17 September 1942, p. 2.
182. Green, *The Warplanes*, p. 351; DM, FA001/904, E. Heinkel an Generalleutnant Kammhuber: He 219, 11 Juli 1942; E. Heinkel an RLM/Dr. Pasewaldt, 23 Juli 1942.
183. DM, FA001/866, Protokoll. Lieferplan 222-Bedarf an Menschen, Raum, Material. Sitzung des Sonderausschusses F3 am 2 November 1942, p. 6.
184. DM, FA001/261, Heinkel an RLM/Vorwald: Arbeiterlage in [sic] ihrer Auswirkung auf das Lieferprogramm, 8 Januar 1943, Anlage 4.
185. DM, FA001/906, Francke: He 219-Anruf des Herrn General Kammhuber, 28 Januar 1943.
186. *Ibid*, EHF/Betriebsdirektion: 0-Serie He 219-Verlagerung Wien-Schwechat, 2 Juni 1943. Unterredung am 28 Mai 1943.
187. DM, FA001/260, Besprechung bei Generalfeldmarschall Milch am 21 Juli 1943, pp. 2–3.
188. DM, FA001/906, Dr. Heinkel: Leutebeschaffung He 219, 22 Juli 1943.
189. *Ibid.*, v. Pfistermeister: Aktennotiz betr. Grossserie He 219, 29 Juni 1943.

190. *Ibid.*, Dr. Heinkel: Flugzeugwerk Budzyn — Anlauf Grossserie He 219, 29 Juni 1943.
191. *Ibid.*, v. Pfistermeister an Heinkel-Wien: He 219 Serie, 16 Juli 1943.
192. *Ibid.*, Maschket: Besprechung — Niederschrift, 27 Juli 1943.
193. *Ibid.*, Meschkat: Direktionsbesprechung in Jenbach am 7/8 August 1943.
194. *Ibid.*, Meschkat: Besprechung am 7 September 1943 im Berliner Büro.
195. *Ibid.*, EHW/Betriebsdirektion: He 219 Fertigung Lagebericht, 28 Oktober 1943.
196. Green, *The Warplanes*, p. 351.
197. DM, FA001/906, Meschkat: Besprechung am 7 September 1943 im Berliner Büro.
198. Green, *The Warplanes*, p. 351.
199. *Ibid.*, pp. 353–354.
200. *Ibid.*, p. 355.
201. DM, LRD/2441, EHG/Wien: He 219 Baureihenübersicht und Flugleistungen, Mai 1944, p. 18.
202. DM, FA001/827, Aktennotiz: Besuch bei Oberstlt. Knemeyer am 12 Juli 1944.
203. LRD/2441, E. Heinkel: Denkschrift zu dem Baumuster He 219, 15 Juni 1944, pp. 3–4.
204. DM, FA001/832, Dr. Heinkel: Jahresschlussgedanken, 29 Dezember 1944.
205. DM, FA001/261, Heinkel an Vorwald: Arbeiterlage, 16 Oktober 1942, p. 2.
206. *Ibid.*, Heinkel an RLM/Vorwald: Arbeiterlage in [sic] ihrer Auswirkung auf das Lieferprogramm, 8 Januar 1943.
207. DM, FA001/261, Heinkel an Vorwald: Arbeiterbedarf, 18 Dezember 1942.
208. DM, FA001/332, Hayn an Francke: Gefolgschaftsaufstellung der Ernst Heinkel AG Wien-Schwechat (Stichtag 1 Mai 1944), 2 Mai 1944, pp. 2–3.
209. Budrass, *Flugzeugindustrie*, p. 791.
210. *Ibid.*, p. 797; Budrass, "Der Schritt," p. 156.
211. Budrass, "Der Schritt," p. 158.
212. DM, FA001/278-281.
213. NARA T83/17/3361119-23 Focke-Wulf an das RLM/GL Planungsamt: Verlagerung einzigartiger Fertigungen, 28 November 1942, Anlagen 7: Belegschafts-Aufstellung.
214. NARA, T83/6/3747376-77, Haberstolz an FW Bremen: Umsetzung ital. Personals nach Deutschland, 29 Januar 1944.
215. Budrass, "Der Schritt," p. 59.
216. DM, FA001/906, Meschkat: Besprechung am 7 September 1943 im Berliner Büro.
217. NARA, M888/8/590, Stenographische Niederschrift der Besprechung beim Reichsmarschall am Donnerstag, dem 28 Oktober 1943, p. 7.
218. NARA, M888/6/334, Stenographischer Bericht über die Besprechung mit den Flotteningenieuren und Oberquartiermeistern unter dem Vorsitz von Generalfeldmarschall Milch am Sonnabend, 25 März 1944, p. 22.
219. NARA T83/16/3360575, Gesamtwerkschutzleitung: Ausländer in der Werkfeuerwehr, 12 April 1944.
220. USSBS, *Aero Engine and Automobile Factories of BMW at Munich (Oberwiesenfeld) and Allach* (n.p., 1945), p. 1.
221. Fröbe, "KZ-Häftlinge," pp. 644–645.
222. Schabel, *Die Illusion*, p. 207.
223. Budrass, *Flugzeugindustrie*, p. 798.
224. Cited in: Schabel, *Die Illusion*, p. 145.
225. Martin Weinmann, ed., *Liste der Unternehmen, die im Nationalsozialismus von der Zwangsarbeit profitiert haben*. Available for download at http://www.zweitausendeins.de/pdf/ZA.pdf.
226. See map in NARA, RG243/6/Box 238, Messerschmitt AG Augsburg, Main Production Lines.

227. NARA, T73/123/3285013, Konzentrationslager Dachau/Kommandantur an Rüstungskommando Augsburg: Brand in der Unterkunftsbaracke des SS-Arbeitslagers Haunstetten b/Augsburg, 19 Februar 1944.
228. NARA, RG243/6/Box 237, Gesamtlageplan Kempten-Kottern; Box 238, Messerschmitt AG Augsburg, Produktions-Stätten, 30 April 1945.
229. Kaienburg, *Die Wirtschaft*, pp. 618–619.
230. Hans Brenner, "Der 'Arbeitseinsatz' in den Außenlagern des KZs Flossenbürg," in *Die nationalsozialistischen Konzentrationslager — Entwicklung und Struktur*, Bd. I, eds. Ulrich Herbert, Karin Orth, and Christoph Dieckmann (Göttingen: Wallstein, 1998), pp. 686–688.
231. Kaienburg, *Die Wirtschaft*, pp. 619, 639–640.
232. Hesse, *Rüstungsindustrie in Leipzig*, Teil II, p. 86.
233. Brenner, "Der 'Arbeitseinsatz,'" p. 688.
234. DM, LRD/6406, Urteil VII ZR 181/65 des Bundesgerichtshofs, 22 Juni 1967, p. 3.
235. Hesse, *Rüstungsindustrie in Leipzig*, Teil II, pp. 96–97.
236. Schmoll, *Messerschmitt-Giganten*, p. 132.
237. BA-B, NS3/3571, Pohl und den RFSS, 14 Juni 1944.
238. Brenner, "Der 'Arbeitseinsatz," pp. 690–691. The Ar 234 exhibition in the Luftwaffe Museum at Gatow dedicates relatively large space to slave labor at Rathnow.
239. NARA, T73/123/3285164, Messerschmitt AG an das Rüstungskommando Augsburg: Gefolgschaftszahlen, Stand 30. Dezember 1943, 19 Januar 1944.
240. BA-MA, RH8/1210, Heeresanstalt Peenemünde: Besichtigung des Häftling-Einsatzes bei den Heinkel-Werken, Oranienburg, am 12 April 1943. It should be noted that as early as summer 1939 Rudolph and his team visited the Oranienburg factory as part of the early planning phase of the Peenemünde production facilities. RH8/1206, Entstehungsgeschichte der FSP, August u. September 1939, p. 86.
241. BA-B, NS19/1542, Milch an Maurer, 13 April 1943.
242. Schelp played a central role in promoting the development of jet engines within the RLM. See NARA, RG331/13d/Box 93, CIOS, Interrogation of Dr. Ing. Helmut Schelp; NARA, RG72/111/Box 2, Helmut Schelp: Survey on Special Engine Development in Germany, 25 May 1945.
243. BA-B, NS19/57, Chef SSFHA an Himmler, 15 September 1943.
244. BA-MA, RL3/38, Stenographischer Bericht über die Entwicklungs-Besprechung unter Vorsitz von Generalfeldmarschall Milch am Freitag, dem 27 August 1943, p. 4638.
245. See, for example, the correspondence between Milch and Himmler in his SS file: NARA, BDC, SSO Zborowski, Helmut A3343/018C and *Der Dienstkalender*, p. 324. From what his colleagues told about him after the war, the Americans summed up: "Nominally head of Rocket Development Department but apparently he was more concerned with his membership of the SS and political affairs and left most of the technical matters to Schneider and Ziegler [rocket engine designer and deputy, respectively]." See NARA, RG331/13d/Box 90, CIOS, Gas Turbine Development by BMW, p. 7.
246. BA-B, NS19/3711, Zborowski an Jüttner/FHA, 2 September 1942. See also Schabel, *Die Illusion*, p. 183.
247. BA-B, NS19/57, Chef SS-FHA an Himmler, 15 September 1943.
248. Loc. cit.
249. Schabel, *Die Illusion*, pp. 49, 238.
250. Smith and Creek, *Me-262*, vol. 2, p. 308.
251. Irving, *The Rise and Fall*, p. 257.
252. NARA, M888/6/2-8, Stenographische Niederschrift der 54 Sitzung der zentralen Planung betreffend Arbeitseinsatz, 1 März 1944, pp. 5–7. The meeting took place at the RLM.
253. NARA, M888/6/2-8, Stenographische Bericht über die Besprechung mit den Flotteningenieuren und Oberquar-

tiermeistern unter dem Vorsitz von Generalfeldmarschall Milch am 25 März 1944, pp. 1–2.
254. NARA, M888/6/548-552, Chef des SS WVHA: Häftlingseinsatz für Zwecke der Luftfahrtindustrie, 21 Februar 1944.
255. Brenner, "Der 'Arbeitseinsatz,'" p. 693. At that time it also became a center of tank engine production with the construction of the giant underground factory "Richard" in Leitmeritz.
256. YVA, TR.2/PS-1584, Himmler an Göring, Einsatz von Häftlingen in der Luftfahrtindustrie, 9 März 1944 (also in NARA, M888/6/544-547).
257. Bernhard Strebel, "Ravensbrück-das zentrale Frauenkonzentrationslager," in *Die nationalsozialistischen Konzentrationslager—Entwicklung und Struktur*, Band II, eds. Ulrich Herbert, Karin Orth, and Christoph Dieckmann (Göttingen: Wallstein, 1998), p. 234.
258. NARA, M888/6/543, Göring an Himmler: Aufstellung der 7 Staffel/ Fliegergruppe z.b.V. 7, 14 Februar 1944.
259. YVA, TR.2/PS-1584, Aufstellung Pohls vom 21 Februar 1944.
260. Ibid., Himmler an Göring, Einsatz von Häftlingen in der Luftfahrtindustrie, 9 März 1944 (also in NARA, M888/6/544-547).
261. Thomas Irmer, "Zwangsarbeit von jüdischen KZ-Häftlingen für die Rüstungsproduktion in der Region Berlin-Brandenburg während der Schlussphase des Zweiten Weltkrieges—die Außenlager Glüwen und Schwarzheide des KZ Sachsenhausen," in *Zwangsarbeit während der NS-Zeit in Berlin und Brandenburg*, ed. Winfried Meyer (Potsdam: Berlin-Brandenburg, 2001), p. 169.
262. Ibid., p. 170.
263. NARA, RG243/6/Box 225, Stenographische Niederschrift der 56 Sitzung der Zentralen Planung am 5 April 1944, pp. 1–13.
264. NARA, M888/6/601, Stenographischer Bericht über die Jägerstab Besprechung am 6 März 1944, p. 18.
265. NARA, M888/6/662-663, Stenographischer Bericht über die Jägerstab Besprechung am Freitag 17 März 1944, pp. 13–14.
266. Ibid., pp. 35–36; Mommsen and Grieger, *Das Volkswagenwerk*, p. 811.
267. NARA, M888/6/662-663, Stenographischer Bericht über die Jägerstab Besprechung am Freitag 17 März 1944, pp. 13–14.
268. NARA, RG243/6/Box 224, Punkte aus den Besprechungen beim Führer am 6 und 7 April 1944, p. 6.
269. NARA, M888/6/880, Stenographischer Bericht der Jägerstab Besprechung am 25 April 1944, p. 10.
270. Irmer, "Zwangsarbeit," p. 170.
271. NARA, M888/6/1280-1282, Stenographischer Bericht über die Jägerstab Besprechung am Freitag, 26 Mai 1944, pp. 32–34.
272. Ibid., p. 81.
273. See, for example, YVA, 0.3/7936, Testimony by Miriam Frank; 0.15E/3080, Testimony by Samuel Abrahamovich.
274. Budrass, *Flugzeugindustrie*, p. 796.
275. Raim, *Die Dachauer*, p. 36.
276. Ibid., pp. 36–37. See also NMT, vol. 2, p. 574; YVA, TR.2/NOKW-359, Jägerstab Besprechung am 27 Juni 1944.
277. NARA, M888/6/1228, Stenographischer Niederschrift der Jägerstab Besprechung am 25 Mai 1944, p. 22.
278. YVA, TR.2/NOKW-359, Stenographischer Bericht über die Jägerstab Besprechung am Dienstag, 27 Juni 1944, p. 31.
279. NARA, M888/6/1624, Stenographischer Bericht über die Jägerstab Besprechung am Mittwoch, dem 7 Juni 1944, p. 13.
280. NMT, vol. 2, p. 574; YVA, TR.2/NOKW-359, Stenographischer Bericht über die Jägerstab Besprechung am Dienstag, 27 Juni 1944, pp. 26–27, 31; Fröbe, "KZ-Häftlinge," p. 668.

281. NARA T73/1/1045147, Auszug aus: Maßnahmen zur Fliegerschadenbeseitigung bei den Hydrierwerken, Sitzung Leuna 16/5/1944.
282. BA-B, R3/3058, Hauptausschuss Triebwerke/Sonderausschuss T3: Monatsbericht Juni 1944. On the problems of Daimler-Benz with its Czech workers, see Gregor, *Daimler-Benz*, pp. 193–194.
283. BA-B, R3/3058, Lagebericht des Werkes Allach, Monat Juli 1944.
284. BA-MA,RL3/9, Stenographischer Bericht über die Jägerstab Besprechung am Freitag, 9 Juli 1944, p. 479.
285. NARA, RG331/13d/Box 87, CIOS, Bayrische Motorenwerke AG Munich-Oberwiesenfeld, p. 13.
286. NARA T83/11/3353822-5, Brief von Dr. Schmelter an Herrn Steinbach vom 6 Juli 1944; Abschrift des Schriftwechsels wegen des Einsatzes des KZ-Lagers Wronke, 26 Mai–7 Juli 1944.
287. NARA, T83/11/353743-6, Direktor Werk Adelheide an Tank, 23 Dezember 1944.
288. NARA, RG243/6/Box 244, Sub Assembly Plant of Junkers Flugzeug und Motoren Werke AG, Leopoldshall, Germany. Junkers Flugzeug- und Motorenwerk AG Leopoldshall-Stassfurt: Beschäftigtenmeldung-Industrie, 31 Januar 1945.
289. Perz, "Der 'Arbeitseinsatz,'" p. 543.
290. Schmoll, *Nest of Eagles*, p. 162.
291. YVA, O.3/7250, Testimony by Peter Erben.
292. Rainer Fröbe, "Der Arbeitseinsatz von KZ-Häftlinge und die Perspektive der Industrie, 1943–1945," in *Europa und der "Reichseinsatz." Auslandische Zivilarbeiter, Kriegsgefangenen und KZ-Häftlinge in Deutschland, 1938–1945*, ed. Ulrich Herbert (Essen: Klartext, 1991), p. 370; Raim, *Die Dachauer*, p. 129.
293. Gregor, *Daimler-Benz*, p. 188.
294. Ibid., p. 123.
295. NARA, T73/132/3295137, Rüstungsinspektion VII: Aktenvermerk. Besprechung bei Fa. Messerschmitt und Sonderausschuss F2 am 7 März 1945.
296. Smith and Creek, *Me-262*, vol. 2, pp. 392–393, 396, 399.
297. USSBS, *Aircraft Division*, Exhibit IV.
298. NARA, RG243/31/Box 1, USSBS: Interview No. 44, Dr. Karl Frydag, Chief of Air Frame Industry, 9 July 1945, p. 13.
299. NARA, RG331/13d/Box 98, CIOS, Survey of Production Techniques used by the German Aircraft Industry, pp. 11–12.
300. Mommsen, *Der Mythos*, p. 10.
301. Fröbe, "KZ-Häftlinge," p. 637.

Chapter 5

1. BBSU, pp. 103–105.
2. NARA, RG243/6/Box 181, Statistical Summery of Eight Air Force Operations, European Theatre, 17 August 1942–8 May 1945, pp. 38–39.
3. NARA, RG243/6/Box 220, Anlage 3 zu Lw. Fü. St.Ic, 6 September 1944.
4. Ibid., Anlage 3 zu Lw. Fü. St.Ic, 12 Januar 1945.
5. NARA, T83/12/3353513, Sch/K/2: Aktennotiz, 22 Juni 1944.
6. Hesse, *Rüstungsindustrie in Leipzig*, Teil I, pp. 94–95.
7. NARA, RG243/31/Box 1, USSBS: Minutes of meeting with Reichsminister Albert Speer, Flensburg, 18 May 1945, pp. 7–8.
8. NARA, RG243/6/Box 218, Air P/W Interrogation Unit (Ninth Air Force Adv.): The Me 262, 12 April 1945, p. 10.
9. NARA, T83/14/3357681, Der Werkluftschutzleiter, Bad-Eilsen: Mitteilung: Tieffliegerangriffe, 13 November 1944.
10. Hermann, *Focke-Wulf Ta 152*, pp. 110–111, 116.

11. Schmoll, *Messerschmitt-Giganten*, p. 153.
12. USSBS, *Gothaer*, p. 7.
13. Alan J. Levine, *The Strategic Bombing of Germany, 1940–1945* (Westport: Praeger, 1992), pp. 163–169.
14. BBSU, pp. 25–26; Craven and Cate, *The Army Air Forces*, vol. 3, pp. 639, 732.
15. Craven and Cate, *The Army Air Forces*, vol. 3, pp. 734–735.
16. BBSU, p. 133.
17. NARA, RG243/6/Box 242, Report of USSBS Field Team No. 43 on Junkers Flugzeug- und Motorenwerk AG, 12–14 May 1945, p. 2; Box 229, Daten laut Kontrollliste für die Untersuchung der Anlagen Focke-Wulf, p. 25.
18. BA-B, R3/3034a, Rüstungsstab: Kurzbericht Flugzeugenbau und Reparatur, 26 Januar 1945.
19. Werner, *Kriegswirtschaft*, p. 165.
20. NARA, RG243/10/Box 1, Headquarters USSBS: Intelligence Notes No. 7, 21 June 1945, p. 6.
21. NARA, RG243/6/Box 242, Report of USSBS Field Team No. 43 on Junkers Flugzeug- und Motorenwerk AG, 12–14 May 1945, p. 2.
22. Hertel, *The German Air Force*, p. 428.
23. NARA T83/11/3353733, Einkaufsbüro Berlin: Aktennotiz für Herrn Prof. Tank, 7 Dezember 1944.
24. BBSU, p. 127.
25. Werner, *Kriegswirtschaft*, pp. 165–166.
26. NARA, M888/8/500-504, Stenographischer Bericht über die Jägerstab Besprechung am Montag, dem 19 Juni 1944, pp. 46–50.
27. NASM, 3984/65, Denkschrift betreffend die Produktion, Entwicklung und Verwendung von Jagdeinsitzern der Firma Messerschmitt, 30 Oktober 1944, p. 8.
28. Schmoll, *Nest of Eagles*, pp. 54–55.
29. NARA, T83/11/3353537, Flugbetriebsleitung Sorau an Tank: Bericht, 22 März 1944.
30. NARA, RG243/180/3392188, Statistische Schnellberichte zur Kriegsproduktion: Rüstungsendfertigung, Stand Februar 1945, p. 13.
31. Boelcke, *Deutschlands Rüstung*, p. 465.
32. NARA, RG243/6/Box231, USSBS, Junkers Aircraft and Aero Engine Works, Dessau, Germany, Exhibits Y3-4, X.
33. NARA, RG243/6/Box 235, Report of Field Team no. 81: Focke-Wulf Flugzeugbau GmbH, Bremen, IIIb.
34. NARA, RG243/6/Box 231, USSBS, Junkers Aircraft and Aero Engine Works, Dessau, Germany, Exhibits Y3-4, X.
35. NARA, RG243/6/247, MIMO: Produktionszahlen 1945.
36. United States Strategic Bombing Survey, *ATG Maschinenbau GmbH, Leipzig, Germany* (n.p., 1945), pp. 12–13.
37. NARA, RG243/6/Box 235, Report of Field Team no. 81: Focke-Wulf Flugzeugbau GmbH, Bremen, IIIc.
38. BA-B, R3/3011, Rumi/Rüstungsstab/Der Hauptwerksbeauftragte Hauptmann Erich Schwarz an den Rüstungsstab/Beauftragter f.d. Flugzeugprogramm O. Lange: Verlagerung, 13 Februar 1945.
39. *Ibid.*, Rumi/Rüstungsstab/Der Hauptwerksbeauftragte Hauptmann Erich Schwarz an den Rüstungsstab/Beauftragter f.d. Flugzeugprogramm O. Lange, 1 Februar 1945.
40. BA-B, R3/3012, Speer an Gauleiter Emil Stuertz, 30 Januar 1945.
41. BA-B, R3/3011, Anlaufbeauftragter in FW Cottbus an Lange: Sonderaktion zur Bergung von Verrichtungen aus den FW Werken Posen-Kreising, 29 Januar 1945.
42. BA-B, R3/3012, Schwarz und die Reichsbahn, 18 Februar 1945.
43. NARA, RG243/6/Box 232,12 Army Group/G-2, Intelligence Report No. EW-Te 4, 26 April 1945.
44. NARA, RG243/6/Box 243, Addenda to Report on Schönebeck Junkers Plant: Dispersal Plant at Tarthun (Underground), 17 May 1945.

45. NARA, RG331/13d/Box90, CIOS, German Aircraft Industry Bremen-Hamburg Area, pp. 14, 16.
46. *Ibid.*, p. 24.
47. USSBS, *Messerschmitt AG. Overall Report*, p. 10.
48. Fröbe, "'Wie bei den alten Ägyptern,'" p. 441.
49. Workers in Messerschmitt's Augsburg factory worked 69-hour weeks at least until May. NARA, T73/123/3284737, Messerschmitt AG Augsburg an den Bezirks Arbeitseinsatz Ingenieur: Erfolgsmeldung für Monat April 1944.
50. NARA, M888/6/905, Stenographischer Bericht der Jägerstab Besprechung am 25 April 1944, p. 35.
51. NARA, T83/15/3358973, Der Betriebsführer, Bad Eilsen: Mitteilung. Arbeitszeit zu Weihnachten und Neujahr, 14 Dezember 1944.
52. NARA, T73/123/3284539, Messerschmitt AG Augsburg an das Rüstungskommando Augsburg, 14 December 1944.
53. NARA, RG331/13d/Box 87, CIOS: Bavarian Motor Works (BMW). A production Survey, pp. 50–51.
54. USSBS, *Messerschmitt AG. Appendix*, Appendix II, p.4.
55. See, for example, NARA, T83/4/3744480, Focke-Wulf: Lohnregelung für Gassen, 1 Oktober 1943.
56. Budrass, "Der Schritt," p. 154; NARA, RG331/13d/Box 86, CIOS: Administration, Plastics, Production Tooling, Spare Parts and Servicing in German Aircraft Industry, pp. 4, 6; NARA, T177/37/3727177, Aktenvermerk: Lohnerhebung in der metallerarbeitende Industrie für das 3 Kalendervierteljahr 1943.
57. NARA, RG243/6/Box 242, Report of USSBS Field Team No. 43 on Junkers Flugzeug- und Motorenwerk AG, 12–14 May 1945, p. 5.
58. NARA, RG243/32/Box 2, USSBS: Interview no. 34, General Galland (Part I), 4 June 1945, p. 10.
59. USSBS, *Aircraft Division*, p. 85; Budrass, "Der Schritt," p. 137.
60. NARA, RG243/10/Box 1, Headquarters USSBS: Intelligence Notes No. 7, 21 June 1945, p. 7.
61. The general prewar practice was to avoid extra hours as far as possible through efficient management in order to cut expenses. See: Junkers Flugzeug- und Motorenwerke AG, *Vier Jahre sozialer Aufbau*, Dessau, 1937, p. 20.
62. Budrass, "Der Schritt," p. 158.
63. BA-B, R3/3693, Dr. Ing. H. Conradis, "Vereinfachung, Entrümpelung, Entfeinerhung," in *Der Condor. Werkzeitschrift der Betriebsgemeinschaft Focke-Wulf Bremen*, 6. Jahrgang/Heft 2, April–September 1944, p. 6.
64. NARA T83/14/3357841, Reichsminister für Rüstung und Kriegsproduktion/Technisches Amt: Massnahmen zur Einsparung von Werkstoffen im Flugzeugzellen- und Triebwerksbau, 29 Oktober 1944.
65. NARA T83/14/3357842-56, Ing. Hermann Tödter: Aktion "Wir sparen Werkstoff," 27 October 1944.
66. DM, FA001/335, EHAG an Tödter: Aktion "50 Km schneller," 5 Februar 1945.
67. NARA, M888/6/43, Stenographische Niederschrift der 54 Sitzung der zentralen Planung betreffend Arbeitseinsatz, 1 März 1944, p. 45.
68. DM, FA001/261, Aktennotiz über Besprechung bei Fl. Oberstabsing. Alpers, RLM, am 17 September 1942, p. 3.
69. P.A. Unbehagen, "Unsere soziale Betreuung," in *20 Jahre Focke-Wulf. Werkzeitschrift der Betriebsgemeinschaft Focke-Wulf*, 6 Jahrgang, Heft 1, Januar–März 1944, p.15. See also NARA, RG331/13d/Box 87, CIOS, Bayrische Motorenwerke AG Munich-Oberwiesenfld, p. 14.
70. NARA, RG331/13d/Box 99, CIOS, Underground Factories in Central Germany, p. 117.
71. Fröbe, "'Wie bei den alten Ägyptern,'" p. 405.
72. NARA, RG331/13d/Box 90, CIOS, German Activities in the French Aircraft Industry, pp. 48–49.

73. NARA T83/6/3747440-42, Aktenvermerk: Besprechung mit Herrn Baurat Elten vom Jägerstab Abteilung Leistungssteigerung in St. Astier am 19–20 Juli 1944.
74. BA-B, R3/1554, Punkte für die Rede des Reichsministers Speer vor den Betriebsführern, Produktionschefs, Hauptwerk- und Werkbeauftragten sowie Firmenvertretern und Hauptausschuss- und Sonderausschussleitern, 24 August 1944.
75. Loc. cit.
76. NARA, T83/14/3357696, Gefolgschaftsabteilung, Bad-Eilsen: Mitteilung: Wochenendeheimfahrten, 13 November 1944.
77. NARA, T83/14/3357702, Der Betriebsführer, Bad-Eilsen: Mitteilung: Einführung der vorläufigen Urlaubssperre, 13 November 1944.
78. DM, FA001/332, Dr. Honegger an E. Heinkel: Kündigungsgründe, 25 August 1943.
79. NARA, M888/6/607-608, Stenographischer Bericht über die Jägerstab Besprechung am 6 März 1944, pp. 24–25.
80. NARA, M888/6/607-608, Stenographischer Bericht über die Jägerstab Besprechung am 6 März 1944, pp. 24–25.
81. NARA, RG72/116/Box 2, Technical Report No. 124-45 A Survey of Production Techniques Used in the German Aircraft Industry, July 1945, p. 9.
82. Schmoll, *Nest of Eagles*, pp. 24–25.
83. Budrass, *Flugzeugindustrie*, p. 779.
84. BA-B, R49/47, Ernst Heinkel AG-Werk Oranienburg, 25 April 1944, p. 3.
85. Kohl, *Auto Union und Junkers*, p. 161.
86. USSBS, *Aero Engine*, Figure I.
87. NARA, T83/4/3744449, Der Reichswirtschaftsminister: Einsatz von Kräfte, die für den Arbeitseinsatz nicht vol zur Verfügung stehen, 17 April 1943.
88. NARA, T177/39/3729793, BMW: FOB Monatsbericht, Werk I München, 7 May 1943.
89. NARA, T83/4/3744726, Focke-Wulf: Beurlaubung werktätiger Frauen während des *Wehrmacht* surlaubes des Ehemannes, 20 November 1943.
90. BA-B, R3/3034b, Rüstungsstab: Kurzbericht Flugzeugenbau und Reparatur, 26 Oktober 1944.
91. Herbert, *Fremdarbeiter*, pp. 299–301.
92. NARA, T83/14/3357703, Der Betriebsführer, Bad-Eilsen: Mitteilung: Ordnungsmassnahmen, 3 September 1944.
93. USSBS, *Focke-Wulf*, p. 8.
94. See, for example, NARA, T83/14/3357681, Der Betriebsführer, Bad-Eilsen: Mitteilung: Fehlzeit nach Flieger-alarm, 1 November 1944.
95. USSBS, *Aero Engine*, Exhibit E.
96. NARA, RG243/6/Box 242, Report of USSBS Field Team No. 43 on Junkers Flugzeug- und Motorenwerk AG at Aschersleben, 5–14 May 1945, p. 5.
97. NARA, RG243/6/Box 241, 13a6 Production Data, p. 2.
98. NARA, RG243/10/Box 1, Headquarters USSBS: Intelligence Notes No. 6, 1 June 1945, p. 1.
99. NARA, RG83/4/3744728, Focke-Wulf: Freistellung von der Arbeit bei Beschädigungen der eigenen Wohnung durch Luftangriffe, 16 September 1943.
100. Gregor, *Daimler-Benz*, p. 161.
101. *Ibid.*, p. 207.
102. NARA, T177/34/3723605, Monats-Bericht für den Verwaltungsrat der JFM, August 1941.
103. NARA, T83/14/3357684, Der Betriebsführer, Bad-Eilsen: Zahnärztliche Behandlung unserer Gefolgschaftsmitglieder, 1 November 1944.
104. NARA, RG243/6/Box 242, Report of USSBS Field Team No. 43 on Junkers Flugzeug- und Motorenwerk AG at Aschersleben, 5–14 May 1945, p. 5.

105. Dr. Warning, "Rasse und Reich," in *Der Condor: Werkzeitschrift der Betriebsgemeinschaft Focke-Wulf Bremen* 5, nos. 1 & 2 (Januar–Februar 1943): p. 8.
106. Budrass, *Flugzeugindustrie*, pp. 802–803.
107. Inge Marszolek und Rene Ott, *Bremen im Dritten Reich. Anpassung, Wiederstand, Verfolgung* (Bremen: Schünemann, 1986), pp. 426–427.
108. *Ibid.*, p. 427.
109. YVA, TR.2/NOKW-416, GL Besprechung vom 26 August 1942.
110. NARA, M888/8/588, Stenographische Niederschrift der Besprechung beim Reichsmarschall Donnerstag, dem 28 Oktober 1943, p. 5.
111. NARA, M888/6/331, Stenographischer Bericht über die Besprechung mit den Flotteningenieuren und Oberquartiermeistern unter dem Vorsitz von Generalfeldmarschall Milch am Sonnabend, 25 März 1944, p. 19.
112. NARA T83/17/3361463, Umlauf für Abt. LK, 6 November 1944 (?).
113. Fröbe, "'Wie bei den alten Ägyptern,'" p. 438.
114. "Nowotny," in *20 Jahre Focke-Wulf: Werkzeitschrift der Betriebsgemeinschaft Focke-Wulf* 6, no. 1 (Januar–März 1944): p. 23.
115. NARA, M888/6/312-321, Stenographischer Bericht über die Besprechung mit den Flotteningenieuren und Oberquartiermeistern unter dem Vorsitz von Generalfeldmarschall Milch am Sonnabend, 25 März 1944, p. 9.
116. See the entire issue of *20 Jahre Focke-Wulf: Werkzeitschrift der Betriebsgemeinschaft Focke-Wulf* 6, no. 1 (Januar–März 1944).
117. BA-B, R3/3034b, Rüstungsstab: Kurzbericht Flugzeugenbau und Reparatur, 11 Oktober 1944.
118. NARA, RG243/6/Box 224, Punkte aus der Besprechung beim Führer am 5 März 1944, p. 2.
119. NARA, M888/6/1153, Stenographischer Niederschrift der 6 Reise des "Unternehmens Hubertus" vom 8 bis 10 Mai 1944, p. 25.
120. NARA, M888/6/1342, Stenographischer Bericht über die Jägerstab Besprechung am Freitag, 26 Mai 1944, p. 103.
121. NARA T83/14/3357741, Der Betriebleiter, Bad Eilsen: Mitteilung. Sonderzuweisung von Lebens- und Genussmitteln, 24 April 1944.
122. See for example: NARA, RG243/6/Box 242, Besprechung des Einsatzstabes am 7 Juli 1944 im FZA.
123. NARA, M888/6/318-9, Stenographischer Bericht über die Besprechung mit den Flotteningenieuren und Oberquartiermeistern unter dem Vorsitz von Generalfeldmarschall Milch am Sonnabend, 25 März 1944, pp. 8–9.
124. NARA, RG243/6/Box 224, Besprechungen im Führerhauptquartier am 22, 23 und 25 Mai 1944, p. 1.
125. NARA T83/17/3361373, Mitteldeutsche Motorenwerk: Bericht über Stand und Entwicklung der Produktion zur Beiratsitzung am 4 Dezember 1944.
126. USSBS, *Messerschmitt AG Augsburg, Germany. Part B* (n.p., 1945), p. 7; Schmoll, *Nest of Eagles*, pp. 62–65.
127. USSBS, *Messerschmitt AG Augsburg*, p. 8.
128. Irving, *The Rise and Fall*, p. 270.
129. Loc. cit.
130. Schmoll, *Nest of Eagles*, p. 66.
131. NARA, T83/14/3357686, Der Werkluftschutzleiter, Bad-Eilsen: Mitteilung: Luftschutzdisziplin, 31 Oktober 1944.
132. NARA, T83/14/3357682, Der Werkluftschutzleiter, Bad-Eilsen: Anordnung: Tragen von hellen Arbeitskitteln während Fliegeralarms, 9 November 1944.
133. Megargee, *Encyclopedia*, vol. 1, Part A, p. 596.
134. BA-B, R3/1749, Extract from minutes of a Rüstungsstab meeting on 6 November 1944.
135. NARA, RG243/31/Box 1, USSBS: Interview no. 7, Mr. Seiler, 16 May 1945, p. 2. Seiler, the chairman of Mes-

serschmitt, was particularly disturbed by this fact and mentioned it several times.

136. NASM, 3984/65, Denkschrift betreffend die Produktion, Entwicklung und Verwendung von Jagdeinsitzern der Firma Messerschmitt, 30 Oktober 1944, p. 17.
137. BA-B, R3/3034a, Rüstungsstab: Kurzbericht Flugzeugenbau und Reparatur, 15 Januar 1945.
138. NARA T73/123/3284525, Rü.Kdo. Augsburg: Aktenvermerk, 30 Dezember 1944.
139. NARA T83/11/3353641-6, Betr.Ltg. Detmold an Kommissar Diehle, Gestapo Bielfeld: Fertigungsablauf der Ta 154 in Detmold und bei dem angeschlossenen Nachbaufirmen, 28 August 1944.
140. NARA, T73/132/3295136-38, Rüstungsinspektion VII: Aktennotiz. Besprechung bei Fa. Messerschmitt und Sonderausschuss F2 am 7 März 1945.
141. Werner, *Kriegswirtschaft*, p. 187.
142. YVA, 0.17/30, Testimony by Eva Bohcher.
143. YVA, 0.3/6463, Testimony by Sidney Birnbaum.
144. YVA, 0.3/7643, Testimony by Itzchak Baldinger.
145. Weitz, "KZ-Häftlinge," p. 171.
146. YVA, 0.3/6463, Testimony by Sidney Birnbaum.
147. Budrass, "Der Schritt," p. 156.
148. Fröbe, "'Wie bei den alten Ägyptern,'" p. 441.
149. YVA, 0.3/7237, Testimony by Liviena Epstein.
150. YVA, 0.3/7250, Testimony by Peter Erben.
151. BA-MA, RH8/1210, Heeresanstalt Peenemünde: Besichtigung des Häftlings-Einsatzes bei den Heinkel-Werken, Oranienburg, am 12 April 1943; BA-B, R50/50, Protokoll über die Gesprächstunde Heinkel, 6 November 1982, p. 2.
152. BA-MA, RH8/1210, Heeresanstalt Peenemünde: Besichtigung des Häftlings-Einsatzes bei den Heinkel-Werken, Oranienburg, am 12 April 1943.
153. See, for example, YVA, 0.3/6617, Testimony by Zehava Fruchter.
154. Fröbe, "'Wie bei den alten Ägyptern,'" p. 437.
155. DM, FA001/277, Leistungsbericht 1942 der Heinkel Flugzeugwerke, p. 33.
156. Fröbe, "'Wie bei den alten Ägyptern,'" p. 441.
157. YVA, 0.3/6906, Testimony by Josef Pinsker.
158. Fröbe, "KZ-Häftlinge," pp. 650–654.
159. Mommsen and Grieger, *Das Volkswagenwerk*, pp. 821–823.
160. Raim, *Die Dachauer*, p. 37.
161. *Ibid.*, p. 126.
162. Fröbe, "'Wie bei den alten Ägyptern,'" p.444.
163. YVA, 0.3/7643, Testimony by Itzchak Baldinger. Regarding Oranienburg, see BA-MA, RH8/1210, Heeresanstalt Peenemünde: Besichtigung des Häftlings-Einsatzes bei den Heinkel-Werken, Oranienburg, am 12 April 1943.
164. Schmoll, *Nest of Eagles*, p. 24.
165. Fröbe, "KZ-Häftlinge," p. 644.
166. YVA, 0.3/3676, Testimony by Rachel Zolf.
167. YVA, 0.3/6782, Testimony by Samuel Yoskowitz.
168. YVA, 0.3/6277, Testimony by Agnes Geva.
169. BA-B, R49/47, Ernst Heinkel AG-Werk Oranienburg, 25 April 1944, p. 1.
170. Gregor, *Daimler-Benz*, p. 214.
171. Fröbe, "'Wie bei den alten Ägyptern,'" p. 438.
172. Schmoll, *Nest of Eagles*, p. 164.
173. YVA, 0.3/8099, Testimony by Tzipora Shwiatowitz.
174. YVA, 0.3/6617, Testimony by Zehava Fruchter.
175. Kohl, *Auto Union und Junkers*, p. 167.
176. YVA, 0.3/6463, Testimony by Sidney Birnbaum.
177. NARA, M888/6/948, Stenographischer Bericht der Jägerstab Besprechung am 2 Mai 1944, p. 18.
178. NARA, M888/6/1624, Stenographischer Bericht über die Jägerstab Besprechung am Mittwoch, dem 7 Juni 1944, p. 13.
179. Strebel, "Ravensbrück," p. 234.
180. NARA, M888/6/1040, Stenographischer Bericht der Besprechungen des Jägerstabes anlässlich der 5 Reise des "Unternehmens Hubertus," 2 und 3 Mai 1944, p. 111.
181. NARA, T83/15/3358404, Gesamt Werkschutz, Focke-Wulf: Dienstanweisung für den nebenamtlichen Werkschutz (NWS).
182. Kohl, *Auto Union und Junkers*, p. 165.
183. NARA, T83/16/3360429, Der Reichsminister für Rüstung und Kriegproduktion an Dienstellen und Firmen: Übergang der Betreuung des Werkschutzes der Luftwaffenindustrie auf den Reichsführer SS, 27 November 1944.
184. NARA, M888/6/1622-1625, Stenographischer Bericht über die Jägerstab Besprechung am Mittwoch, dem 7 Juni 1944, pp. 11–14.
185. NARA, M888/8/487-488, Stenographischer Bericht über die Jägerstab Besprechung am Montag, dem 19 Juni 1944, pp. 34–35.
186. NARA, RG243/6/Box 243, Besprechung des Einsatzstabes am 7 Juli 1944 im FZA.
187. Schmoll, *Nest of Eagles*, p. 164.
188. Brenner, "Der 'Arbeitseinsatz,'" pp. 688–689; Hesse, *Rüstungsindustrie in Leipzig*, Teil II, p. 97.
189. Brenner, "Der 'Arbeitseinsatz,'" p. 692.
190. YVA, O.3/6447, Testimony by Hanna Sternlicht.
191. YVA, O.15E/1404, Testimony by Eva Grunfeld.
192. YVA, O.15E/1424, Testimony by Samuel Kramer.
193. Schmoll, *Nest of Eagles*, p. 24.
194. Megargee, *Encyclopedia*, vol. 1, Part A, p. 596.
195. Quoted in Herbert, *Hitler's Foreign Workers*, p. 249.
196. Quoted in *Ibid.*, p. 250.
197. Fröbe, "'Wie bei den alten Ägyptern,'" p. 408.
198. Perz, "Der 'Arbeitseinsatz,'" p. 549.
199. Fröbe, "'Wie bei den alten Ägyptern,'" p. 434.
200. *Ibid.*, p. 436.
201. Perz, "Der 'Arbeitseinsatz,'" pp. 543, 548.
202. NARA, RG243/6/Box 243, JFM/FZA: Besprechung Rüstungsstab am 28 März 1945.
203. NARA, RG331/13d/Box 99, CIOS, Underground Factories in Central Germany, pp. 84–85.
204. Pöhlmann, Markus, "Die Stadt Augsburg im Bombenkrieg 1939–1945," in http://www.bombenkrieg.historicum-archiv.net/themen/augsburg.html.
205. NARA, T73/123/3284918, Messerschmitt AG Augsburg: Aktenvermerk. Messerschmitt Tagesangriff vom 16 März 1944, 17 März 1944.
206. Vajda and Dancey, *German Aircraft Industry*, p. 208. Other sources report 400 to 500 inmates killed during the raid. Magergee, *Encyclopedia of Camps and Ghettos*, vol. 1, Part B, p. 1333.
207. NARA T83/14/3357768, Der Betriebleiter, Bad Eilsen: Mitteilung. Tagesalarm, 21 Dezember 1943.
208. See an example from Fallersleben in: YVA, 0.15E/2651, Testimony by Bela Friedländer.
209. Schmoll, *Nest of Eagles*, p. 63.
210. YVA, O.69/310, Testimony by Mark Stern.
211. BA-B, R50/50, Protokoll über die Gesprächsrunde Heinkel, 6 November 1982, p.6.
212. NARA, RG243/6/Box 251, Report of USSBS Field Team No. 43 on Aero Engine Factory of Junkers Flugzeug- und Motorenwerke Köthen, Germany, p. VIII.
213. NARA, RG243/31/Box 1, USSBS: Interview N0.36. Interrogation of Officers of the Luftwaffe, 21 June 1945, p. 6.
214. Price, *The Last Year*, p. 15.
215. Henryk Grygiel, "Hungern für Hitler: Erinnerungen an die Zwangsarbeit bei Focke-Wulf," in *Hungern für Hitler: Erinnerungen polnischer Zwangsarbeitern im Deutschen Reich 1940-1945*, ed. Christoph U. Schminck-Gustavus (Reinbek: Rowohlt, 1984), p. 163.
216. NARA, M888/8/512-513, Stenographischer Bericht über die GL Besprechung am Dienstag, dem 5 Mai 1942, pp. 36–37. The discussion with Himmler took place on 19 May 1942.

217. NARA, M888/6/331, Stenographischer Bericht über die Besprechung mit den Flotteningenieuren und Oberquartiermeistern unter dem Vorsitz von Generalfeldmarschall Milch am Sonnabend, 25 März 1944, p. 19.
218. YVA, O.69/310, Testimony by Mark Stern.
219. YVA, O.3/3676, Testimony by Rachel Zolf.
220. NARA T83/14/3357764, Der Betriebleiter, Bad Eilsen: Mitteilung. Mittagspause der Ausländer, 31 Dezember 1943.
221. NARA, M888/13/010-011, M888/13/010, Adolf Janko an Dr. F. Bergold, n.d.; Eidesstattliche Versicherung, Adolf Barthelmess, 28 Januar 1947. See also: Schmoll, *Messerschmitt-Giganten*, pp. 107–108.
222. Kohl, *Auto Union und Junkers*, p. 170.
223. NARA, M888/6/931, Stenographischer Bericht der Jägerstab Besprechung am 2 Mai 1944, pp. 4–5.
224. NARA, M888/8/495-496, Stenographischer Bericht über die Jägerstab Besprechung am Montag, dem 19 Juni 1944, pp. 41–42.
225. Fröbe, "'Wie bei den alten Ägyptern,'" p. 398.
226. Budrass, *Flugzeugindustrie*, pp. 802–803.
227. Chief doctor Honegger of Heinkel's Heidefeld factory was responsible for 1,000 inmates. See DM, FA001/332, Dr. Honegger an E. Heinkel: Kündigungsgründe, 25. August 1943, p. 6.
228. Grygiel, "Hungern," p. 152.
229. YVA, O.3/7643, Testimony by Itzchak Baldinger.
230. BA-B, R3/3328, Sozialer Leistungsbericht der Bayerischen Motoren Werke AG, 1942/1943.
231. NARA, T83/6/3744489, Focke-Wulf: Einstellöhne für Ausländer in Cottbus und Sorau (mit Ausnahme der Ostarbeiter), 5 November 1943.
232. NARA, T83/6/3747470, RLM, GL/F3 an Messerschmitt AG: Spesenregelung für französische Konstrekteure in Augsburg, 5 November 1943.
233. NARA, T83/6/3747472, Aussenstelle Chatillon: Umsetzung französischer Konstruktionsbürogruppen, 16 Oktober 1943.
234. NARA, T83/15/3358949, Mitteilung: Weihnachtsgratifikation, 4 Januar 1945.
235. NARA, T73/123/3284540-41, Bekanntmachung des Betriebsführers Nr.120: Arbeitszeit zwischen Weihnachten und Neujahr, 8 Dezember 1944.
236. NARA, M888/6/1345, Stenographischer Bericht über die Jägerstab Besprechung am Freitag, 26 Mai 1944, p. 106.
237. NARA, T177/34/3723607, Monats-Bericht für den Verwaltungsrat der JFM, August 1941.
238. DM, FA001/260, Tagung der Luftwaffen-Industrie-Beauftragten und der Abt. Leiter Luftwaffe, in Berlin am 20 und 21 August 1942. According to this report 50 percent of the workers were foreigners.
239. BA-MA, RH8/1210, Heeresanstalt Peenemünde: Besichtigung des Häftlings-Einsatzes bei den Heinkel-Werken, Oranienburg, am 12 April 1943.
240. Budrass, "Der Schritt," p. 158.
241. Loc. cit; YVA, TR.2/PS-1584, Himmler an Göring, Einsatz von Häftlingen in der Luftfahrtindustrie, 9 März 1944 (also in NARA, M888/6/544-547).
242. NARA, RG331/13d/Box 98, CIOS, Survey of Production Techniques used by the German Aircraft Industry, p. 6.
243. BA-MA, RH8/1210, Heeresanstalt Peenemünde: Besichtigung des Häftlings-Einsatzes bei den Heinkel-Werken, Oranienburg, am 12 April 1943.
244. YVA, O.3/7643, Testimony by Itzchak Baldinger.
245. YVA, O.3/7158, Testimony by Ruth Tetarko.
246. Schmoll, *Messerschmitt-Giganten*, p. 112.
247. Gregor, *Daimler-Benz*, p. 213.
248. Megargee, *Encyclopedia*, vol. 1, Part A, p. 595.
249. YVA, O.3/4745, Testimony by Miriam Giveon.
250. BA-B, R3/3328, Sozialer Leistungsbericht der Bayerischen Motoren Werke AG, 1942/1943.
251. YVA, O.3/7379, Testimony by Miriam Alter.
252. YVA, O.3/6617, Testimony by Zehava Fruchter.
253. NARA, T73/132/3294832, Messerschmitt AG an das Rüstungskommando Augsburg: Zusätzliche Verpflegung bei 72-stündiger Arbeitszeit, 9 Mai 1944.
254. YVA, O.3/6906, Testimony by Josef Pinsker.
255. YVA, O.3/4745, Testimony by Miriam Giveon.
256. Fröbe, "KZ-Häftlinge," p. 646.
257. Mommsen and Grieger, *Das Volkswagenwerk*, p. 822.
258. Irmer, "Zwnagsarbeit," p. 173.
259. YVA, O.3/7643, Testimony by Itzchak Baldinger.
260. Megargee, *Encyclopedia*, vol. 1, Part A, pp. 596–597.
261. YVA, O.3/6782, Testimony by Samuel Yoskowitz.
262. Brenner, "Der 'Arbeitseinsatz,'" pp. 699–700.
263. YVA, O.3/6463, Testimony by Sidney Birnbaum.

Chapter 6

1. The most comprehensive book published so far about the He 162 is Robert Forsyth and Eddie J. Creek, *Heinkel He 162 from Drawing Board to Destruction: The Volksjäger* (Hersham: Ian Allan, 2008).
2. OKL/Gen. Qu. 6 Abt.: Anlage. Flugzeuglage Monatsmeldung Januar 1945. A copy kindly provided by Manfred Boehme via Richard Eger. On the poor combat performance of the Me 262, see Price, *The Last Year*, pp. 176–177; Jeffrey Ethell and Alfred Price, *World War II Fighting Jets* (Annapolis: Naval Institute, 1996), pp. 23–50.
3. DM, FA001/827, Giese: Abschrift von FS Nr. 22181 vom 7 September 1944; Schabel, *Die Illusion*, pp. 248–249; NASM, 8131/303, Knemeyer: The Development of the 8-162, 12 July 1945.
4. DM, FA001/827, Giese: Abschrift von FS Nr. 22181 vom 7 September 1944.
5. DM, FA001/924, E. Heinkel an Göring, 11 Juli 1944; DM, FA001/261, Heinkel an Knemeyer, 6 Juli 1944; DM, FA001/924, E. Heinkel: Denkschrift zu den Strahlbomber-Baumustern He 343 und Ju 287, 16 Juni 1944.
6. Richard J. Smith and Eddie J. Creek, *Jet Planes of the Third Reich* (Boylston: Monogram, 1982), p. 239.
7. DM, FA001/827, Dr. Heinkel: P 1073, 12 Juli 1944.
8. *Ibid.*, Aktennotiz: Besuch bei Oberstlt. Knemeyer am 12 Juli 1944.
9. *Ibid.*, Francke: Anlauf des P 1073, 11 September 1944.
10. *Ibid.*, Francke: Ablauf der Besprechungen im Amt vom 12 bis 14 September 1944.
11. Loc. cit.
12. DM, FA001/827, Francke an Schüngel: Vollmacht für P 1073, 15 September 1944; Francke: Mitteilung für Herrn Prof. Dr. Heinkel, Betr. P 1073, 13 September 1944.
13. DM, FA001/827, Francke: Ablauf der Besprechungen im Amt vom 12 bis 14 September 1944.
14. NASM, 2055/638, Günther: Geschichte und Erfahrung der 162, 6 Juli 1945, p. 1.
15. DM, FA001/827, Günter: Bemerkungen zu der Stellungnahme von Prof. Messerschmitt zum Volksjäger 162, o.D.
16. NASM, 3984/65, Dipl.Ing. Bölkow an Brigadeführer von Schultze-Tratzigg: Thema Volksjäger, 25 Oktober 1944.
17. Schabel, *Die Illusion*, pp. 249–250.
18. DM, FA001/827, Protokoll. He 500 Attrappenbesichtigung am 22 September 1944.
19. NARA, RG243/6/Box 224, Besprechung beim Führer am 21–23 September 1944, p. 1.
20. NARA, RG72/116/Box 2, Technical Report No. 101-45. Bavarian Motor Works (BMW)—A Production Survey, June 1945, p. 51; Schabel, *Die Illusion*, pp. 250–251.
21. Alfred Hiller, *Heinkel He 162 Volksjäger: Entwicklung, Produktion, Einsatz* (Wien: Verlag Alfred Hiller, 1984), p. 30.

22. DM, FA001/827, Francke: He 162, 27 September 1944.
23. Hiller, *Heinkel He 162*, p. 32.
24. BA-B, R3/2598, Der Führer: Aktion Hochleistungsflugzeuge, 12 Oktober 1944.
25. DM, FA001/827, Betriebsführung/Hr: Gewaltaktion 162, 30 Oktober 1944.
26. DM, FA001/832, E. Heinkel: Verlust von 162 Zeichnungen, 2 November 1944.
27. NASM, 2668/611, Meyer: Lagebericht für Fertigung Heinkel Rostock, 27 Januar 1945.
28. *Ibid.*, Francke an Kessler und Lucht: 162, 6 Februar 1945.
29. NARA, RG243/6/Box 227, Punkte aus der Besprechung beim Führer am 26 Februar 1945, p. 2; Vajda and Dancey, *German Aircraft Industry*, p. 99.
30. Schabel, *Die Illusion*, p. 248.
31. NARA, RG243/31/Box 1, HAF/P J: Abschrift vom Flugzeug-Programm. Gegenüberstellung LP227/LP228, 16 März 1945, attached to USSBS: Interview No. 44, Dr. Karl Frydag, Chief of Air Frame Industry, 9 July 1945.
32. DM, FA001/827, Francke: Attrappenbesichtigung durch GdJ, 7 Oktober 1944.
33. *Ibid.*, Prof. Messerschmitt an Dr. Heinkel: Stellungnahme zum Projekt Volksjäger, 23 Oktober 1944, p. 1.
34. NASM, 3984/65, Denkschrift betreffend die Produktion, Entwicklung und Verwendung von Jagdeinsitzern der Firma Messerschmitt, 30 Oktober 1944, p. 15.
35. Schabel, *Die Illusion*, p. 249.
36. Dealing with this power struggle and its aftermath is beyond the focus of this research. For a comprehensive overview, see Smith and Creek, *Me-262*, vol. 3, pp. 510–511, 528–535.
37. Forsyth and Creek, *Heinkel He 162*, p. 26.
38. DM, LRD/2442, Vier-Wochen Schulung der Geräteführer, o.D.
39. NASM, 3391/16, Technische Direktion: Bericht über die wichtigsten Entwicklungsfragen in Wien (Stand Jahresende 1944), 29 Dezember 1944, p. 4.
40. *Ibid.*, Technische Direktion: Julia-Stillegung, 21 Februar 1945.
41. DM, FA001/827, Francke: 162 Kurzbericht vom 20 Oktober 1944.
42. NASM, 3391/16, Technische Direktion: 162 — Wochenbericht, 17 November 1944.
43. DM, FA001/832, Francke: 162 Besprechung bei General Fliegerausbildung am 30 Oktober 1944.
44. Although the aircraft was also known as "Salamander," after the first flight Heinkel emphasized that this was only a code name for the wings and nothing else. NASM, 3391/16, Techn. Direktion, 7 Dezember 1944. It was also known as *Spatz* (Sparrow), but this was almost certainly an unofficial nickname, because this name does not appear in official documents related to the aircraft.
45. Smith and Creek, *Jet Planes*, p. 239.
46. See, for example, DM, FA001/827, Prof. Messerschmitt an Dr. Heinkel: Stellungnahme zum Projekt Volksjäger, 23 Oktober 1944.
47. BA-B, R3/1749, Extract from minutes of a Rüstungsstab meeting on 3 October 1944.
48. Hiller, *Heinkel He 162*, p. 20.
49. Rüstungsstab: Semmelbericht–21 Oktober 1944, published *ibid.*, p. 70; Peter Petrick, "Das Schulflugzeug fürs letzte Aufgebot," *Jet and Prop* 4 (1994): p. 54.
50. Hiller, *Heinkel He 162*, p. 72; Hainz Nowarra, *Der Volksjäger He 162* (Friedberg: Podzun, 1984), p. 13.
51. DM, FA001/335, Francke: 162 Wochenbericht, 24 November 1944.
52. Hiller, *Heinkel He 162*, p. 88.
53. NASM, 3391/16, Technische Direktion: 162-Wochenbericht, 7 Dezember 1944.
54. DM, FA001/832, 162, 27 Dezember 1944.
55. NASM, 3391/16, Francke an OKW/Filmstab/Oberst Kallab, 13 Dezember 1944.
56. Ethell and Price, *World War II Fighting Jets*, p. 156.
57. DM, FA001/832, EHAG: 162 Flugerprobung, 15 Dezember 1944.
58. DM, FA001/832, Protokoll Nr.10147 über eine in Wien am 12 Dezember 1944 stattgefundene Besprechung über 162 Tragflächen.
59. NASM, 3391/16, Technische Direktion: 162 Wochenbericht vom 7–22 Dezember 1944.
60. DM, FA001/837, Prof. Heinkel an Francke: 162 Metallfläche, 2 Januar 1945.
61. DM, FA001/335, Flugbericht. 162 M-2 am 22 Dezember 1944 von Fl.Stabsing. Bader.
62. NASM, 3391/16, Technische Direktion: Personal TD, 18 November 1944.
63. Heinz Mankau, "Die Heinkel He 280," *Jet and Prop* 6 (2004): pp. 50–51.
64. Beauvais, *Flugerprobungsstellen*, p. 138.
65. NARA, T321/59/806953, OKL/Gen. Qu: Aufstellung E-Kdo. 162, 9 Januar 1945.
66. NARA, RG72/116/Box17, Narrative Report No. 286-45, 4 October 1945, p. 3.
67. DM, FA001/838, Francke: 162, 15 Februar 1945; Smith and Creek, *Jet Planes*, p. 247.
68. DM, FA001/838, Francke: Zwischenbericht Flügeltankraum 162, 9 März 1945.
69. The normal V (*Versuch*: experiment) designation for prototypes was replaced at the end of December 1944 by the designation M (*Muster*: type).
70. NASM, 2668/611, Francke: Aktennotiz über Telefongespräch mit Gen. Dir. Frydag am 19 Februar, 1945.
71. *Ibid.*, Technische Direktion: 162 Besuch GdJ und KdE am 10 Februar 1945.
72. DM, FA001/335, Francke: 162 Wochenbericht vom 23–28 Januar 1945.
73. *Ibid.*, Francke: 162 Wochenbericht vom 5–12 Februar 1945.
74. NASM, 2668/611, EHW: Besprechung am 12. Februar 1945 in der Angermayergasse.
75. NASM, 2668/611, Heinkel: 162, 9 Februar 1945.
76. *Ibid.*, Technische Direktion: 162 Wochenbericht vom 13–18 Februar 1945; 3391/16, Wochenbericht in der Zeit vom 11–18 Februar.
77. NASM, 3391/16, Francke: Telefongespräch mit Gen.Dir Frydag am 20 Februar 1945.
78. NASM, 3391/16, Technische Direktion: 162 Wochenbericht vom 19–25 Februar 1945.
79. *Ibid.*, Technische Direktion: 162 Wochenbericht vom 26 Februar–4 März 1945.
80. DM, FA001/335, Flugabteilung-Peter: Erfahrungsaustausch mit Flugerprobung Arado und Nachfliegen Ar 234 im Hinblick auf 162, 6 November 1944.
81. DM, FA001/832, E. Heinkel: Brandgefahr 162, 28 November 1944.
82. NASM, 3391/16, Technische Direktion: 162 Wochenbericht vom 5–11 März 1945.
83. NARA, RG72/116/Box 17, Narrative Report No. 286-45, 4 October 1945, p. 3.
84. Ethell and Price, *World War II Fighting Jets*, p.162.
85. Eric Brown, "Mastering Heinkel's Minimus," *Air Enthusiast* (June 1972): p. 300.
86. NARA, 2709/145, 162 Fertigung Beprechung am 27 September 1945 in Wien-Fichtegasse.
87. NARA, RG243/32/Box 2, USSBS: Karl Frydag and Dr. Ernst Heinkel, Jet-Fighter He 162, n.d, p. 3.
88. Michael Feodrowitz, "Heinkel-Werke im Untergrund. Ein Bericht über die Produktionsstandorte der He 162 Volksjäger," *Jet and Prop* 1 (2006): p. 21.
89. Megargee, *Encyclopedia*, vol. 1, Part B, pp. 958–959.

90. DM, FA001/827, Betriebsführung/Hr: Gewaltaktion 162, 30 Oktober 1944.
91. *Ibid.*, Francke: Anlauf der P 1073, 11 September 1944.
92. NASM, 2668/611, EHAG: Anlaufbericht, 15 November 1944, Anlage 10, 12.
93. Megargee, *Encyclopedia*, vol. 1, Part B, pp. 959–960.
94. NASM, 2668/611, EHAG: Anlaufbericht, 15 November 1944, Anlage 10, 12.
95. *Ibid.*, EHAG: Anlaufbericht, 15 December 1944.
96. Hiller, *Heinkel He 162*, p. 88.
97. Budrass, "Der Schritt," p. 159.
98. Megargee, *Encyclopedia*, vol. 1, Part B, p. 963. The unverified source is Konzentrationslager "Schwechat I"–"Santa," in www.geheimprojekte.at.
99. DM, FA001/832, Francke: 162 Wochenbericht, 12 November 1944.
100. Feodrowitz, "Heinkel-Werke," pp. 22–23; Wagner, *Produktion des Todes*, pp. 207–208.
101. DM, FA001/832, Francke: 162, 8 November 1944.
102. *Ibid.*, Protokoll Nr. 31144a: Anlauf Baumuster 162 bei Organisation Direktor Kunze, Besprechung am 4 Dezember 1944 in Wernigrode.
103. DM, FA001/827, Francke: 162-Besprechung am 14 Oktober 1944.
104. DM, FA001/832, Protokoll über die Besprechung am 15 November 1944 mit Generalkommissar Kessler im Ausbildungswesen.
105. It should be noted that wooden parts were used in later models of the Me 109 and FW 190 fighters. These were restricted, however, mainly to control surfaces, tails and flaps. See: BA-B, R3/3787, Werkschrift: Fw 190 und Ta 152 Bau- und Reparaturanleitungen für Holzbauteile, Februar 1945.
106. Hiller, *Heinkel He 162*, p. 63.
107. DM, FA001/832, Protokoll über die Besprechung am 15 November 1944 mit Generalkommissar Kessler im Ausbildungswesen.
108. Feodrowitz, "Heinkel-Werke," p. 24.
109. NASM, 2668/611, 5 und 6 Bericht über den Fertigungsstand des Baumusters 100 in Rostock, 5 Februar 1945.
110. NARA, RG243/6/Box 245, Addenda to Report on Schönebeck Junkers Plant: Dispersal Plant at Tarthun (Underground), 17 Mai 1945.
111. NARA, RG331/13d/Box 93, CIOS, Junkers Aircraft and Engines Facilities, pp. 8, 13–14; RG243/6/Box243, JFM/FZA: Rüstungsstab Besprechung am 28 März 1945.
112. NARA, RG243/6/Box 244, Sub Assembly Plant of Junkers Flugzeug und Motoren Werke AG, Leopoldshall, Germany.
113. DM, FA001/335, Francke, 23 Januar 1945; NARA, RG243/32/Box 2, USSBS: Karl Frydag and Dr. Ernst Heinkel, Jet-Fighter He 162, n.d, p. 5.
114. NARA, RG72/116/Box 2, Technical Report No. 101-45. Bavarian Motor Works (BMW)—A Production Survey, June 1945, p. 10; DM, FA001/825, EHAG Wien: Beschreibung Baumuster 162, 15 Oktober 1944, p. 10.
115. NARA, RG243/6/Box 222, USSBS/Aircraft Division to Commander S.P. Johnston: Interrogation of Georg Rickhey, 6 June 1945, p. 4 ff.
116. NARA, RG72/116/Box 2, Technical Report No. 101-45. Bavarian Motor Works (BMW)—A Production Survey, June 1945, p. 50.
117. DM, FA001/827, Besprechung He 162 am 2 October 1944.
118. Allen, *The Business*, pp. 242–244.
119. NARA, RG331/13d/Box 96, CIOS, Plastics and Wooden Parts in German Aircraft, Appendix A.
120. BA-B, R3/3862, Organisation May: Schalenentwicklung 162 Bauart Esslingen.
121. NARA, RG331/13d/Box96, CIOS, Plastics and Wooden Parts in German Aircraft, p. 7.

122. DM, FA001/827, Francke: Anruf Gen.Ing. Herrmann, TLR, am 22 October 1944.
123. BA-B, NS19/2056, Chef SS-HA an RFSS, Btr. Flugzeugprogramm 162, 9 November 1944.
124. NASM, 3391/16, Technische Direktion: Bericht über die wichtigsten Entwicklungsfragen in Wien (Stand Jahresende 1944), 29 Dezember 1944, p. 4.
125. *Ibid.*, Technische Direktion: Entscheidungen Gen. Dir. Frydag, 30 Januar 1945.
126. NASM, 3391/16, Technische Direktion: 162 Holzfertigung, 22 November 1944.
127. *Ibid.*, Technische Direktion an Frydag: 162 Fertigung, 25 November 1944.
128. DM, FA001/837, Meschkat: Bericht über die bestehenden Mängel der Holzfertigung, 5 Januar 1945.
129. *Ibid.*, Prof. Heinkel: Bericht von Herr Maschket über einen von mir veranlassten Besuch bei den Holzfirmen, 5 Januar 1945.
130. NASM, 2668/611, Meyer: Lagebericht für Fertigung Heinkel Rostock, 27 Januar 1945.
131. NASM, 3391/16, Francke an Kessler: 162, 23 Januar 1945.
132. Hiller, *Heinkel He 162*, pp. 88–89.
133. NARA, RG72/111/Box 3, S. Günther and Hohbach, He 162 Report No. 2. Performances with Jumo 004 (Heinkel Report), October 1946, p. 3.
134. NASM, 2668/611, Nachbau-Besprechung am 2 Februar 1945 in HWO Baumuster 162.
135. *Ibid.*, Technische Direktion: Entscheidungen Gen. Dir. Frydag, 30 Januar 1945.
136. *Ibid.*, EHW Betriebsdirektion: Umrüstung 162, 15 Februar 1945.
137. *Ibid.*, Technische Direktion: 162 Wochenbericht vom 29 Januar–4 Februar 1945.
138. NASM, 2668/611, Technische Direktion: 162 Besuch GdJ und KdE am 10 Februar 1945.
139. NASM, 3391/16, Technische Direktion: 162 Auslieferung der Flugzeuge für Ltn. Hachtel, 21 März 1945.
140. NASM, 2668/611, Nachbau-Besprechung am 2 Februar 1945 in HWO Baumuster 162.
141. NASM, 3391/16, Francke: 162 Schleuse Halle 47, 26 Februar 1945.
142. *Ibid.*, Technische Direktion: 162 Auslieferung der Flugzeuge für Ltn. Hachtel, 21 März 1945.
143. *Ibid.*, Technische Direktion: 162 Wochenbericht vom 26 Februar–4 März 1945.
144. *Ibid.*, Technische Direktion an Prof. Heinkel: Bericht, 28 März 1945.
145. Smith and Creek, *Jet Planes*, p. 302.
146. NARA, RG243/10/Box 1, Headquarters USSBS: Intelligence Notes No. 7, 21 June 1945, p. 7.
147. Feodrowitz, "Heinkel-Werke," p. 22.
148. NASM, 2668/611, EHW: Besprechung am 12 Februar 1945 in der Angermayergasse.
149. *Ibid.*, Nachbau-Besprechung am 2 Februar 1945 in HWO Baumuster 162.
150. NARA, RG243/32/Box 2, USSBS: Karl Frydag and Dr. Ernst Heinkel, Jet-Fighter He 162, n.d, p. 5.
151. Werner, *Kriegswirtschaft*, p. 165.
152. NASM, 2668/611, EHW: Besprechung am 12 Februar 1945 in der Angermayergasse. It should be noted that the plane was designed from the start to carry either of these guns. See DM, FA001/825, EHAG Wien: Baubeschreibung Baumuster 162, 15 Oktober 1944, p. 16.
153. NASM, 2668/611, Francke an Gen. Komm. Kessler: 162-EZ 42, 24 Februar 1945.
154. NARA,RG243/6/Box 243, JFM/FZA: Besprechung Rüstungsstab am 28 März 1945.
155. NASM, 3391/16, Technische Direktion: Julia, Besprechung am 5 März 1945
156. DM, FA001/332, Heinkel an den Reichverteidi-

gungskommissar des Reichsgaues Wien: Liquidierung des Werkes Wien, 14 April 1945.
   157. Megargee, *Encyclopedia*, vol. 1, Part B, p. 960.
   158. Hiller, *Heinkel He 162*, p. 89.
   159. DM, FA001/336, Schaubergwerk, "Seegrotte" Hinterbrühl.
   160. NARA, RG243/32/Box 2, USSBS: Karl Frydag and Dr. Ernst Heinkel, Jet-Fighter He 162, n.d, pp. 5–6.
   161. NASM, 3391/16, Francke: 162 Rümpfe für Linktrainer, 9 März 1945.
   162. Schabel, *Die Illusion*, p. 284.
   163. NARA, RG 243/6/Box 244, Junkers Bernburg: Beantwortung des am 9 Mai 1945 besprochenen Fragebogen, 11 Mai 1945, p. 13.
   164. Beauvais, *Flugerprobungsstelen*, p. 288.
   165. Smith and Creek, *Jet Planes*, p. 312.
   166. NASM, 3391/16, Technische Direktion: 162 Wochenbericht vom 26 Februar–4. März 1945.
   167. *Ibid.*, Technische Direktion: 162 Wochenbericht vom 12–18 März 1945.
   168. *Ibid.*, Francke: 162 Auslieferung der Flugzeuge für Ltn. Hachtel, 21 März 1945.
   169. *Ibid.*, Francke an Heinkel: Bericht, 28 März 1945.
   170. DM, FA001/832, Francke: 162 Wochenbericht vom 23–30 Dezember 1944.
   171. NARA, RG72/116/Box 14, Technical Report No. 550-45 History of German Jet Engine Developments, October 1945, p. 16.
   172. Smith and Creek, *Jet Planes*, p. 312.
   173. G. Hanf, "Ich flog den He 162," *Flugzeug* 6 (1992), p. 17.
   174. Hiller, *Heinkel He 162*, p. 82.
   175. *Ibid.*, p. 76.
   176. *Ibid.*, p. 80.
   177. Ethell and Price, *World War II Fighting Jets*, pp. 159–161.
   178. NARA, RG72/116/Box 2, Technical Report No. 101-45. Bavarian Motor Works (BMW)—A Production Survey, June 1945, pp. 21–22, Figures 15–17.
   179. DM, FA001/335, Das Ende der Ernst Heinkel AG in Wien im Jahr 1945, pp. 2–3.
   180. Butler, *War Prizes*, pp. 89–90, 268.
   181. DM, FA001/840, Royal Aircraft Establishment, Farnborough: Note on the performance in flight of the German jet-propelled aircraft Messerschmitt 262, Heinkel 162 and Arado 234, October 1945.
   182. NARA, RG243/32/Box 2, USSBS: Interview no. 10, Dr. Karl Frydag and Mr. Heinkel, 19 May 1945, p. 7.
   183. NASM, 2497/714, Messerschmitt AG: Baustunden-Aufwand Me 262, 25 Januar 1945.
   184. NASM, 3429/672, JFM: Aufwand-Wirkung-Lebensdauer (EF126), Dezember 1944.

Conclusion

   1. Budrass, "Zwischen Unternehmen und Luftwaffe," p. 160.
   2. NARA, RG331/13d/Box95, CIOS, Messerschmitt Company's Design and Development Department Oberammergau, p. 2. A report composed by British experts.
   3. Holley, "A Detroit Dream," pp. 587–592.
   4. Kenneth P. Werrell, *Blankets of Fire: U.S. Bombers Over Japan During World War II* (Washington, D.C.: Smithsonian Institution Press, 1996), pp. 56–74.
   5. *Ibid.*, pp. 74–83.
   6. BA-B, R3/1959, RLM/GL/C-Fertg.XI: Bericht zur Fertigungsführung: Leistungssteigerung durch Betriebskonzentrazion und Typenverminderung im Zellenbau, 30 Juni 1944.
   7. *Ibid.*, Das Planungsamt im Rumi, Rüstungsendfertigung. Deutsches Reich und Feindmächte, September 1944.
   8. NARA, T73/132/3295136-37, Rüstngsinspektion VII: Aktennotiz. Besprechung bei Fa. Messerschmitt und Sonderausschuss F2 am 7 März 1945.
   9. Mommsen, *Der Mythos*, p. 26.
   10. NARA, RG331/13d/Box 95, CIOS, Messerschmitt Company's Design and Development Department Oberammergau, p. 2. Report composed by British experts.
   11. See especially Budrass et al., "Demystifying"; Mommsen, *Der Mythos*, p. 12.

Bibliography

Primary Sources

Bundesarchiv-Berlin (BA-B):
 NS19 — Persönlicherstab Reichsführer SS.
 R3 — Reichsministerium für Bewaffnung und Munition.
 R50 — Organisation Todt.

Bundesarchiv-Militärarchiv, Freiburg (BA-MA):
 RH8II — Heeresversuchsanstalt Peenemünde.
 RL3 — Generalluftzeugmeister.

Deutsches Museum (DM):
 FA001 — Heinkel Sammlung.
 LRD — Luft- und Raumfahrt Dokumentation.

Library of Congress (LOC):
 Heinkel Werkzeitung.
 Der Propeller.
 Wir arbeiten bei Junkers. Ein Bildbericht vom praktischen Sozialismus eines Industriewerkes im Kampf um das neue Europa. München, 1943.

National Air and Space Museum (NASM):
 Captured German and Japanese Air Technical Documents Collection.
 Junkers Flugzeug- und Motorenwerke AG. *Vier Jahre sozialer Aufbau.* Dessau, 1937.
 Junkers Nachrichten.
 USSBS reports.

U.S. National Archives and Records Administration (NARA):
 M888 — Milch Trial.
 RG72 — U.S. Naval Technical Mission in Europe and Translated BuAer Captured German Records.
 RG238 — The Nuremberg Trials and Other War Crime Trials.
 RG242 — Captured German Records.
 RG243 — USSBS Records.
 RG319 — Records of the Army Staff Assistant Chief of Staff (G-2).
 RG331 — CIOS reports.
 T73 — Armaments Ministry.
 T83 — German Firms.
 T177 — Reichsluftfahrtministerium.

Yad Vashem Archive (YVA):
 O.3 — Testimonies.
 O.15E — Testimonies.
 O.17 — Testimonies.
 O.69 — Testimonies.
 TR.2 — Nuremberg Documents.

Published Sources

Boelcke, Willi. *Deutschlands Rüstung im Zweiten Weltkrieg: Hitlers Konferenzen mit Albert Speer 1942–1945.* Frankfurt am Main: Athenaion, 1969.

Forschungsstelle für Zeitgeschichte in Hamburg, eds. *Der Dienstkalender Heinrich Himmlers 1941/1942.* Hamburg: Christians, 1999.

Hertel, Walter. *The German Air Force: Aircraft Procurement.* Karlsruhe: Studiengruppe Geschichte des Luftkrieges, 1955.

Moll, Martin. *Führer-Erlasse 1939–1945.* Stuttgart: Franz Steiner, 1997.

The Strategic Air War Against Germany, 1939–1945: Report of the British Bombing Survey Unit (BBSU). London: Frank Cass, 1998.

Trials of War Criminals Before the Nuremberg Military Tribunals Under Control Council, October 1946–April 1949. Washington, D.C.: U.S. Government Printing Office, 1949–1953.

United States Strategic Bombing Survey. *Aero Engine and Automobile Factories of BMW at Munich (Oberwiesenfeld) and Allach.* N.p., 1945.

———. *Aircraft Division Industry Report*, 2nd edition. N.p., 1947.

———. *Airframes Plant Report NO.1. Junkersflugzeug und Motorenwerk, Dessau, Germany.* N.p., 1945.

———. *ATG Maschinenbau GmbH, Leipzig, Germany.* N.p., 1945.

———. *The Effect of Strategic Bombing on the German War Economy.* Washington, D.C., 1945.

———. *Erla Maschinenwerke GmbH, Mockau, Germany.* N.p, 1947.

———. *Focke-Wulf Aircraft Plant Bremen*, 2nd edition. N.p., 1947.

———. *Gerhard Fieseler Werke GmbH, Kassel, Germany.* N.p., 1945.

———. *Gothaer Waggonfabrik AG, Gotha, Germany*, 2nd edition. N.p., 1947.

———. *Messerschmitt AG Augsburg, Germany. Appendices I–III.* N.p. 1945.

_____. *Messerschmitt AG Augsburg, Germany. Overall Report.* N.p., 1945.

_____. *Messerschmitt AG Augsburg, Germany. Part B.* N.p., 1945.

Secondary Literature

Abelshauser, Werner. "Germany: Guns, Butter and Economic Miracles." In *The Economics of World War II: Six Great Powers in International Comparison*, edited by Mark Harrison, pp. 122–176. Cambridge: University Press, 1998.

Air Ministry (Great Britain). *The Rise and Fall of the German Air Force, 1933–1945.* New York: St. Martin's Press, 1983.

Albrecht, Ulrich. "Artefakte des Fanatismus: Technik und nationalsozialistische Ideologie in der Endphase des Dritten Reiches." *Informationsdienst Wissenschaft and Frieden* 7 (1989): Heft 4, pp. 21–28.

Allen, Michael Thad. *The Business of Genocide: The SS, Slave Labor, and the Concentration Camps.* London: Chapel Hill, 2002.

_____. "The Puzzle of Nazi Modernism: Modern Technology and Ideological Consensus in an SS Factory at Auschwitz." *Technology and Culture* 37 (1996): pp. 527–71.

Bajohr, Frank. "Dynamik und Disparität: Die nationalsozialistische Rüstungsmobilisierung und die Volksgemeinschaft." In *Volksgemeinschaft: Neue Forschungen zur Gesellschaft des Nationalsozialismus*, edited by Frank Bajohr and Michael Wildt, pp. 78–93. Frankfurt/M: Fischer, 2009.

Beauvais, Heinrich, Heinrich Kössler, Max Mazer, and Christoph Regel. *Flugerprobungsstellen bis 1945.* Bonn: Bernard and Graefe, 1998.

Beyerchen, Alen. "Rational Means and Irrational Ends: Thoughts on the Technology of Racism in the Third Reich." *Central European History* 30 (1997): pp. 386–402.

Boog, Horst. "Der angloamerikanische strategische Luftkrieg über Europa und die deutsche Luftverteidigung." In *Das Deutsche Reich und der Zweite Weltkrieg*, Bd. 6, edited by MGFA, pp. 429–565. München: Dt. Verl.-Anst., 1990.

Braun, Hans-Joachim. "Aero-engine Production in the Third Reich." *History of Technology* 14 (1992): pp. 1–15.

_____. "Krieg der Ingenieure? Technik und Luftkrieg 1914 bis 1945." In *Erster Weltkrieg, Zweiter Weltkrieg: Krieg, Kriegerlebnis, Kriegserfahrung in Deutschland*, edited by Bruno Thoß and Hans-Erich Volkmann, pp. 193–210. Paderborn: Schöningh, 2002.

Brenner, Hans. "Der 'Arbeitseinsatz' in den Aussenlagern des KZs Flossenbürg." In *Die nationalsozialistischen Konzentrationslager—Entwicklung und Struktur*, Bd. I, edited by Ulrich Herbert, Karin Orth and Christoph Dieckmann, pp. 682–706. Göttingen: Wallstein, 1998.

Brown, Eric. "Mastering Heinkel's Minimus." *Air Enthusiast*, June 1972, pp. 295–300.

Brunzel, Ulrich. *Hitler's Treasures and Wonder Weapons.* Zella-Mehlis: H. Jung, 1997.

Buchheim, Christoph. "Introduction: German Industry in the Nazi Period." In *German Industry in the Nazi Period*, edited by Christoph Buchheim, pp. 11–26. Stuttgart: Franz Steiner, 2008.

Budrass, Lutz. "Arbeitskräfte können ausreichlich vorhandenen jüdischen Bevölkerung gewonnen werden: Das Heinkel-Werk in Budzyn 1942–1944." *Jahrbuch für Wirtschaftsgeschichte* 1 (2004): pp. 41–64.

_____. *Flugzeigindustrie und Luftrüstung in Deutschland 1918–1945.* Düsseldorf: Droste, 1998.

_____. "Hans Joachim Pabst von Ohain: Neue Erkenntnisse zu seiner Rolle in der nationalsozialistischen Rüstung." In *Technikgeschichte kontrovers: Zur Geschichte des Fliegens und des Flugzeugbaus in Mecklenburg-Vorpommern*, edited by Friedrich-Ebert-Stiftung, Landesbüro Mecklenburg-Vorpommern, pp. 52–69. Beiträge zur Geschichte Mecklenburg-Vorpommern, Bd. 13, Schwerin, 2007.

_____. "Der Schritt über die Schwelle: Ernst Heinkel, das Werk Oranienburg und der Einstieg in die Beschäftigung von KZ-Häftlingen." In *Zwangsarbeit während der NS-Zeit in Berlin und Brandenburg*, edited by Winfried Meyer, pp. 129–162. Potsdam: Berlin-Brandenburg, 2001.

_____. "Zur Heinkel-Ausstellung." *Zeitgeschichte Regional* 2, no. 6 (2002): pp. 91–96.

_____. "Zwischen Unternehmen und Luftwaffe: Luftfahrtforschung im Dritten Reich." In *Rüstungsforschung im Nationalsozialismus*, edited by Helmut Maier, pp. 142–182. Göttingen: Wallstein, 2002.

Budrass, Lutz, and Manfred Grieger. "Die Moral der Effizienz: Die Beschäftigung von KZ-Häftlingen am Beispiel des Volkswagenwerkes und der Henschel Flugzeugwerke." *Jahrbuch für Wirtschaftsgeschichte*, 1992–93, pp. 89–136.

Budrass, Lutz, Jonas Scherner, and Jochen Streb. "Demystifying the German 'Armament Miracle' During World War II: New Insights from the Annual Audits of German Aircraft Producers." *Economic Growth Center Discussion Papers*, No. 905. Yale University, January 2005.

Butler, Phil. *War Prizes: An Illustrated Survey of German, Italian and Japanese Aircraft Brought to Allied Countries During and After the Second World War.* Leicester: Midland Counties, 1994.

Ciesla, Burghard. "German High Velocity Aerodynamics and Their Significance for the U.S. Air Force 1945–52." In *Technology Transfer Out of Germany After 1945*, edited by Matthias Judt and Burghard Ciesla, pp. 93–106. Amsterdam: Harwood, 1996.

Coates, Steve. *Helicopters of the Third Reich.* Hersham: Classic Publications, 2002.

Corum, James. "Defeat of the Luftwaffe, 1935–1945." In *Why Air Forces Fail: The Anatomy of Defeat*, edited by Robin Higham and Stephan J. Harris, pp. 203–226. Lexington: Kentucky University Press, 2006.

Craven, Wesley Frank, and James Lea Cate, eds. *The Army Air Forces in World War II.* Washington, D.C.: Office of Air Force History, 1983.

Eiber, Ludwig, and Jan Reemtsma, eds. *Deutsche Wirt-*

schaft: Zwangsarbeit von KZ-Häftlingen für Industrie und Behörden. Hamburg: VSA-Verlag, 1991.

Engel, Reinhard, and Joanna Radzyner. *Sklavenarbeit unterem Hakenkreuz: Die verdrängte Geschichte der österreichischen Industrie.* Wien-München: Deuticke, 1999.

Ethell, Jeffrey, and Alfred Price. *World War II Fighting Jets.* Annapolis: Naval Institute, 1996.

Feodrowitz, Michael. "Heinkel-Werke im Untergrund: Ein Bericht über die Produktionsstandorte der He 162 'Volksjäger.'" *Jet and Prop* 1 (2006): pp. 20–25.

Ferguson, Robert G. "One Thousand Planes a Day: Ford, Grumman, General Motors and the Arsenal of Democracy." *History and Technology* 21, no. 2 (June 2005): pp. 149–175.

Fings, Karola. *Krieg, Gesellschaft und KZ. Himmlers SS-Baubrigaden.* Paderborn: Schöningh, 2005.

Forsyth, Robert, and Eddie J. Creek. *Heinkel He 162 From Drawing Board to Destruction: The Volksjäger.* Hersham: Ian Allan, 2008.

Frei, Norbert. "Wie Modern war der Nationalsozialismus?" *Geschichte und Gesellschaft* 19 (1993): pp. 367–87.

Freund, Florian. "Der Entscheidungsprozess zum Einsatz von KZ-Häftlingen in der Raketenrüstung." In *Zwangsarbeit und die unterirdische Verlagerung,* edited by Torsten Hess, pp. 20–36. Berlin: Westkreuz, 1994.

_____, and Bertrand Perz. *Das KZ in der Serbenhalle: Zur Kriegsindustrie in Wiener Neustadt.* Vienna: Verlag für Gesellschaftskritik, 1987.

Fröbe, Rainer. "Der Arbeitseinsatz von KZ-Häftlinge und die Perspektive der Industrie, 1943–1945." In *Europa und der "Reichseinsatz": Ausländische Zivilarbeiter, Kriegsgefangenen und KZ-Häftlinge in Deutschland, 1938–1945,* edited by Ulrich Herbert, pp. 351–383. Essen: Klartext, 1991.

_____. "KZ-Häftlinge als Reserve qualifizierter Arbeitskraft: Eine späte Entdeckung der deutschen Industrie und ihre Folgen." In *Die nationalsozialistischen Konzentrationslager — Entwicklung und Struktur,* Band II, edited by Ulrich Herbert, Karin Orth, and Christoph Dieckmann, pp. 636–681. Göttingen: Wallstein, 1998.

_____. "'Wie bei den alten Ägyptern': Die Verlegung des Daimler-Benz Flugmotorenwerk Genshagen nach Obrigheim am Necker 1944/45." In *Das Daimler-Benz-Buch: Ein Rüstungskonzern im "Tausendjährigen Reich,"* edited by Hamburger Stiftung für Sozialgeschichte des 20. Jahrhunderts, pp. 392–470. Nördlingen: Delphi, 1987.

Gall, Lothar, and Manfred Pohl, eds. *Unternehmen im Nationalsozialismus.* München: Beck, 1998.

Gersdorff, Kyrill von, and Kurt Grasmann. *Flugmotoren und Strahltriebwerke: Entwicklungsgeschichte der deutschen Luftfahrtantriebe von den Anfängen bis zu den internationalen Gemeinschaftsentwicklungen.* Koblenz: Bernard and Graefe, 1985.

Geyer, Michael. *Deutsche Rüstungspolitik 1860–1980.* Frankfurt a/M: Suhrkamp. 1984.

Green, William. *The Augsburg Eagle: Messerschmitt Bf 109.* London: Jane's, 1980.

_____. *The Warplanes of the Third Reich.* London: Jane's, 1970.

Gregor, Neil. *Daimler-Benz in the Third Reich.* New Haven: Yale University Press, 1998.

Grieger, Manfred. "'Vernichtung durch Arbeit' in der deutschen Rüstungsindustrie." In *Vernichtung durch Fortschritt am Beispiel der Raketenproduktion im Konzentrationslager Mittelbau,* edited by Torsten Hess and Thomas Seidel, pp. 35–62. Bonn: Westkreuz, 1995.

Griehl, Manfred. *Last Days of the Luftwaffe: German Luftwaffe Combat Units 1944–45.* Barnsley: Frontline, 2009.

Grube-Lieblich, Renate. *"...und morgen war Krieg!": Arado Flugzeugwerke GmbH Wittenberg 1936–1945.* Wittenberg: Grube-Lieblich, 1995.

Grygiel, Henryk. "Hungern für Hitler: Erinnerungen an die Zwangsarbeit bei Focke-Wulf." In *Hungern für Hitler: Erinnerungen polnischer Zwangsarbeitern im Deutschen Reich 1940–1945,* edited by Christoph U. Schminck-Gustavus. Reinbek: Rowohlt, 1984.

Hachtmann, Rüdiger. *Industriearbeit im Dritten Reich: Untersuchungen zu den Lohn- und Arbeitsbedingungen in Deutschland 1933–1945.* Göttingen: Vandenhoeck and Ruprecht, 1989.

Hallion, Richard P. "Doctrine, Technology and Air Warfare: A Late Twentieth Century Perspective." *Airpower Journal*l, no. 2 (1987): pp. 16–27.

Handel, Michael. "Numbers Do Count: The Question of Quality Versus Quantity." *Journal of Strategic Studies* 4 (1981): pp. 225–260.

Hanf, G. "Ich flog den He 162." *Flugzeug* 6 (1992): pp. 13–18.

Herbert, Ulrich. *Fremdarbeiter: Politik und Praxis des "Ausländer-Einsatzes" in der Kriegswirtschaft des Dritten Reiches.* Berlin: Dietz, 1986.

_____. *Hitler's Foreign Workers: Enforced Foreign Labor in Germany Under the Third Reich.* Cambridge: University Press, 1997.

Hermann, Dietmar. *Focke-Wulf Ta 152: The Story of the Luftwaffe's Last Variant High-Altitude Fighter.* Atglen: Schiffer, 1999.

Hesse, Klaus. *1933–1945 Rüstungsindustrie in Leipzig,* 2 Teile. Leipzig: Eigenverlag, 2000.

Hiller, Alfred. *Heinkel He 162 "Volksjäger": Entwicklung, Produktion, Einsatz.* Wien: Verlag Alfred Hiller, 1984.

Hirschel, Ernst-Heinrich, Horst Prem, and Madelung Gero, eds. *Luftfahrtforschung in Deutschland.* Bonn: Bernard and Graefe, 2001.

Holley, Irving Brinton, Jr. "A Detroit Dream of Mass-produced Fighter Aircraft: The XP-75 Fiasco." *Technology and Culture* 28 (1987): pp. 578–593.

Hölsken, Heinz Dieter. *Die V-Waffen: Entstehung, Propaganda, Kriegseinsatz.* Stuttgart: Deutsche Verlags Anstalt, 1984.

Homze, Edward M. *Arming the Luftwaffe: The Reich Air Ministry and the German Aircraft Industry 1919–1939.* Lincoln: University of Nebraska, 1976.

_____. *Foreign Labor in Nazi Germany.* Princeton: Princeton University Press, 1967.

_____. "The German MIC." In *War, Business and*

World Military-Industrial Complexes, edited by Benjamin Franklin Cooling, pp. 51–83. New York: Kennikat Press, 1981.

Hoppe, Joseph. "Fernsehen als Waffe: Militär und Fernsehen in Deutschland 1935–1950." In *Ich diente nur der Technik: Sieben Karrieren zwischen 1940 und 1950*, edited by Museum für Verkehr und Technik, pp. 53–88. Berlin: Nicolai, 1995.

Horten, Reimar, and Peter F. Selinger. *Nurflügel: Die Geschichte der Horten-Flugzeuge 1933–1960*. Graz: Weishaupt, 1983.

Irmer, Thomas. "Zwangsarbeit von jüdischen KZ-Häftlingen für die Rüstungsproduktion in der Region Berlin-Brandenburg während der Schlussphase des Zweiten Weltkrieges — die Aussenlager Glüwen und Schwarzheide des KZ Sachsenhausen." In *Zwangsarbeit während der NS-Zeit in Berlin und Brandenburg*, edited by Winfried Meyer, pp. 163–192. Potsdam: Berlin-Brandenburg, 2001.

Irving, David. *The Rise and Fall of the Luftwaffe: The Life of Field Marshal Erhard Milch*. Boston: Little, 1974.

Jahnke, Karl Heinz. *Ernst Heinkel und die Stadt Rostock: Eine Dokumentation*. Rostock: Ingo Koch, 2002.

Jones, David R. "From Disaster to Recovery: Russia's Air Forces in Two World Wars." In *Why Air Forces Fail: The Anatomy of Defeat*, edited by Robin Higham and Stephan J. Harris, pp. 261–286. Lexington: Kentucky University Press, 2006.

Kaienburg, Hermann. "KZ-Haft und Wirtschaftsinteresse: Das Wirtschaftsverwaltungshauptamt der SS als Leitungszentrale der Konzentrationslager und der SS-Wirtschaft." In *Konzentrationslager und deutsche Wirtschaft 1939–1945*, edited by Hermann Kaienburg, pp. 29–60. Opladen: Leske and Budrich, 1996.

———. *Die Wirtschaft der SS*. Berlin: Metropol, 2003.

Kaiser, Johann B. *Die Geschichte des Messerschmitt-Flugzeugbaus*. Köln: Dt. Forschungs- u. Versuchsanst. für Luft- u. Raumfahrt, 1975.

Kay, Anthony L. *German Jet Engine and Gas Turbine Development 1930–1945*. Shrewsbury: Airlife, 2002.

Koehler, Dieter H. *Ernst Heinkel — Pionier der Schnellflugzeuge: Eine Biographie*. Bonn: Bernard and Graefe, 1999.

Kohl, Peter, and Peter Bessel. *Auto Union und Junkers: Die Geschichte der Mitteldeutschen Motorenwerke GmbH Taucha 1935–1948*. Wiesbaden: Franz Steiner, 2003.

Koziol, Michael Sylvester. *Rüstung, Krieg und Sklaverei: Der Fliegerhorst Schwaebisch-Hall-Hessental und das Konzentrationslager. Eine Dokumentation*. Sigmaringen: J. Thorbecke Verlag, 1986.

Krag, Bernd. "Erfahrung bei der Entwicklung und Erprobung der ersten Strahlflugzeuge mit Pfeilflügeln." In *Die Pfeilflügelentwicklung in Deutschland bis 1945*, edited by Hans-Ulrich Meier, pp. 303–364. Bonn: Bernard and Graefe, 2006.

Kranzhoff, Jörg Armin. *Arado: History of an Aircraft Company*. Atglen: Schiffer, 1997.

Kreis, John F., ed. *Piercing the Fog: Intelligence and Army Air Forces Operation in World War II*. Washington, D.C.: AFHandMP, 1996.

Kroener, Bernhard R. "General Heldenklau: Die 'Unruh-Kommission' im Strudel polykratischer Desorganisation (1942–1944)." In *Politischer Wandel, organisierte Gewalt und nationale Sicherheit. Beiträge zur neueren Geschichte Deutschlands und Frankreichs. Festschrift für Klaus-Jürgen Müller*, edited by Ernst-Willi Hansen, Gerhard Schreiber, and Bernd Wegner, pp. 269–285. München: Oldenbourg, 1995.

Kugler, Anita. "Airplanes for the Führer: Adam Opel AG as Enemy Property, Model War Operation and General Motors Subsidiary, 1939–1945." In *Working for the Enemy: Ford, General Motors, and Forced Labor in Germany*, edited by Reinhold Billstein, pp. 33–82. Oxford: Berghahn, 2000.

Levine, Alan J. *The Strategic Bombing of Germany, 1940–1945*. Westport: Praeger, 1992.

Lewis, David L. "'They may save our honor, our hopes — and our necks': Michigan's Damnedest Colossus" *Michigan History* (September–October 1993). http://www.michiganhistorymagazine.com/extra/willow_run/willow_run.html

Longerich, Peter. *Heinrich Himmler: Biographie*. München: Siedler, 2008.

Madsaac, David. *Strategic Bombing in World War Two: The Story of the United States Strategic Bombing Survey*. New York: Garland, 1976.

Mankau, Heinz. "Die Heinkel He 280." *Jet and Prop* 6 (2004): pp. 44–52.

Marssolek, Inge, and Rene Ott. *Bremen im Dritten Reich: Anpassung, Wiederstand, Verfolgung*. Bremen: Schünemann, 1986.

McLarren, Robert. "Captured Tunnel Advances U.S. Research." *Aviation Week* 51, no. 1 (August 1949).

Megargee, Geoffrey P., ed. *The United States Holocaust Memorial Museum Encyclopedia of Camps and Ghettos, 1933–1945*, vols. 1–2. Bloomington: Indiana University Press, 2009.

Mierzejewski, Alfred C. *The Collapse of the German War Economy, 1944–1945: Allied Air Power and the German National Railway*. Chapel Hill: University of North Carolina Press, 1988.

Milward, Alan S. "Arbeitspolitik und Produktivität in der deutschen Kriegswirtschaft unter vergleichendem Aspekt." In *Kriegswirtschaft und Rüstung 1939–1945*, edited by Friedrich Forstmeier and Hans-Erich Volkmann, pp. 73–91. Düsseldorf: Droste, 1977.

———. *War, Economy and Society, 1939–1945*. Berkeley: University of California, 1979.

Mommsen, Hans. *Der Mythos von der Modernität: Zur Entwicklung der Rüstungsindustrie im Dritten Reich*. Essen: Klartext, 1999.

———, and Manfred Grieger. *Das Volkswagenwerk und seine Arbeiter im Dritten Reich*. Düsseldorf: Econ, 1996.

Morrow, John H. "Defeat of the German and Austro-Hungarian Air Forces." In *Why Air Forces Fail: The Anatomy of Defeat*, edited by Robin Higham and

Stephan J. Harris, pp. 99–134. Lexington: Kentucky University Press, 2006.

Müller, Klaus W., and Willy Schilling. *Deckname Lachs: Die Geschichte der unterirdischen Fertigung der Me 262 im Walpersberg bei Kahla 1944/45*. Zella-Mehlis: Heinrich-Jung-Verlagsgesellschaft, 1995.

Müller, Peter. *Das Bunkergelände im Mühldorfer Hart*. Mühldorf a. Inn: Kreismuseum, 2000.

Müller, Rolf-Dieter. "Albert Speer und die Rüstungspolitik im totalen Krieg." In *Das Deutsche Reich und der Zweite Weltkrieg*, Bd. 5/2, edited by MGFA. München: Dt. Verl.-Anst., 1999.

Murray, Williamson. *Strategy for Defeat: The Luftwaffe 1933–1945*. Maxwell AFB: Air University Press, 1983.

Naasner, Walter. *Neue Machtzentren in der deutschen Kriegswirtschaft 1942–1945*. Boppard: Harald Boldt, 1994.

Neufeld, Michael J. "Overcast, Paperclip, Osoaviakhim: Looting and the Transfer of German Military Technology." In *The United States and Germany in the Era of the Cold War, 1945–1990*, edited by Detlef Junker, pp. 197–203. Cambridge: University Press, 2004.

———. "Rocket Aircraft and the 'Turbojet Revolution': The Luftwaffe's Quest for High-Speed Flight, 1935–1939." In *Innovation and the Development of Flight*, edited by Roger D. Launius, pp. 207–233. College Station: Texas A&M, 1999.

———. *The Rocket and the Reich: Peenemünde and the Coming of the Ballistic Missile Era*. New York: Harvard University Press, 1995.

Noian, Mary. *Visions of Modernity: American Business and the Modernization of Germany*. Oxford: Clarendon, 1994.

Nowarra, Hainz. *Der "Volksjäger" He 162*. Friedberg: Podzun, 1984.

Overy, Richard J. *The Air War 1939–1945*. New York: Stein and Day, 1980.

———. "German Aircraft Production 1939–1942." Diss., Cambridge, 1978.

———. *Goering*. London: Routledge and Kegan, 2000.

———. *War and Economy in the Third Reich*. Oxford: Clarendon, 1994.

Perz, Bertrand. "Der Arbeitseinsatz im KZ Mauthausen." In *Die nationalsozialistischen Konzentrationslager — Entwicklung und Struktur*, Band II, edited by Ulrich Herbert, Karin Orth, and Christoph Dieckmann, pp. 533–557. Göttingen: Wallstein, 1998.

———. "Politisches Management im Wirtschaftskonzern. Georg Meindl und die Rolle des Staatskonzerns Steyr-Daimler-Puch bei der Verwirklichung der NS-Wirtschaftsziele in Österreich." In *Konzentrationslager und deutsche Wirtschaft 1939–1945*, edited by Hermann Kaienburg, pp. 95–113. Opladen: Leske and Budrich, 1996.

Petrick, Peter. "Das Schulflugzeug für's letzte Aufgebot." *Jet and Prop* 4 (1994): pp. 54–55.

Pohl, Hans, Stephanie Habeth, and Beate Brüninghaus, eds. *Die Daimler Benz AG in den Jahren 1933–1945. Eine Dokumentation*. Stuttgart: ZUG Beihefte, 1986.

Price, Alfred. *The Last Year of the Luftwaffe: May 1944 to May 1945*. London: Arms and Armour, 1991.

Raim, Edith. *Die Dachauer KZ-Außenkommandos Kaufering und Mühldorf: Rüstungsbauten und Zwangsarbeit im letzten Kriegsjahr 1944/1945*. Landsberg a. Lech: Landsberger Verl.-Anst., 1992.

Ransom, Stephan, and Hans-Hermann Cammann. *Me 163 Rocket Interceptor*, 2 vols. Crowborough: Classic Publications, 2001.

Ritchie, Sebastian. *Industry and Air Power: The Expansion of British Aircraft Production 1935–1941*: London: Frank Cass, 1997.

Schabel, Ralf. *Die Illusion der Wunderwaffen: Die Rolle der Düsenflugzeuge und Flugabwehrraketen in der Rüstungspolitik des Dritten Reiches*. München: Oldenbourg, 1994.

Schmidt, Matthias. *Albert Speer: Das Ende eines Mythos. Speers wahre Rolle im Dritten Reich*. München: Netzeitung, 1982.

Schmoll, Peter. *Messerschmitt-Giganten und der Fliegerhorst Obertraubling 1936–1945*. Regensburg: MZ Buchverlag, 2002.

———. *Nest of Eagles: Messerschmitt Production and Flight-Testing at Regensburg 1936–1945*. Hersham: Ian Allen, 2010.

Schulte, Jan Erik. *Zwangsarbeit und Vernichtung: Das Wirtschaftsimperium der SS. Oswald Pohl und das SS-Wirtschafts-Verwaltungshauptamt, 1933–1945*. Paderborn: Schöningh, 2001.

Schweitzer, Arthur. *Big Business in the Third Reich*. Bloomington: Indiana University Press, 1977.

Siegel, Tilla, and Thomas von Freyberg. *Industrielle Rationalisierungsmuster unter dem Nationalsozialismus*. Frankfurt a/M: Campus, 1991.

Smith, Richard J., and Eddie J. Creek. *Arado 234 Blitz*. Sturbridge: Monogram, 1992.

———, and ———. *Jet Planes of the Third Reich*. Boylston: Monogram, 1982.

———, and ———. *Me 262*, 4 vols. Burgess Hill: Classic Publication, 1998.

Speer, Albert. *Der Sklavenstaat: Meine Auseinandersetzung mit der SS*. Berlin: Ullstein, 1981.

Spoerer, Mark. *Zwangsarbeit unter dem Hakenkreuz: Ausländische Zivilarbeiter, Kriegsgefangene und Häftlinge im Deutschen Reich und im besetzten Europa 1939–1945*. München: Deutsche Verlags-Anstalt, 2001.

Staerck, Christopher, and Paul Sinnot. *Luftwaffe: The Allied Intelligence Files*. Washington D.C.: Brassey's, 2002.

Stasjulevics, Heiko. *Gotha, die Fliegerstadt*. Gotha: URANIA, 2001.

Stilla, Ernst. *Die Luftwaffe im Kampf um die Luftherrschaft: Entscheidende Einflussgrössen bei der Niederlage der Luftwaffe im Abwehrkampf im Westen und über Deutschland im Zweiten Weltkrieg unter besonderer Berücksichtigung der Faktoren "Luftrüstung," "Forschung und Entwicklung" und "Human Ressourcen."* Bonn: Unpublished diss., 2005.

Strebel, Bernhard. "Ravensbrück — das zentrale Frauenkonzentrationslager." In *Die nationalsozialistis-*

chen Konzentrationslager — Entwicklung und Struktur, Band II, edited by Ulrich Herbert, Karin Orth, and Christoph Dieckmann, pp. 215–258. Göttingen: Wallstein, 1998.

Stubner, Helmut. *Der Kampfflugzeug Heinkel 177 Greif und seine Weiterentwicklung.* Zürich: Eurodoc, 2005.

Suchenwirth, Richard. *Historical Turning Points in the German Air Force War Effort* (USAF Historical Studies, No. 189): New York, 1968.

Trischler, Helmut. "Historische Wurzeln der Grossforschung: Die Luftfahrtforschung vor 1945." In *Grossforschung in Deutschland*, edited by Margit Szöllösi-Janye and Helmut Trischler, pp. 23–37. Frankfurt/M: Campus, 1990.

———. *Luft- und Raumfahrtforschung in Deutschland 1900–1970: Politische Geschichte einer Wissenschaft.* Frankfurt/M: Campus, 1992.

Vajda, Ferenc A., and Peter Dancey. *German Aircraft Industry and Production 1933–1945.* Bath: SAE, 1998.

Van der Vat, Jan. *The Good Nazi: The Life and Lies of Albert Speer.* New York: Weidenfeld and Nicholson, 1997.

Vernaleken, Christoph, and Martin Handig. *Junkers Ju 388: Entwicklung, Erprobung und Fertigung des letzten Junkers-Höhenflugzeugs.* Ochsenfurt-Hohestadt: Aviatic, 2003.

Wagner, Jan-Christian. *Produktion des Todes: Das KZ Mittelbau-Dora.* Göttingen: Wallstein, 2004.

Wagner, Wolfgang. *Kurt Tank: Focke-Wulf's Designer and Test Pilot.* Atglen: Schiffer, 1998.

Webster, Charles, and Noble Frankland. *The Strategic Air Offensive Against Germany 1939–1945*, 4 vols. London: NandM Press, 1961.

Weitz, Birgit. "Der Einsatz von KZ-Häftlinge und jüdischen Zwangsarbeitern bei der Daimler-Benz AG (1941–1945). Ein Überblick." In *Konzentrationslager und deutsche Wirtschaft 1939–1945*, edited by Hermann Kaienburg, pp. 169–198. Opladen: Leske and Budrich, 1996.

Werner, Constanze. *Kriegswirtschaft und Zwangsarbeit bei BMW.* München: Oldenbourg, 2006.

Werrell, Kenneth P. *Blankets of Fire: U.S. Bombers Over Japan During World War II.* Washington, D.C.: Smithsonian Institution Press, 1996.

———. "The Strategic Bombing of Germany in World War II: Costs and Accomplishments." *The Journal of American History* 73, no. 3 (1986): pp. 702–713.

Wiederholt, Thorsten. *Gerhard Fieseler — eine Karriere: Ein Wirtschaftsführer im Dienste des Nationalsozialismus.* Kassel: Jenior, 2006.

Williamson, Gordon. *U-Boat Bases and Bunkers 1941–1945.* Oxford: Osprey, 2003.

Zeitlin, Jonathan. "Flexibility and Mass Production at War: Aircraft Manufacture in Britain, the United States and Germany." *Technology and Culture* 36 (1995): pp. 46–79.

Zilbert, Edward R. *Albert Speer and the Nazi Ministry of Arms: Economic Institutions and Industrial Production in the German War Economy.* London: Fairleigh Dickinson, 1981.

Zofka, Zdenek. "Allach — Sklaven für BMW. Zur Geschichte eines Aussenlagers des KZ Dachau." *Dachauer Heft* 2, pp. 68–78. München: DTV, 1993.

Newspapers and Journals

Heinkel Werkzeitung.

Junkers Flugzeug- und Motorenwerke AG. *Vier Jahre sozialer Aufbau.* Dessau, 1937.

Junkers Nachrichten.

Der Propeller.

Werkszeitschrift der Betriebsgemeinschaft Focke-Wulf Flugzeugbau GmbH.

Wir arbeiten bei Junkers: Ein Bildbericht vom praktischen Sozialismus eines Industriewerkes im Kampf um das neue Europa. München, 1943.

Index

Abterode 128
Adelheide 190
Aero 47
Aerodynamische Versuchsanstalt (AVA) 106–107, 242
AGO *see* Apparatebau GmbH Oschersleben
Ainring 238
Air France 46
Airbus 49, 106
Allach 112, 120, 123, 128, 161–162, 168, 178–179, 185, 189, 217, 221, 258
Allen, Michael T. 2, 75
Allgemeine Elektrizität Gesellschaft 10
Allgemeine Transportanlagen-Gesellschaft (ATG) 10, 48, 90, 128, 170
ALS *see* Amme Luther Sack
Alt Lönnewitz 100, 111
Altona 128
Altrohlau 203
Amme Luther Sack (ALS) 254–255
Amsterdam 53
"Anhydrit" 124, 127
Apparatebau GmbH Oschersleben (AGO) 56, 58, 98, 110, 131, 137
Ar 96 47
Ar 196 48, 89
Ar 234 65, 69, 77, 99–102, 111, 150, 154–155, 182, 233, 241
Ar 296 46
Ar 396 47
Arado 10, 12, 25–26, 46, 48, 58, 61, 65, 98–103, 106, 114, 148–150, 164, 167, 169, 177, 182, 185, 213, 217, 219, 224–225, 227, 232–234, 239, 247
Argus 65, 92, 153
Arnold, Henry H. ("Hap") 56, 266
AS 10 47
Aschersleben 79, 170, 211, 223, 251
Askania 92, 131–132, 153, 207
ATG *see* Allgemeine Transportanlagen-Gesellschaft
Augsburg 16, 22, 26, 54, 57–59, 63, 78, 99, 101, 118, 136–137, 155–156, 179–180, 182, 185, 193, 205, 217, 226, 235

Auschwitz 91, 144, 185, 187–189, 220, 232–233
AVA *see* Aerodynamische Versuchsanstalt
Avia 47

B-17 Flying Fortress 56, 215
B-24 Liberator 31–32
B-29 Superfortress 266
Baatz 231
Babelsberg 25, 213
Bad Eilsen 44, 91, 101, 110
Bad Gendersheim 258
Bader, Paul 244
Bartensleben 131–132, 207
Barth 177, 185, 222–224, 251
Bayerische Motoren Werke *see* BMW
Beaverbrook, Lord (Max Aitken) 85
Bell 266
Bergen-Belsen 224
"Bergkristall-Esche II" 32, 134–135, 190, 219
Berlin 10, 18, 56, 76, 114, 128, 146, 176, 213, 243, 255, 258–259
Bernburg 57, 79, 177, 214, 237, 251, 255–257, 261
"Berta" 251
Bilfinger 121, 123
Birkenholz 214–215
Birkenwerder 187
Birmingham 30
Blankenburg 258
Blohm & Voss 10, 45, 106, 128, 204, 238–240, 247
Blomberg, Werner von 8
Blume, Walter 100
BMW (Bayerische Motoren Werke) 38, 41, 46, 60, 67, 112, 114, 120–121, 123, 126, 128, 131, 138, 148–149, 160–162, 168, 177–178, 183–185, 189, 200, 205–206, 210–211, 217, 221, 252, 255–256
BMW 003 61, 63, 65, 127, 178, 183–184, 237, 247, 249, 252, 256
BMW 132 46
BMW 801 38, 45, 68, 161, 178
BMW P.3302 64, 184

Böblingen 74
Bock 204
Boeing 266
Bordeaux 42, 121, 208
Brandenburg 25, 65, 114, 128, 148, 164, 167–168
Braun, Wernher von 62
Braunschweig 242
Breda 48–49
Breguet 45–46
Bremen 16, 21–22, 25, 36, 55, 110, 147, 149, 190, 212–213, 230
Brenner, Hans 185
Breslau 90, 184
Brown, Eric 247
Bü 181 47
Buchenwald 180, 182, 185–186, 220–221, 232, 234
Budapest 188
Budrass, Lutz 4, 25, 100, 163, 263
Budweis 234
Budzyn 171–172, 175, 177, 232–233
Burgau (aka "Kuno I" and "Kiesweg I") 83
Büssing 132
BV 45–46, 144, 155, 106

Cambeis, Walter 111
Casablanca 5
Castle Bromwich 30, 109
Cham 154
Chatillon sur Bagneaux 42–43, 45, 167
Chemnitz 115
Chicago 135
Christian, Eckhard 245
Cincinnati 135
Consolidated 35
Cottbus 36, 79, 195, 200, 203, 230
Cox, Sebastian 85
Cracow 132
Creil 120
Crewe 109

Dachau 128, 137, 168, 178–180, 185, 191, 221, 232, 234
Dähne, Paul-Heinrich 259–260
Daimler-Benz 26, 38, 124–125,

128, 132, 149, 171–172, 192, 204, 208, 211, 213, 219–220, 225–226, 230–231, 233
Darmstadt 105
DB 601 47
DB 603 68
DB 605 47, 128, 132, 268
Dedelsdorf 103
Degenkolb, Gerhard 78, 84
Demuth, Emil 247, 259
Derby 109
Dernau 233
De Schelde 47
Dessau 16–18, 22, 24, 29, 106, 123, 141, 196, 204, 231
DESt *see* Deutsche Erde- und Steinwerke
Detmold 90, 217
Detroit 31, 135
Deutsche Erde- und Steinwerke (DESt) 134, 180, 182, 186, 190
Deutsche Versuchsanstalt für Luftfahrt (DVL) 245
Deutsche Forschungsinstitut für Segelflug (DFS) 103, 105–106, 238, 244
DFS *see* Deutsche Forschungsinstitut für Segelflug
"Diana II" 136
Diesing, Ulrich 83, 201, 239
Dilg, Adolf 227
Do 24 47–48
Do 335 77, 104, 128, 130–131, 170, 241
Do 635 83
Dornier 47–48, 128, 176, 185, 225, 229
Dorsch, Franz Xaver 80, 122–123, 134, 189
Dresden 256
Düsseldorf 251
DVL *see* Deutsche Versuchsanstalt für Luftfahrt
Dynamit AG 91

E395 102
E580 239
Egeln 251
Eger 87
EHAG *see* Heinkel
Eigruber, August 161
Eisenhower, Dwight David 199
"Emil" 180
Erfurt 90, 252
Erla Maschinenwerk 10, 16–17, 33, 53, 57–58, 98, 115, 134, 180, 182, 185, 197, 215, 224, 229
Eschenlohe 127, 141
Esslingen 252–253, 255
Ezian 83

Fallersleben 30–31, 65, 91, 128, 153–154, 166, 220
Farman 176
Farnborough 261
Fi 103/V-1 31, 65–66, 77, 79, 87, 91, 92–93, 100, 122–123, 128–129, 150, 153–154, 186, 220, 233
Fi 156 12, 46, 118

Fiat 48–49
Fieseler 12, 16, 41, 65, 92–93, 98, 114, 153–154, 211
Finkenwerder 106
Flöha 182
Flossenbürg 180, 182, 185, 221–222, 224, 227–229, 234
Flugmotorenwerke Reichshof 171
Flushing 47
Focke-Achgelis 88, 148
Focke-Wulf 10, 16, 21–22, 25, 35–36, 41–45, 47–49, 55, 58, 72, 90–92, 97–98, 100–101, 109, 121, 124, 141, 146–149, 152, 167, 172, 176–177, 190, 195, 197, 200, 202–203–206, 208, 210–213, 215–217, 228, 230, 248, 259, 264
Fokker 48, 53
Ford 30–32, 35, 145
Ford, Henry 28, 30, 32
Fort Worth 35
Francke, Carl 174, 239–241, 244–246, 249, 253–255, 259
Frankfurt/M 55
Franz, Anselm 204
"Franz II" 128
Freiberg 65, 182, 217, 224–225, 232, 234
Freiburg (Silesia) 90
Friedrichroda 128
Friedrichshafen 122
Fröbe, Rainer 193
Frydag, Karl 71, 83, 93, 114, 149, 156, 171, 193, 200, 240, 253–254, 257, 261
Full, Gerhard 246
FW 58 12
FW 189 42, 46–47
FW 190 28, 35, 42–44, 46, 48–49, 55–56, 58, 67–68, 89, 93, 97–98, 110, 121, 124, 137, 141, 170, 178, 190, 196, 202, 206, 213, 221, 227–228
FW 200 42, 44, 89
FW 206 42
FW 300/Ta 400 42, 44, 48–49

G-55 49
Gabel 142
Gablingen 185
Galland, Adolf 69, 82, 213, 240–243
Gassen 203
Gauting 115, 118, 141, 268
Geilenberg, Edmund 140
Geist 239
General Motors 30, 265
Genshagen 26, 125, 128, 192, 204, 211, 219–220, 230–231
Georgii, Walter 105
Giese, Gerhard 238
Glasgow 109
Glöwen 233
Gnôme et Rhône 41, 46, 53, 121
"Goldfisch" 128, 132, 208, 213, 219, 225–226
Gollob, Gordon 82, 242, 255, 259
Göring, Hermann 3, 8–9, 14–15,

34, 46, 55, 68, 72, 74, 77, 79–87, 91, 93–94, 96, 99, 103, 106–108, 111, 122–124, 131, 134, 137–138, 149–150, 152, 158–161, 168–169, 172, 175, 184–186, 192, 231, 238–243, 247, 268
Gotha 113–114
Gothaer Waggonfabrik 10, 91, 98, 113–114, 128, 155, 248
Göttingen 106–107, 242
Gross Rosen 233
Gross Schierstadt 142
Günther, Siegfried 102, 238, 261
Gusen 134–135, 191
Gusen II 134, 219

Haberstolz 48
Hachtel, August 255, 259–260
Halberstadt 57, 170, 226, 251
Halle 149
Hamburg 55–56, 93, 105–106, 110, 128, 150, 204, 245
Hamburgische Schiffbau Versuchsanstalt 105, 245
Hänsslein 183
Harthof 160
Haunstetten 180, 185, 226
Hayn, Karl 163–164, 168, 173–176, 231, 239
He 59 12
He 111 12, 72, 94, 159, 164, 170–171, 173, 175
He 162 66, 94, 103–104, 207, 226, 236ff
He 177 13–14, 36, 72, 86–88, 90, 94, 96, 102–104, 128–129, 149–150, 159, 163–165, 167–172, 174–176, 193, 227, 231, 238, 264, 266
He 219 72, 91, 147, 172–175, 246
He 274 45, 176
He 280 169, 244
He 343 102–103, 238
Heidelberg 128, 132
Heidfeld 224, 246, 248–249, 254–255
Heiligenrode 128
Heimkehle 124, 128, 132
Heinkel (EHAG) 10, 13–14, 16, 18, 35, 41, 61, 66, 72, 76, 102–104, 132, 146–148, 157–159, 163–165, 167–177, 179, 182–185, 193, 204, 206–207, 209, 214, 218–222, 224, 226, 231–232, 234, 238ff, 264
Heinkel, Ernst 1–2, 10–11, 18, 21, 25, 60, 69, 164–165, 169, 171, 173–175, 238–240, 244, 246–247, 253, 255–256, 258, 261, 264
Helling 248
Hemelingen 204
Henschel 10, 15, 18, 101, 122, 126–128, 148, 155, 160, 171, 222, 232
Herget, Wilhelm 192
Hertel, Heinrich 100, 108
HeS 30 76
Himmler, Heinrich 75–76, 80, 122–124, 158, 162–164, 183–187,

212, 222, 228, 231, 234
Hinterbrühl 248
Hitler, Adolf 2-3, 7-9, 65, 71-72, 77, 79-81, 83-87, 89, 93-94, 96, 99-100, 102, 104, 108, 111, 116, 120-122, 124-126, 133-134, 138, 141, 149-152, 154, 158, 160, 162, 165, 185, 187, 212, 214-215, 229, 240-241, 243, 245
Ho IX 107
Homze, Edward 4
Horgau 118
Horní Bříza 234
Horten 106-107
Horten brothers 106-107
Hs 117 127
Hs 129 15, 41
Hs 293 101
Huchting 204
Hugo Danger 204

IG Farben 18
Ihlefeld, Herbert 260
Industriekontor 125
Inglewood 35
Innsbruck 141
Ishikawagima NA-20 61

Jena 137
Johangeorgenstadt 182, 224
Ju 52 37, 46, 48, 177
Ju 86 29
Ju 87 48, 97
Ju 88 13, 15, 27, 30-31, 36, 57, 68, 128, 153, 164, 170, 202, 221
Ju 188 15, 68, 128, 153, 172
Ju 252 37
Ju 287 94, 102-104, 106, 128-129
Ju 288 36, 74
Ju 352 37, 248
Ju 388 15, 45, 48, 50, 68, 93, 170, 175
Ju 488 45
"Julius" 248-249
Jumo see Junkers Motorenwerk
Jumo 004 38, 61, 64, 111, 124, 126-127, 155, 202, 204, 221, 226, 232, 247, 255, 267
Jumo 213 124, 126-127, 202
Junkers 2, 10, 16-18, 22-27, 29, 31, 37, 41, 48, 50, 57, 68, 74-75, 100, 102-103, 106, 108, 112-114, 123, 126, 128, 140-142, 149, 156, 158-159, 170, 175, 185-186, 190, 196, 200, 202, 204, 207, 211-212, 214, 220-221, 223-224, 231, 233-234, 237, 242, 248, 251, 254-255, 258, 260, 264
Junkers Motorenwerk (Jumo) 18, 41, 60, 90, 111, 113-114, 124, 126, 128, 155, 160, 202, 204, 221-222, 227-228, 232
Jüttner, Hans 183-184

Kaether, Willi 48
Kahla 84, 137-139, 141, 216
Kaltenbrunner, Ernst 161
Kammhuber, Josef 85, 245

Kammler, Hans 80-81, 84-85, 95, 122, 124-125, 140, 188, 220, 225, 229, 234, 258
"Karl" 251
Kassel 16, 65, 93, 116, 153-154, 160
Kaufering 118, 133, 136
Keller, Alfred 243, 253, 259
Kematen 130, 133, 135, 141, 149
Kempten 185
Kessler, Phillip 240-241, 244, 249, 254-256, 258
Kl 35 12, 47
Klagenfurt 251
Kleinrath, Kurt 192
Klemm 74, 98
Klemm, Hans 74
Kloth, Albert 78
Knemeyer, Siegfried 100, 102, 237-239
"Knurrhahn" 224
Koppenberg, Heinrich 23, 74
Kosin, Rüdiger 101-103, 106
Köthen 126
Kottern 150, 156, 180, 182, 185
Kreipe Werner 240, 242
Krome 133
Krupp 10
Kudicke, Helmuth 244
"Kuno I/Kiesweg I" (Burgau) 83, 267
"Kuno II" 267

Lage 44
Landsberg 136, 258, 261
Landshut 102, 113
Langdalze 234
Lange, Otto 188, 203
Langenstein 141
Langenwerke 98, 221
"Languste" 243-244, 248-251, 253-254, 258
Lärz 245
Latécoère 45
Le Mans 53
Lechfeld 130, 155
Leck 260-261
Lehmann, Walther 101, 103, 106
Leipheim 89, 117, 133, 142, 153, 179, 182, 198, 267
Leipzig 16-18, 57-58, 114-115, 180, 185, 196, 215
Leipzig-Möckau 57
Leitmeritz 175, 234
"Leo" 127, 192, 234
Leonberg 127-128, 134, 141, 192, 234
Leonhard Moll 137
Leopoldshall 190, 200, 251
Les Mureaux 42
Letov 47
LFA see Luftfahrtforschungsanstalt Hermann Göring
LFM see Luftfahrtforschungsanstalt München
Lichtenwört 173
Lioré et Olivier 120
Lippisch, Alexander 245
"Lisa" 248, 258

London 87
Lucht, Roluf 79, 100, 108, 117-118, 239-241
"Ludwig II" 128, 251
Ludwigslust 255, 260
Luftfahrtforschungsanstalt Hermann Göring (LFA) 242
Luftfahrtforschungsanstalt München (LFM) 105-106
Lufthansa 8-9, 42, 253-255
Lukesch, Dieter 69
Lysia Gora 172

Magdeburg 18, 111, 114, 126, 251
Mahnke, Franz 183
"Malachit" 226
Mansfeld 148, 159
Marienburg 36, 55, 177, 211, 215
Marienehe 18, 21, 167-168, 172, 174, 241, 251, 254-255, 260
Marietta 266
Markirsch 128
Markkleeberg 221-222, 228, 233
Matford Ford 41
"Maulwurf" 251
Maurer, Gerhard 174, 183
Mauthausen 134-135, 161, 174, 180, 182, 186, 191, 226, 232, 234, 248, 258
May, Kurt 252-253, 255
Me 108 42-43, 46
Me 109 12-13, 16, 24, 27-28, 33, 36-37, 42, 47, 49, 55-56, 58, 66-67, 86, 93, 98-99, 115-116, 118, 121, 132, 156, 180, 182, 191, 196, 205, 209, 224, 229, 239, 251, 267
Me 110 128
Me 163 62-63, 65, 66, 74, 77, 89, 103, 242-243
Me 208 46
Me 209 86, 151-152, 157
Me 210 13-14, 42, 49-50, 86, 90, 168
Me 262 4, 32, 36, 47, 58, 62-65, 67, 69, 77-80, 83-84, 86, 93, 95-97, 99-100, 104-105, 118-119, 126-127, 130, 133-136, 138, 142, 150-156, 179, 184, 191-193, 197, 198, 200, 216, 219, 234, 236, 239, 241, 243, 247, 254, 256, 258, 260-261, 264, 266
Me 264 83
Me 323 41, 105, 174
Me 410 15, 54, 205
Mechanik GmbH 221
Mechanische Werkstätte Neubrandenburg 185
Meimershausen 124
Meindl, Georg 161
Messerschmitt 14, 16, 22, 26-27, 32, 37, 41-43, 48, 54-58, 65, 69, 78, 80, 83-84, 86, 89, 99, 101, 105, 111, 114, 117-119, 127-128, 130, 134-136, 138, 141, 148-150, 153, 155-157, 167, 169, 176-177, 179-182, 184-185, 190-193, 201, 205, 209, 212, 215-217, 221-222, 224, 226-233, 239,

245, 252, 258, 264–267
Messerschmitt, William (Willy)
  14, 66, 78, 86, 120, 150–152,
  157, 168, 179, 192, 241
Meteor 247
Mettenheim 137
Metz 187
Mielec 132, 170–172, 174, 214, 218,
  221, 227–228, 230, 232, 234
"Miessgeldingen" 119
Milbertshofen 162
Milch, Erhard 8–9, 11, 14–15, 28,
  30–31, 34, 37, 40–41, 43–44,
  71–74, 76–83, 85–86, 89–91,
  93, 95–96, 99, 108, 111, 123–124,
  126, 129, 131, 133, 147, 149–150,
  152, 154–158, 161–162, 164–169,
  171–174, 177, 183–188, 212, 215,
  228, 264–265
Mimetall 98, 248
MIMO see Mitteldeutsche Motorenwerke
Mitteldeutsche Motorenwerke
  (MIMO) 16–18, 144, 147, 149,
  202, 215, 225, 229, 231
Mitteldeutsche Stahlwerke 10
Mittelwerk/Mittelbau-Dora 65,
  80, 122–129, 140, 142–143, 202,
  227, 233, 249, 251–252, 258
Mooyer, Otto 158, 165
Morane-Saulnier 43
Morris, William (Lord Nuffield)
  29–30
Morris Motors 29
Moscow 160
Mosquito 91
Mühldorf am Inn 137, 235
Mülhausen 114
Müller, Karl Christian 192
Müller, Max Adolf 75–76
Mülsen-St. Michelin 182, 229
Munich 118, 120, 127–128, 137,
  149, 160, 162, 168, 179, 200,
  210–211, 225, 256, 258
Munich-Riem 261
Murray, Williamson 15

NDW see Norddeutschen
  Dornierwerke
Nebel 83
Neckargerach 225
Neuaubing 185
Neubrandenburg 185, 251
Neuburg 197
Neufeld, Michael J. 75
Neuhausen 197
Neusollstedt 124
Neustadt 257
Neustadt-Glewe 251
Niederorschel 220–221, 233
Niedersachswerfen 122
Nobel 156
Norddeutschen Dornierwerke
  (NDW) 98
Nordhausen 122, 128, 144, 231,
  249
"Nordwerk" 126–127, 226–227,
  249
Nordwerke 126

North American 35
Northrop, Jack 107
Nowotny, Walter 213
Nürnberg 91, 158

Oberammergau 69, 100, 105, 182,
  267
Obergreinau 127
Oberottmarshausen 133
Oberpfaffenhofen 229
Obersalzberg 79, 82, 93
Obertraubling 58, 105, 111, 116,
  118, 120, 202, 229, 232
Oels 31
Offingen 83
Offingen Metallbau 83
Ohain, Hans von 1–2, 11, 76
Opel 29
Oranienburg 10, 18, 21, 107, 128,
  146, 159, 163–165, 167–171, 176–
  177, 179, 183–184, 187, 193, 209,
  219, 221, 227, 231, 253, 255
Oranienburg-Germendorf 35,
  163–164, 167, 170–171, 226, 232
Oschersleben 56–58, 79, 110, 131,
  137
Overlach 78
Overy, Richard 4, 12, 27, 86

P-51 Mustang 56, 69, 99, 198
P-75 265
P 211 239
P 1063 238
P 1068 102–103
P 1073 238–239, 248
P 1077 "Julia" 242, 253, 257
Paris 40–42, 44, 46, 50, 120–121,
  167
Peenemünde 62, 93, 122
Peenemünde-West 106
Perchim 251, 259–260
Peres, Werner 207
Perigeux 208
Peter, Gotthold 244, 247
Petersen, Edgar 245
Peugeot 91
Pfistermeister, Hermann von 171,
  173
Piaggio 48–49
Plaszow 233
Ploemintz 128
Pohl, Oswald 162–164, 174, 182,
  187
Polansky & Zöllner 137
Popp, Franz Josef 46, 121
Porsche, Ferdinand 187
Posen 36, 58, 90–91, 176–177, 190,
  195, 203, 256
Prague 160
Prandtl, Ludwig 106
Prenzlau 159
Price, Alfred 227
PZL 170–171

Raim, Edith 188
Randstein 241
Rathenow 25–26, 182, 233
Rathsdamnitz 141
Ravensbrück 185, 219, 222, 224

Raxwerke 122
Rebstein 182
"Rebstock" 233
Rechlin 8, 37, 82, 95, 106, 183,
  244–245
Regensburg 16, 27, 32, 37, 55–56,
  65, 99, 111, 116–118, 127, 180,
  182, 190, 202, 209, 212, 215,
  221, 224, 227, 229, 232, 267
Regensburg-Prüfening 58
Reichsmarschall Hermann Göring
  (REIMAHG) 137–139
REIMAHG see Reichsmarschall
  Hermann Göring
Reitsch, Hanna 243
Reperaturwerk Erfurt 90
Revensburg 131
Rhone 121
"Richard" 234
Riesenfeld 160
Rimpl, Herbert 18
"Robert" 251
Rochlitz 221
Rolls-Royce 109
Rostock 1, 10–11, 16, 18, 21, 165,
  168, 170, 172, 174, 177, 185, 207,
  239, 241, 248, 251, 254, 258–
  260
Rottleberode 124
Rudolph, Arthur 183
Rzeszow 170–172, 222, 232–233

S-199 47
Sachsenhausen 163–164, 168–169,
  183–184, 219, 232–233
St. Aegyd 75
St. Astier 43, 121, 208
St. Georgen 134, 190
Salzburg 258
"Salzwerke" 251
Sandbach 27
"Santa" 249
Sauckel, Fritz 137–138, 150, 155–
  156, 165–167, 185, 216
Saur, Karl Otto 43, 45, 50, 77–
  79, 81–85, 87, 92–93, 104, 112–
  113, 126, 133–134, 138, 140, 143,
  150, 155–156, 163, 186, 188, 190,
  203, 209, 239–240, 243, 253,
  258–259, 268
Savoia 49
Savoia-Marchetti 49
Sawatzki, Alwin 249
Schabel, Ralf 4, 62, 85
Schaberger 174
Schaffer-Linz 257
Schaler 183
Schelp, Helmuth 183
"Schildkröte" 249
Schirach, Baldor von 240
Schlatt 76
Schmelter, Fritz 155, 157, 188
Schmidt, Eberhardt 216
Schmidt, Rudolf 260
Schmiele, Arthur 220
Schmoll, Peter 96
Schönebeck 185, 248, 251
Schönefeld 15, 127
Schuchardt 48

Index

Schüngel 258
Schwäbisch Hall 119, 128, 267
Schwechat 170, 172–174, 177, 244, 246, 248–249
Schweinfurt 56, 58, 240
Schwichtenberg 190
"Seelachs" 130
Seiler, Friedrich Wilhelm 78, 155, 192
Si 204 46–47
SIAI *see* Società Idrovolanti Alta Italia
Siebel 149, 170, 248
Siemens-Schuckert 92, 185
Simpson, Homer 4
SIPA *see* Société Industrielle Pour l'Aéronautique
SNCAN *see* Société Nationale des Constructions Aéronautiques du Nord
SNCASO *see* Société Nationale des Constructions Aéronautiques du Sud-Ouest
Società Idrovolanti Alta Italia (SIAI) 48
Société Industrielle Pour l'Aéronautique (SIPA) 46
Société Nationale des Constructions Aéronautiques du Nord (SNCAN) 42–43, 46
Société Nationale des Constructions Aéronautiques du Sud-Ouest (SNCASO) 42–45, 121, 167, 208
Solle 124
Sommer, Erich 69
Sommer, Josef 230
Sorau 22–23, 36, 79, 195, 202–203, 230
Southhampton 109
Spandau 160
Speer, Albert 2, 38, 72–74, 76–78, 80–86, 94, 108, 116, 121–123, 125, 133, 140, 143, 150, 152, 154–156, 163, 167, 184, 188–189, 196, 203, 211, 241, 258, 268
Spitfire 99, 109
Springen 131, 206, 252
Stalingrad 149, 157, 166
Stassfurt 126, 128, 131, 190, 251, 256
"Stauffen" 119–120
Steeb, Erwin 260
Sternberg 235
Stettin 79
Steyr 58, 161
Steyr-Daimler-Puch 161

Stobbe-Dethleffsen, Carl 133
Strasbourg 41, 114
Streitwieser, Anton 248, 258
Stuttgart 57, 252, 255
Süd Deutsche Bremse Werk 128
Sulzhayn 76
Supermarine 109

Ta 152 42, 45, 48–49, 67–69, 92–93, 126, 134, 137, 190, 197, 203, 216, 241, 261
Ta 154 81, 89–93, 97, 108, 124, 172, 174, 190, 217, 244, 248–249, 252
Ta 183 97
Ta 254 97
Tank, Kurt 42–44, 48, 90, 97, 100–101, 110, 146, 152, 212–213
"Tarthun" 204, 251, 256
Taucha 16–18, 202, 215
Tautz 259
Technique de Chatillon 42, 44
Tedder, Arthur William 199
Telfs 141
Tempelhof 80
Tempest 260
Thiedemann, Richard 50, 251
Tiercelet 91, 123, 187, 220, 233
TL-300 76
Todt, Fritz 3, 72
Tödter, Hermann 207
Toulouse 45
Trautloff, Hannes 213
Trebbin 243
Treiber 124
Trischler, Helmut 104
Turin 49
Tutow 58

Überlingen 131
Udet, Ernst 9, 11–15, 28, 31, 37, 42, 47, 52, 71–72, 74, 79, 85–86, 161–162, 264, 268
Ulm 118

V-2/A-4 62, 65, 75, 77, 80, 84, 92–93, 100, 121–122, 128, 133, 151, 153–154, 182–183
V-3 77, 142
VDM *see* Vereinigte Deutsche Metallwerke
Venusberg 224
Vereinigte Deutsche Metallwerke (VDM) 55, 74, 110
Vereinigte Ostwerke 170
Vienna 111, 170, 172–174, 177, 209, 224, 238–246, 248–249, 253–260
Vienna-Florisdorf 248
Voigt, Richard 240
Volkswagen 29, 30–31, 65, 91, 93, 123, 128, 153–154, 166, 187, 220, 233
Vorwald, Wolfgang 168

Wahl, Karl 78
"Walnuss I" 136
Wanke 247
Warnemünde 10, 25–26, 227
Warning, Herbert 25, 212, 230
Warsitz, Erich 11
Wedemeyer, Georg 118, 246
Weingut Betriebsgesellschaft 136, 191–192
"Weingut I" 137
"Weingut II" 118, 136–137, 191
Weiss, Martin Gottfried 179, 191
Wendel, Fritz 69, 151
Werner, Constanze 144
Werner, Wilhelm 38, 123
Weser Flugzeugbau 10, 98, 155, 182
Wesserling-Urbès 233
Wieliczka 132, 234
Wiener-Neustadt 24, 55, 122, 212
Wiener Neustädter Flugzeugwerke (WNF) 24, 55, 111, 116, 251, 254–255
WIFO *see* Wirtschaftliche Forschungsgesellschaft
Wilhelm Gustloff Werke 98, 137
Willow Run 30–32, 35, 145
Wilson, Samuel G. 135
Wirtschaftliche Forschungsgesellschaft (WIFO) 122, 124, 143
Wittenberg 26, 65, 182, 193, 219
WNF *see* Wiener Neustädter Flugzeugwerke
Woffleben 127
Wronki 190
Wuppertal 91

Ypsilanty 31

Zborowski, Helmuth Graf von 184
Zeppelin Lufschiffbau 122
Ziereis, Franz 174
Zipprich, Erich 38
Zittau 39, 111, 155–156, 232–233
Zlin 47

www.ingramcontent.com/pod-product-compliance
Ingram Content Group UK Ltd.
Pitfield, Milton Keynes, MK11 3LW, UK
UKHW050702160426
5217IPUK00038B/2026